APPLIED ELECTROMAGNETICS

Using QuickField™ and MATLAB®

License, Disclaimer of Liability, and Limited Warranty

APPLIED ELECTROMAGNETICS

Using QuickField™ and MATLAB®

By

James R. Claycomb, PhD

JONES AND BARTLETT PUBLISHERS
Sudbury, Massachusetts
BOSTON TORONTO LONDON SINGAPORE

FIREWAL MEDIA
(An Imprint of Laxmi Publications Pvt. Ltd.)
BANGALORE • CHENNAI • COCHIN • GUWAHATI • HYDERABAD
JALANDHAR • KOLKATA • LUCKNOW • MUMBAI • RANCHI
NEW DELHI

Published by :
FIREWAL MEDIA
(An Imprint of Laxmi Publications Pvt. Ltd.)
113, Golden House, Daryaganj,
New Delhi-110002

Phone : 011-43 53 25 00
Fax : 011-43 53 25 28

www.laxmipublications.com
info@laxmipublications.com

Economy Edition *Edition* : 2010

OFFICES

© **Bangalore**	080-26 61 15 61	© **Chennai**	044-24 34 47 26
© **Cochin**	0484-237 70 04, 405 13 03	© **Guwahati**	0361-254 36 69, 251 38 81
© **Hyderabad**	040-24 65 23 33	© **Jalandhar**	0181-222 12 72
© **Kolkata**	033-22 27 43 84	© **Lucknow**	0522-220 95 78
© **Mumbai**	022-24 91 54 15, 24 92 78 69	© **Ranchi**	0651-221 47 64

FAP-3041-395-APPLIED ELECTROMAGNETIC-CLA **C— 613 -09-09**

Printed at : Print-o-Pack, Delhi.

TABLE OF CONTENTS

INTRODUCTION

Applied Electromagnetics Using QuickField™ and MATLAB® is intended as an introductory level textbook for teaching computer-based electricity, magnetism and multiphysics. The text is easily accessible to advanced undergraduates and beginning graduate students in physics and engineering. Many exercises and demonstrations may be implemented using QuickField and MATLAB in a traditional introductory level physics course. This second audience will benefit from the visualization of electric and magnetic field distributions and force calculations without a working knowledge of the finite element method or potential theory.

QuickField is a window-based, Finite Element Method (FEM) software package that supports Electrostatics, DC and AC conduction, Magnetostatics, AC and Transient Magnetics, Steady State and Transient Heat Transfer and Stress Analysis problem types. Models are created in a 'point-and-click' CAD environment, where material properties and boundary conditions are assigned. Automatic mesh generation and post processing are fast and user-friendly. Solutions to most problems in the textbook can be displayed in a matter of seconds after the model has been created.

The textbook is packaged with a companion CD with a student version of the software capable of solving all the problems in the text. Additional examples are included with the software. The student version of QuickField may also be downloaded from the Tera Analysis website at www.quickfield.com. The user's guide and demonstration videos are also available on the website.

Application-based examples in the text and on the website include the calculation of currents in biological tissue under electrical stimulation, superconducting magnetic shielding, magnetic levitation, electromagnetic nondestructive testing as well as the motion of charged particles in electric fields. Multiphysics applications include coupled stress, electromagnetic and thermal analysis.

Students taking a course in electromagnetic theory usually concentrate mostly on analytical techniques, e.g., solving differential equations and boundary value problems. Unfortunately, students often come away with a limited understanding of how electromagnetic fields behave. Computer modeling serves to bridge this understanding gap in that it enables visualization of electric and magnetic fields and electrical currents and therefore builds an intuitive and qualitative understanding that is not readily gained in manipulating complex analytical expressions.

Analytical methods developed in this text concentrate on separation of variables, conformal mapping, and Laplace transform techniques. Numerical finite difference and Monte Carlo methods are also introduced with examples in MATLAB. Comparison of numerical solutions with theory helps establish confidence in numerical methods and builds experience in establishing the reliability of computational results and the applicability of theoretical approximations.

The book includes extensive problem sets that facilitate computer-based learning of electromagnetics and the application of QuickField and MATLAB illustrating some of the basic concepts in electromagnetic theory such as Gauss' Law and Ampere's Law. The exercises are designed to allow user selection of different parameters, dimensions, material properties, and initial conditions. Tables of physical properties and characteristic dimensions of engineering materials and biological materials in living cells and the human body are included in Appendices 4 and 5 for the reader's convenience. The reader is encouraged to conform, modify, and extend these exercises according to his or her own interests.

Chapter 1 introduces mathematical preliminaries and MATLAB concurrently with additional MATLAB examples in Appendix 1. The vector analysis component of Chapter 1 provides simple MATLAB examples calculating vector dot and cross products. The divergence, curl, gradient, and Laplacian are also calculated in different coordinate systems. The Laplace Transform introduced in Chapter 1 is used in chapters on transient magnetics, thermal analysis, stress analysis, and electrical circuit modeling.

Analytical and computational methods of solving Laplace and Poisson's equations are developed in Chapter 2. Readers wishing to jump directly into QuickField may begin with Chapter 3 "A Walk Through QuickField." This chapter will get the reader started simulating simple electrostatic and magnetostatics problems in QuickField with step-by-step visual instructions for plotting electric and magnetic fields, creating contour graphs, and calculating integral values.

Chapters 4 through 10 cover electrostatics, magnetostatics, time-harmonic magnetics, transient magnetics, superconductivity, alternating and direct current flow. Chapters 11 and 12 cover thermal and stress analysis and multiphysics examples with coupled heat transfer, stress and electromagnetic coupling. Applications include space capsule atmospheric reentry simulations that couple thermal and stress analysis as well as modeling the temperature distribution resulting from current flow in a fuel cell.

The text concludes with Chapter 13 on passive electrical circuits. QuickField includes a CAD-based electrical circuit simulator that simulates circuits with AC or transient time dependence. Applications include filter circuits and equivalent circuit models of neurons and cells under electrical stimulations.

Chapter 1

MATHEMATICAL PRELIMINARIES

In This Chapter

- Vector Analysis
- Vector Derivatives
- The Scalar Laplacian
- The Vector Laplacian
- Integral Theorems
- Integral Transforms
- The Dirac Delta Function
- Special Functions

1.1 VECTOR ANALYSIS

A scalar physical quantity is described by a numerical factor with units. In general, the numerical factor will depend on the system of units we choose. Examples of scalars include mass, temperature, and time. Two scalars may be added or subtracted provided they have the same units. For example, it would be meaningless to add scalars of differing units such as temperature and time. Scalars of differing units may be multiplied or divided to form a new scalar, however, such as the average density obtained as the ratio of an object's mass to its volume.

A vector physical quantity is described by a numerical factor with units and a specified direction. Examples of vectors include electric and magnetic fields and forces that have a specified direction in space. Vectors are usually designated with boldfaced symbols such as **A** in printed work or with an arrow $\vec{A}$ in handwritten work, such as on a chalkboard or on notebook paper.

Vector Addition

A three-dimensional vector **A** may be expressed in Cartesian coordinates as

$$\mathbf{A} = A_x\mathbf{i} + A_y\mathbf{j} + A_z\mathbf{k}, \tag{1.1}$$

where **i**, **j**, and **k** are unit vectors in the x-, y-, and z-directions, respectively. A_x, A_y, and A_z are the components of **A** along each coordinate axis. Given the vector $\mathbf{B} = B_x\mathbf{i} + B_y\mathbf{j} + B_z\mathbf{k}$, the vector sum or difference $\mathbf{A} \pm \mathbf{B}$ may be written as

$$\mathbf{A} \pm \mathbf{B} = (A_x \pm B_x)\mathbf{i} + (A_y \pm B_y)\mathbf{j} + (A_z \pm B_z)\mathbf{k}. \tag{1.2}$$

Vector addition is commutative,

$$\mathbf{A} + \mathbf{B} = \mathbf{B} + \mathbf{A}. \tag{1.3}$$

In MATLAB vectors are expressed as matrices

```
x=[1 2 3] % specifies the row vector x
y=[2 3 4] % specifies the row vector y
x'        % gives the column vector with x entries above
y'        % gives the column vector with y entries above
x+y       % gives the sum of vectors x and y specified above
```

The magnitude or length of the vector **A** is given by the square root of the sum of squares of each vector component

$$|\mathbf{A}| = \sqrt{A_x^2 + A_y^2 + A_z^2}. \tag{1.4}$$

Dot Product

The dot product, also known as the scalar product, between two vectors **A** and **B** results in a scalar quantity that depends on both the magnitude and angle θ between each vector or

$$\mathbf{A} \cdot \mathbf{B} = |\mathbf{A}| \cdot |\mathbf{B}| \cos(\theta). \tag{1.5}$$

The dot product is zero between orthogonal vectors and is maximal between parallel vectors and is written in terms of the unit vectors

$$\mathbf{A}\cdot\mathbf{B} = \left(A_x\mathbf{i} + A_y\mathbf{j} + A_z\mathbf{k}\right)\cdot\left(B_x\mathbf{i} + B_y\mathbf{j} + B_z\mathbf{k}\right) \quad (1.6)$$

Since $\mathbf{i}\cdot\mathbf{i} = \mathbf{j}\cdot\mathbf{j} = \mathbf{k}\cdot\mathbf{k} = 1$ and $\mathbf{i}\cdot\mathbf{j} = \mathbf{i}\cdot\mathbf{k} = \mathbf{j}\cdot\mathbf{k} = 0$ we have that

$$\mathbf{A}\cdot\mathbf{B} = A_xB_x + A_yB_y + A_zB_z. \quad (1.7)$$

The magnitude of a vector may also be expressed as the dot product with itself square rooted

$$|\mathbf{A}| = \sqrt{\mathbf{A}\cdot\mathbf{A}}. \quad (1.8)$$

The dot product is commutative, that is,

$$\mathbf{A}\cdot\mathbf{B} = \mathbf{B}\cdot\mathbf{A}. \quad (1.9)$$

The dot product may also be distributed across a sum

$$\mathbf{A}\cdot(\mathbf{B}+\mathbf{C}) = \mathbf{A}\cdot\mathbf{B} + \mathbf{A}\cdot\mathbf{C}. \quad (1.10)$$

The dot product may be evaluated in MATLAB as follows

```
x=[1 2 3]     % specifies the row vector x
y=[2 3 4]     % specifies the row vector y
z=[3 4 5]     % specifies the row vector z
sqrt(x*x′)    % gives the magnitude of the vector x
x*y′          % gives the dot product between vectors x and  y
dot(x,y)      % also gives the dot product between x and y
x*(y+z)′      % gives the dot product between x and (y+z)
x*y′+x*z′     % also gives the dot product between x and (y+z)
dot(x,y+z)    % also gives the dot product between x and (y+z)
```

Cross Product

The cross product between two vectors, **A** and **B**, also known as the vector product, results in a third vector that depends on the magnitude and direction of **A** and **B**. The magnitude of the cross product is given by

$$|\mathbf{A}\times\mathbf{B}| = |\mathbf{A}|\cdot|\mathbf{B}|\sin(\theta). \quad (1.11)$$

The cross product is zero between parallel vectors and is maximal between orthogonal vectors. In terms of the unit vectors, the cross product may be expressed as a determinant

$$\mathbf{A}\times\mathbf{B}=\begin{vmatrix}\mathbf{i} & \mathbf{j} & \mathbf{k}\\ A_x & A_y & A_z\\ B_x & B_y & B_z\end{vmatrix}=(A_yB_z-B_yA_z)\mathbf{i}-(A_xB_z-B_xA_z)\mathbf{j}+(A_xB_y-B_xA_y)\mathbf{k} \tag{1.12}$$

expanded across the top row. Note that the vector formed by **A** × **B** is perpendicular to both **A** and **B** so that **A** · (**A** × **B**) = **B** · (**A** × **B**) = 0. Interchanging the last two rows in the determinate we can see that **A** × **B** = –**B** × **A**.

The cross product may be calculated in MATLAB as

```
x=[1 2 3]          % specifies the row vector x
y=[2 3 4]          % specifies the row vector y
cross(x,y)         % gives the cross product between x and y
dot(x,cross(x,y)) % gives the dot product between x and cross(x,y)
```

Triple Vector Product

The vector formed by the cross product of three vectors **A, B,** and **C** is frequently encountered in vector analysis

$$\mathbf{A}\times\mathbf{B}\times\mathbf{C}=\mathbf{B}(\mathbf{A}\cdot\mathbf{C})-\mathbf{C}(\mathbf{A}\cdot\mathbf{B}). \tag{1.13}$$

A mnemonic for remembering the triple cross identity is the "**BAC CAB**" rule, where the brackets go on the back.

```
x=[1 2 3] % specifies the row vector x
y=[2 3 4] % specifies the row vector y
z=[3 4 5] % specifies the row vector z
cross(x,cross(y,z)) % gives the triple cross product
                    % between x, y and z
```

1.2 VECTOR DERIVATIVES

Gradient of a Scalar Field

The gradient, or del operator, is a directional derivative defined as

$$\nabla=\frac{\partial}{\partial x}\mathbf{i}+\frac{\partial}{\partial y}\mathbf{j}+\frac{\partial}{\partial z}\mathbf{k}. \tag{1.14}$$

Given the scalar field $S(x, y, z)$, we have that the gradient of a scalar field is a vector quantity

$$\nabla S = \frac{\partial S}{\partial x}\mathbf{i} + \frac{\partial S}{\partial y}\mathbf{j} + \frac{\partial S}{\partial z}\mathbf{k}. \tag{1.15}$$

In orthogonal curvilinear coordinates, the gradient of a scalar $S(u_1, u_2, u_3)$ may be written as

$$\nabla S = \frac{1}{g_{11}^{1/2}}\frac{\partial S}{\partial u_1}\mathbf{e}_1 + \frac{1}{g_{22}^{1/2}}\frac{\partial S}{\partial u_2}\mathbf{e}_2 + \frac{1}{g_{33}^{1/2}}\frac{\partial S}{\partial u_3}\mathbf{e}_3. \tag{1.16}$$

The unit vectors $(\mathbf{e}_1, \mathbf{e}_2, \mathbf{e}_3)$ point along each coordinate axis. The line element may be written as

$$ds^2 = dx^2 + dy^2 + dz^2 = g_{11}du_1^2 + g_{22}du_2^2 + g_{33}du_3^2. \tag{1.17}$$

The metric coefficients g_{ii} are related to the Cartesian coordinates

$$g_{ii} = \left(\frac{\partial x}{\partial u_i}\right)^2 + \left(\frac{\partial y}{\partial u_i}\right)^2 + \left(\frac{\partial z}{\partial u_i}\right)^2 \tag{1.18}$$

so that $g_{11} = g_{22} = g_{33} = 1$ in Cartesian coordinates.

The gradient of a scalar field may be numerically computed in MATLAB using the following syntax:

```
% gradient of a 1-D scalar field F
Fx = gradient(F)
% gradient of a 2-D scalar field G
[Gx, Gy] = gradient(G)
% gradient of a 3-D scalar field H
[Hx, Hy, Hz] = gradient(H)
```

Divergence of a Vector Field

Given the vector field $\mathbf{V}(x, y, z) = V_x(x, y, z)\mathbf{i} + V_y(x, y, z)\mathbf{j} + V_z(x, y, z)\mathbf{k}$, the divergence of $\mathbf{V}$ is defined as the dot product of the del operator with $\mathbf{V}$ or,

$$\nabla \cdot \mathbf{V} = \left(\frac{\partial}{\partial x}\mathbf{i} + \frac{\partial}{\partial y}\mathbf{j} + \frac{\partial}{\partial z}\mathbf{k}\right) \cdot \left(V_x\mathbf{i} + V_y\mathbf{j} + V_z\mathbf{k}\right). \tag{1.19}$$

This gives

$$\nabla \cdot \mathbf{V} = \left(\frac{\partial V_x}{\partial x} + \frac{\partial V_y}{\partial y} + \frac{\partial V_z}{\partial z}\right). \tag{1.20}$$

In orthogonal curvilinear coordinates the divergence of a vector $\mathbf{V}(u_1, u_2, u_3)$ is

$$\nabla \cdot \mathbf{V} = \frac{1}{g^{1/2}} \left\{ \frac{\partial}{\partial u_1} \left[\left(\frac{g}{g_{11}} \right)^{1/2} V_1 \right] + \frac{\partial}{\partial u_2} \left[\left(\frac{g}{g_{22}} \right)^{1/2} V_2 \right] + \frac{\partial}{\partial u_3} \left[\left(\frac{g}{g_{33}} \right)^{1/2} V_3 \right] \right\}, \tag{1.21}$$

where $g = \begin{vmatrix} g_{11} & g_{12} & g_{13} \\ g_{21} & g_{22} & g_{23} \\ g_{31} & g_{32} & g_{33} \end{vmatrix} = g_{11} g_{22} g_{33}.$

The divergence of a vector field may be numerically computed in MATLAB using the following syntax:

```
% divergence of a 3-D vector field
div = divergence (X, Y, Z, U, V, W)

% U, V, and W are each 3-D numeric arrays corresponding to
% x, y and z-field components

% divergence of a 3-D vector field is also given by
div = divergence (U, V, W)
% where [X Y Z] =meshgrid(1:n,1:m,1:p) and [m, n, p] = size(U)

% divergence of a 2-D vector field
div = divergence (X, Y, U, V)

% divergence of a 2-D vector field is also given by
div = divergence (U, V)
%where [X Y ] =meshgrid(1:n,1:m) and [m,n] = size(U)
```

Curl of a Vector Field

The curl of a vector field is defined as the cross product of the del operator with $\mathbf{V}$

$$\nabla \times \mathbf{V} = \left(\frac{\partial}{\partial x}\mathbf{i} + \frac{\partial}{\partial y}\mathbf{j} + \frac{\partial}{\partial z}\mathbf{k} \right) \times \left(V_x \mathbf{i} + V_y \mathbf{j} + V_z \mathbf{k} \right). \tag{1.22}$$

This may be expressed as the determinate

$$\nabla \times \mathbf{V} = \begin{vmatrix} \mathbf{i} & \mathbf{j} & \mathbf{k} \\ \frac{\partial}{\partial x} & \frac{\partial}{\partial y} & \frac{\partial}{\partial z} \\ V_x & V_y & V_z \end{vmatrix} = \left(\frac{\partial V_z}{\partial y} - \frac{\partial V_y}{\partial z} \right)\mathbf{i} - \left(\frac{\partial V_z}{\partial x} - \frac{\partial V_x}{\partial z} \right)\mathbf{j} + \left(\frac{\partial V_y}{\partial x} - \frac{\partial V_x}{\partial y} \right)\mathbf{k}. \tag{1.23}$$

In curvilinear coordinates the curl may be expressed in determinate form as

$$\nabla \times \mathbf{V} = \begin{vmatrix} \mathbf{e}_1 (g_{11}/g)^{1/2} & \mathbf{e}_2 (g_{22}/g)^{1/2} & \mathbf{e}_3 (g_{33}/g)^{1/2} \\ \dfrac{\partial}{\partial u_1} & \dfrac{\partial}{\partial u_2} & \dfrac{\partial}{\partial u_3} \\ g_{11}^{1/2} V_1 & g_{22}^{1/2} V_2 & g_{33}^{1/2} V_3 \end{vmatrix}. \tag{1.24}$$

The curl of a vector field may be computed numerically in MATLAB using the following syntax:

```
% curl of a 3-D vector field
[curlx, curly, curlz, cav] = curl (X, Y, Z, U, V, W)
% U, V, and W are each 3-D numeric arrays corresponding to
% x, y and z-field components

% curl of a 3-D vector field is also given by
[curlx, curly, curlz, cav] = curl (U, V, W)
% where [X Y Z] =meshgrid(1:n,1:m,1:p) and [m, n, p] = size(U)

% curl of a 2-D vector field
[curlx, curly, cav] = curl (X, Y, U, V)

% curl of a 2-D vector field is also given by
[curlx, curly, cav] = curl (U, V)
% where [X Y ] =meshgrid(1:n,1:m) and [m,n] = size(U)
```

1.3 THE SCALAR LAPLACIAN

The Laplacian of a scalar field is defined as the dot product of the del operator with the gradient of the scalar field $\nabla \cdot \nabla S(x, y, z)$. In curvilinear coordinate systems the scalar Laplacian may be written as

$$\nabla^2 S = \frac{1}{g^{1/2}} \left\{ \frac{\partial}{\partial u_1} \left(\frac{g^{1/2}}{g_{11}} \frac{\partial S}{\partial u_1} \right) + \frac{\partial}{\partial u_2} \left(\frac{g^{1/2}}{g_{22}} \frac{\partial S}{\partial u_2} \right) + \frac{\partial}{\partial u_3} \left(\frac{g^{1/2}}{g_{33}} \frac{\partial S}{\partial u_3} \right) \right\}. \tag{1.25}$$

Cartesian Coordinates

In Cartesian coordinates the scalar Laplacian has the simple form

$$\nabla^2 S = \left(\frac{\partial}{\partial x} \mathbf{i} + \frac{\partial}{\partial y} \mathbf{j} + \frac{\partial}{\partial z} \mathbf{k} \right) \cdot \left(\frac{\partial S}{\partial x} \mathbf{i} + \frac{\partial S}{\partial y} \mathbf{j} + \frac{\partial S}{\partial z} \mathbf{k} \right) = \frac{\partial^2 S}{\partial x^2} + \frac{\partial^2 S}{\partial y^2} + \frac{\partial^2 S}{\partial z^2}. \tag{1.26}$$

The Laplacian of a scalar field may be numerically computed in MATLAB using the following syntax:

```
% Laplacian of a 2-D scalar field U
L = del2(U)   % returns Laplacian/4
% Laplacian of a N-D scalar field U
-L. = del2(U)  % returns Laplacian/N where N is ndims(U)
```

We now consider the scalar Laplacian in several orthogonal coordinate systems.

Cylindrical Coordinates

The cylindrical coordinates (r, z, ϕ) are related to the Cartesian coordinates

$$\begin{aligned} x &= r\cos\phi \\ y &= r\sin\phi, \\ z &= z \end{aligned} \tag{1.27}$$

where $r \geq 0$, $0 \leq \phi < 2\pi$, $-\infty < z < \infty$. The coordinate surfaces are circular cylinders of constant r, half-planes parallel to the z-axis and planes perpendicular to the z-axis

$$\begin{aligned} x^2 + y^2 &= r^2 \\ \tan\phi &= \frac{y}{x}. \\ z &= \mathrm{z} \end{aligned} \tag{1.28}$$

The Laplacian $\nabla^2 S(r, z, \phi)$ is written as

$$\frac{1}{r}\frac{\partial}{\partial r}\left(r\frac{\partial S}{\partial r}\right) + \frac{\partial^2 S}{\partial z^2} + \frac{1}{r^2}\frac{\partial^2 S}{\partial \phi^2}. \tag{1.29}$$

Bicylindrical Coordinates

The bicylindrical coordinates (α, β, z) are related to the Cartesian coordinates

$$\begin{aligned} x &= \frac{a\sinh\alpha}{\cosh\alpha - \cos\beta} \\ y &= \frac{a\sin\beta}{\cosh\alpha - \cos\beta}, \\ z &= z \end{aligned} \tag{1.30}$$

where $-\infty < \alpha < \infty$, $0 \leq \beta \leq 2\pi$, $-\infty < z < \infty$. The coordinate surfaces are circular cylinders parallel to the z-axis and planes perpendicular to the z-axis described by

$$\begin{aligned} &x^2 + \left(y - a\cot\beta\right)^2 = a^2 \csc^2 \beta \\ &\left(x - a\coth\alpha\right)^2 + y^2 = a^2 \text{csch}^2 \alpha. \\ &z = z \end{aligned} \tag{1.31}$$

Figure 1.1 shows the projection of bicylindrical coordinate surfaces in the x-y plane plotted in MATLAB with the generating M-file code.

The Laplacian $\nabla^2 S(\alpha, \beta, z)$ in bicylindrical coordinates is written as

$$\frac{\left(\cosh\alpha - \cos\beta\right)^2}{a^2}\left[\frac{\partial^2 S}{\partial\alpha^2} + \frac{\partial^2 S}{\partial\beta^2}\right] + \frac{\partial^2 S}{\partial z^2}. \tag{1.32}$$

Elliptic Cylindrical Coordinates

The elliptic cylindrical coordinates (α, β, z) are related to the Cartesian co-ordinates

$$\begin{aligned} &x = a\cosh\alpha\cos\beta \\ &y = a\sinh\alpha\sin\beta, \\ &z = z \end{aligned} \tag{1.33}$$

where $0 \le \alpha$, $0 \le \beta \le 2\pi$, $-\infty < z < \infty$. The coordinate surfaces are elliptic (α = const.) and hyperbolic (β = const.) cylinders sharing the same focal line (confocal) with planes perpendicular to the z-axis

$$\begin{aligned} &\left(\frac{x}{a\cosh\alpha}\right)^2 + \left(\frac{y}{a\sinh\alpha}\right)^2 = 1 \\ &\left(\frac{x}{a\cos\beta}\right)^2 - \left(\frac{y}{a\sin\beta}\right)^2 = 1. \\ &z = \mathrm{z} \end{aligned} \tag{1.34}$$

The Laplacian $\nabla^2 S(\alpha, \beta, z)$ in elliptic cylindrical coordinates is written as

$$\frac{1}{a^2\left(\sinh^2\alpha + \sin^2\beta\right)}\left[\frac{\partial^2 S}{\partial\alpha^2} + \frac{\partial^2 S}{\partial\beta^2}\right] + \frac{\partial^2 S}{\partial z^2}. \tag{1.35}$$

ON THE CD Color images for many of the figures may be found on the CD-ROM.

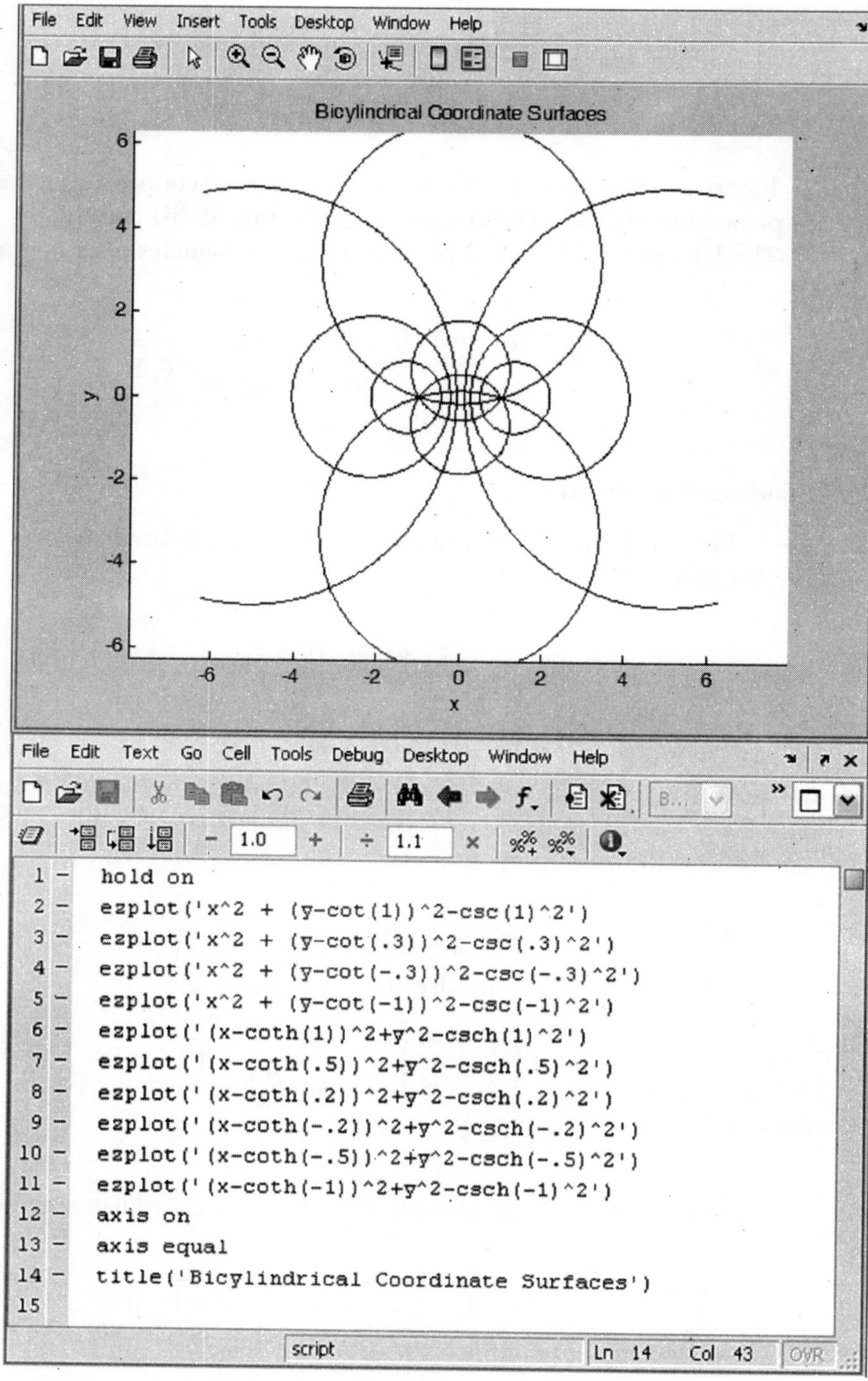

FIGURE 1.1 **Bicylindrical coordinate surface projections plotted in MATLAB with M-file code box (requires Symbolic Toolbox).**

Spherical Coordinates

The spherical coordinates (r, θ, ϕ) are related to the Cartesian coordinates

$$\begin{aligned} x &= r\sin\theta\cos\phi \\ y &= r\sin\theta\sin\phi, \\ z &= r\cos\theta \end{aligned} \tag{1.36}$$

where $r \geq 0$, $0 \leq \theta \leq \pi$, $0 \leq \phi < 2\pi$. The coordinate surfaces are spheres of constant r, cones of constant θ, and half-planes of constant ϕ parallel to the z-axis

$$\begin{aligned} &x^2 + y^2 + z^2 = r^2 \\ &\tan\theta = \frac{\sqrt{x^2 + y^2}}{z}. \\ &\tan\phi = \frac{y}{x} \end{aligned} \tag{1.37}$$

The Laplacian $\nabla^2 S(r, \theta, \phi)$ is written as

$$\frac{1}{r^2}\frac{\partial}{\partial r}\left(r^2\frac{\partial S}{\partial r}\right) + \frac{1}{r^2\sin\theta}\frac{\partial}{\partial\theta}\left(\sin\theta\frac{\partial S}{\partial\theta}\right) + \frac{1}{r^2\sin^2\theta}\frac{\partial^2 S}{\partial\phi^2}. \tag{1.38}$$

Oblate Spheroidal Coordinates

The oblate spheroidal coordinates (ξ, η, ϕ) are related to the Cartesian coordinates

$$\begin{aligned} x &= a\cosh\xi\cos\eta\cos\phi \\ y &= a\cosh\xi\cos\eta\sin\phi, \\ z &= a\sinh\xi\sin\eta \end{aligned} \tag{1.39}$$

where $\xi \geq 0$, $0 \leq \eta \leq \pi$, $0 \leq \phi < 2\pi$. The coordinate surfaces are oblate spheroids (ξ = const.) and hyperboloids of one sheet revolved about the z-axis (η = const.) and half-planes (ϕ = const.) parallel to the z-axis

$$\begin{aligned} &\left(\frac{x}{a\cosh\xi}\right)^2 + \left(\frac{y}{a\cosh\xi}\right)^2 + \left(\frac{z}{a\sinh\xi}\right)^2 = 1 \\ &\left(\frac{x}{a\sin\eta}\right)^2 + \left(\frac{y}{a\sin\eta}\right)^2 - \left(\frac{z}{a\cos\eta}\right)^2 = 1. \\ &\tan\phi = \frac{y}{x} \end{aligned} \tag{1.40}$$

Figure 1.2 shows the projection of oblate spheroidal coordinate surfaces in the x-z plane plotted in MATLAB with the generating M-file code.

The Laplacian $\nabla^2 S(\xi, \eta, \phi)$ in oblate spheroidal coordinates is written as

$$\frac{1}{a^2\left(\sinh^2\xi+\sin^2\eta\right)}\left[\frac{1}{\cosh\xi}\frac{\partial}{\partial\xi}\left(\cosh\xi\frac{\partial S}{\partial\xi}\right)+\frac{1}{\cos\eta}\frac{\partial}{\partial\eta}\left(\cos\eta\frac{\partial S}{\partial\eta}\right)\right]$$
$$+\frac{1}{a^2\cosh^2\xi\cos^2\eta}\frac{\partial^2 S}{\partial\phi^2}. \tag{1.41}$$

Prolate Spheroidal Coordinates

The prolate spheroidal coordinates (ξ, η, ϕ) are related to the Cartesian coordinates

$$\begin{aligned} x &= a\sinh\xi\sin\eta\cos\phi \\ y &= a\sinh\xi\sin\eta\sin\phi, \\ z &= a\cosh\xi\cos\eta \end{aligned} \tag{1.42}$$

where $\xi \geq 0$, $0 \leq \eta \leq \pi$, $0 \leq \phi < 2\pi$.

The coordinate surfaces are prolate spheroids ($\xi =$ const.) and hyperboloids of two sheets revolved about the z-axis ($\eta =$ const.) and half-planes ($\phi =$ const.) parallel to the z-axis

$$\begin{aligned} &\left(\frac{x}{a\sinh\xi}\right)^2+\left(\frac{y}{a\sinh\xi}\right)^2+\left(\frac{z}{a\cosh\xi}\right)^2=1 \\ &-\left(\frac{x}{a\sin\eta}\right)^2-\left(\frac{y}{a\sin\eta}\right)^2+\left(\frac{z}{a\cos\eta}\right)^2=1. \\ &\tan\phi=\frac{y}{x} \end{aligned} \tag{1.43}$$

Figure 1.3 shows the projection of prolate spheroidal coordinate surfaces in the x-z plane plotted in MATLAB with the generating M-file code.

The Laplacian $\nabla^2 S(\xi, \eta, \phi)$ in prolate spheroidal coordinates is written as

$$\frac{1}{a^2\left(\sinh^2\xi+\sin^2\eta\right)}\left[\frac{1}{\sinh\xi}\frac{\partial}{\partial\xi}\left(\sinh\xi\frac{\partial S}{\partial\xi}\right)+\frac{1}{\sin\eta}\frac{\partial}{\partial\eta}\left(\sin\eta\frac{\partial S}{\partial\eta}\right)\right]$$
$$+\frac{1}{a^2\sinh^2\xi\sin^2\eta}\frac{\partial^2 S}{\partial\phi^2}. \tag{1.44}$$

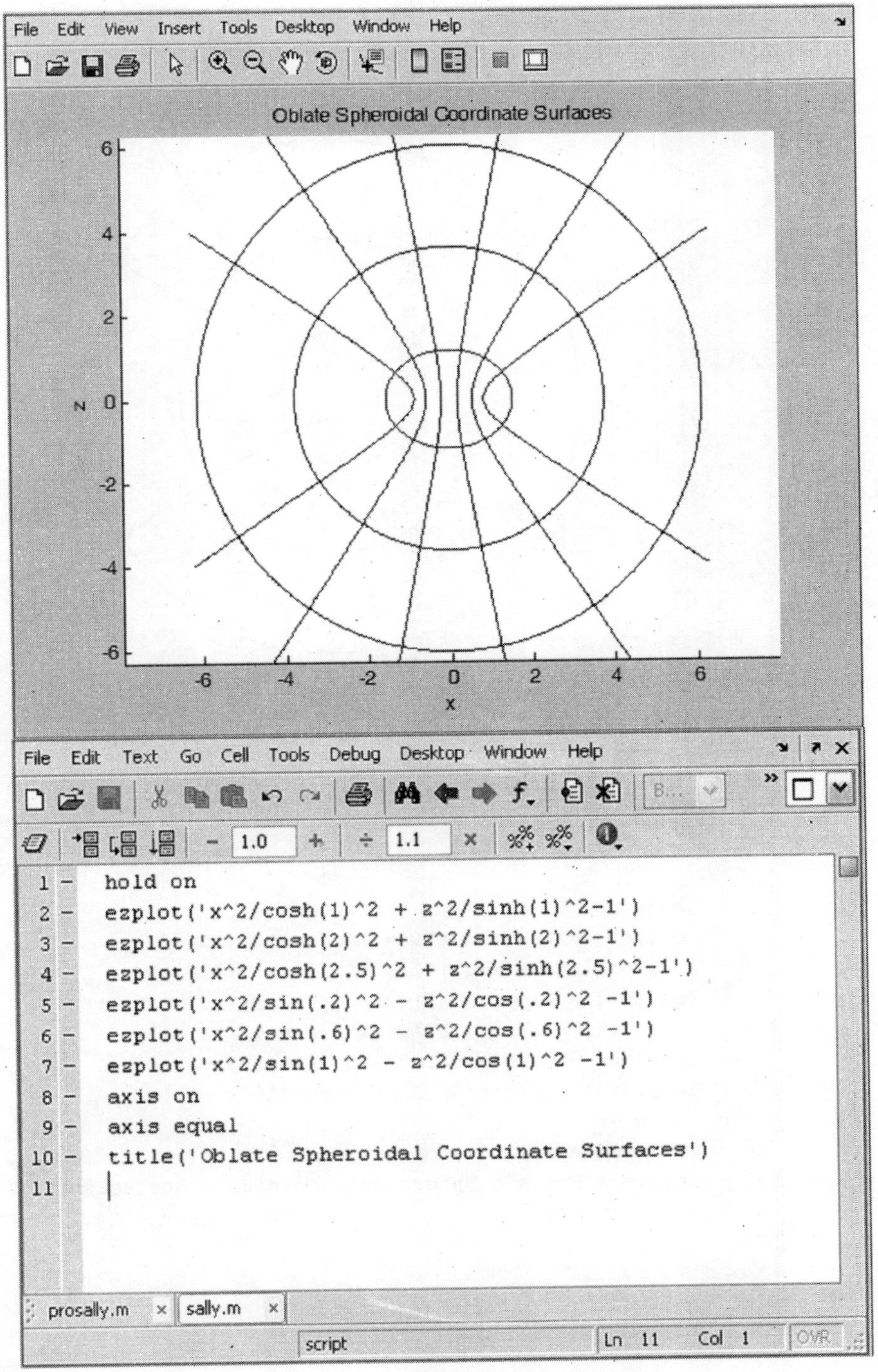

FIGURE 1.2 **Projection of oblate spheroidal coordinate surfaces in the *x-z* plane plotted in MATLAB with the M-file code box (requires Symbolic Toolbox).**

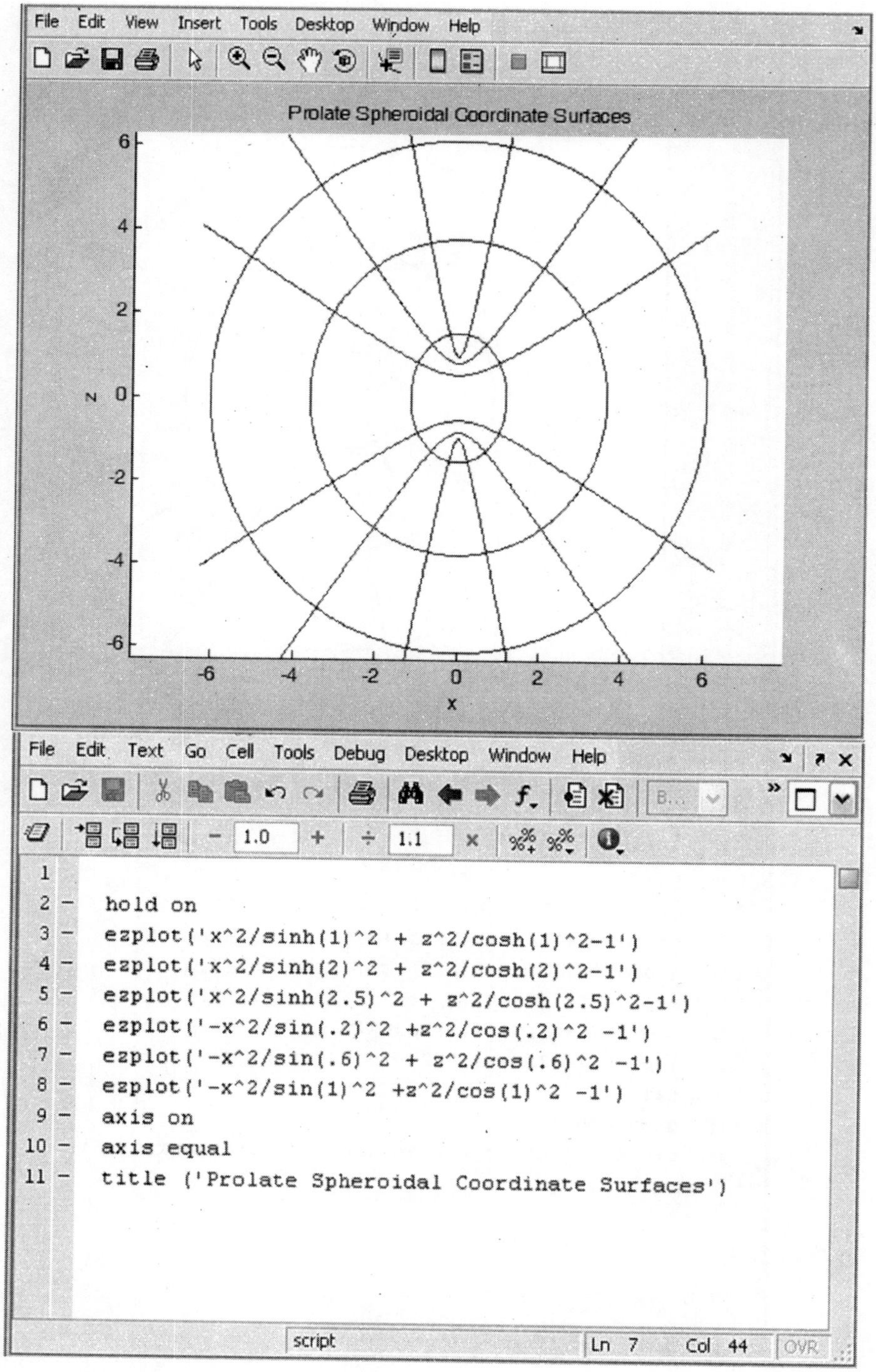

FIGURE 1.3 **Prolate spheroidal coordinate surface projections plotted in MATLAB with the M-file code box (requires Symbolic Toolbox).**

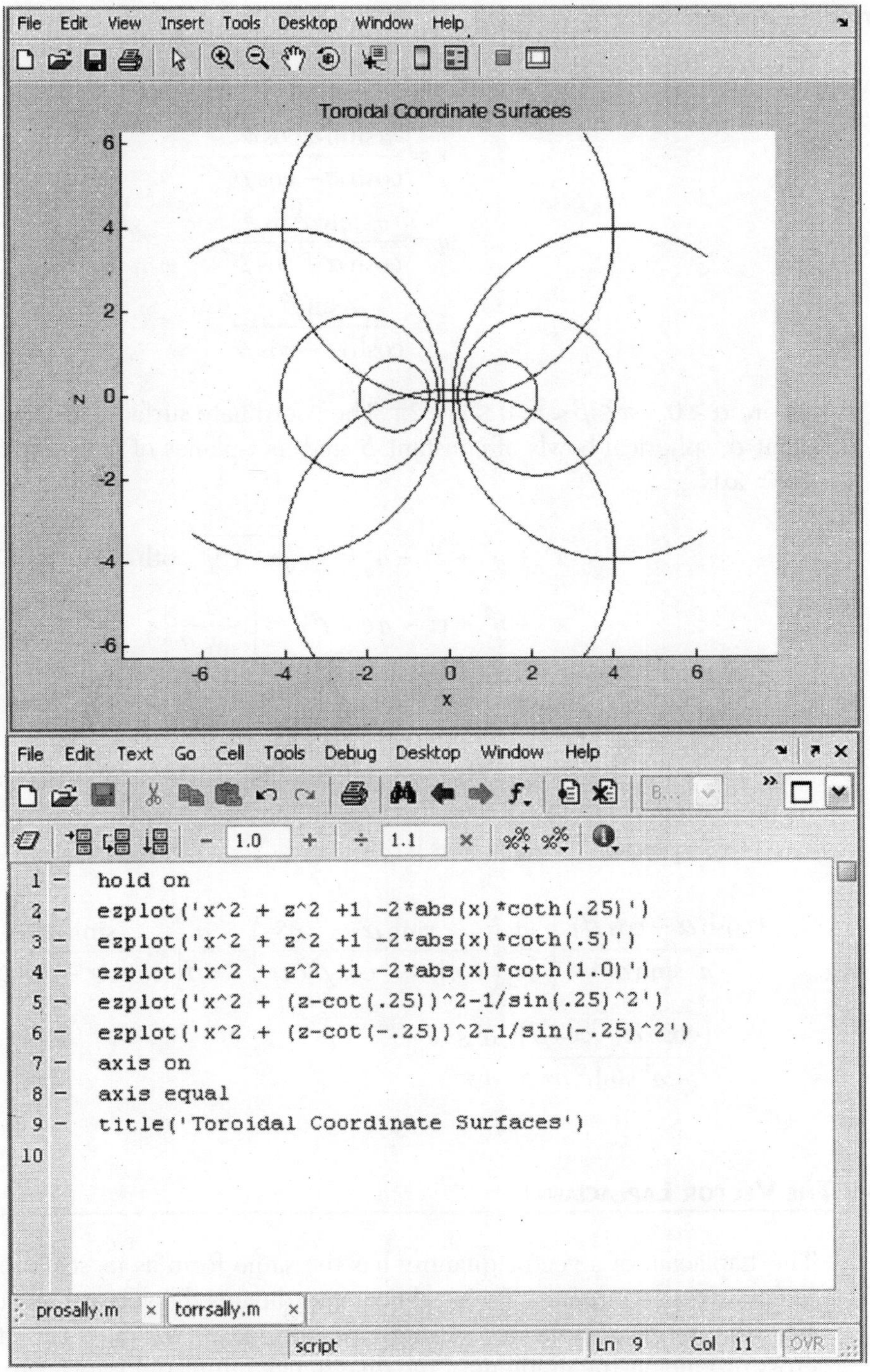

FIGURE 1.4 **Toroidal coordinate surface projections plotted in MATLAB with the M-file code box (requires Symbolic Toolbox).**

Toroidal Coordinates

The toroidal coordinates (α, β, ϕ) are related to the Cartesian coordinates

$$
\begin{aligned}
x &= \frac{a \sinh \alpha \cos \phi}{\cosh \alpha - \cos \beta} \\
y &= \frac{a \sinh \alpha \sin \phi}{\cosh \alpha - \cos \beta}, \\
z &= \frac{a \sin \beta}{\cosh \alpha - \cos \beta}
\end{aligned}
\tag{1.45}
$$

where $\alpha \geq 0$, $-\pi \leq \beta < \pi$, $0 \leq \phi < 2\pi$. The coordinate surfaces are toroids of constant α, spherical bowls of constant β, and half-planes of constant ϕ parallel to the z-axis

$$
\begin{aligned}
& x^2 + y^2 + z^2 + a^2 = 2a\sqrt{x^2 + y^2} \coth \alpha \\
& x^2 + y^2 + (z - a \cot \beta)^2 = \left(\frac{a}{\sin \beta} \right)^2. \\
& \tan \phi = \frac{y}{x}
\end{aligned}
\tag{1.46}
$$

Figure 1.4 shows the projection of toroidal coordinate surfaces in the x-z plane plotted in MATLAB with the generating M-file code.

The Laplacian $\nabla^2 S(\alpha, \beta, \phi)$ is written as

$$
\begin{aligned}
& \frac{(\cosh \alpha - \cos \beta)^3}{a^2 \sinh \alpha} \left[\frac{\partial}{\partial \alpha} \left(\frac{\sinh \alpha}{\cosh \alpha - \cos \beta} \frac{\partial S}{\partial \alpha} \right) + \frac{\partial}{\partial \beta} \left(\frac{\sinh \alpha}{\cosh \alpha - \cos \beta} \frac{\partial S}{\partial \beta} \right) \right] \\
& + \frac{(\cosh \alpha - \cos \beta)^2}{a^2 \sinh^2 \alpha} \frac{\partial^2 S}{\partial \phi^2}.
\end{aligned}
\tag{1.47}
$$

1.4 THE VECTOR LAPLACIAN

The Laplacian of a vector quantity has the same form as its scalar counterpart in Cartesian coordinates. In curvilinear coordinates the vector Laplacian is considerably more complicated. Using equation (1.13) we may write the double cross product as the gradient of the divergence minus the Laplacian

$$
\nabla \times \nabla \times \mathbf{V} = \nabla(\nabla \cdot \mathbf{V}) - \nabla^2 \mathbf{V}. \tag{1.48}
$$

The Laplacian may then be expressed as

$$\nabla^2\mathbf{V} = \nabla(\nabla\cdot\mathbf{V}) - \nabla\times\nabla\times\mathbf{V}. \tag{1.49}$$

In curvilinear coordinates

$$\begin{aligned}\nabla^2\mathbf{A} = \mathbf{e}_1&\left\{\frac{1}{g_{11}^{1/2}}\frac{\partial\gamma}{\partial u_1} + \left(\frac{g_{11}}{g}\right)^{1/2}\left[\frac{\partial\Gamma_2}{\partial u_3} - \frac{\partial\Gamma_3}{\partial u_2}\right]\right\}\\ + \mathbf{e}_2&\left\{\frac{1}{g_{22}^{1/2}}\frac{\partial\gamma}{\partial u_2} + \left(\frac{g_{22}}{g}\right)^{1/2}\left[\frac{\partial\Gamma_3}{\partial u_1} - \frac{\partial\Gamma_1}{\partial u_3}\right]\right\}\\ + \mathbf{e}_3&\left\{\frac{1}{g_{33}^{1/2}}\frac{\partial\gamma}{\partial u_3} + \left(\frac{g_{33}}{g}\right)^{1/2}\left[\frac{\partial\Gamma_1}{\partial u_2} - \frac{\partial\Gamma_2}{\partial u_1}\right]\right\},\end{aligned} \tag{1.50}$$

where the Christoffel symbols

$$\Gamma_1 = \frac{g_{11}}{g^{1/2}}\left\{\frac{\partial}{\partial u_2}\left(g_{33}^{1/2}V_3\right) - \frac{\partial}{\partial u_3}\left(g_{22}^{1/2}V_2\right)\right\} \tag{1.51}$$

$$\Gamma_2 = \frac{g_{22}}{g^{1/2}}\left\{\frac{\partial}{\partial u_3}\left(g_{11}^{1/2}V_1\right) - \frac{\partial}{\partial u_1}\left(g_{33}^{1/2}V_3\right)\right\} \tag{1.52}$$

$$\Gamma_3 = \frac{g_{33}}{g^{1/2}}\left\{\frac{\partial}{\partial u_1}\left(g_{22}^{1/2}V_2\right) - \frac{\partial}{\partial u_2}\left(g_{11}^{1/2}V_1\right)\right\} \tag{1.53}$$

and the factor γ is the divergence of **V** according to equation (1.21).

1.5 INTEGRAL THEOREMS

Divergence Theorem

Gauss' divergence theorem related the volume integral of the divergence of a vector field **E** to the integral of the vector **E** dotted into the surface normal and evaluated over any surface bounding the volume v

$$\iiint_{\text{vol}} \nabla\cdot\mathbf{E}\,dv = \iint_{\text{surf}} \mathbf{E}\cdot\mathbf{n}\,da. \tag{1.54}$$

Stokes' Theorem

Stokes' theorem relates the line integral of a vector field **E** around a closed path Γ to the surface integral evaluated over any capping surface bound by the contour

$$\iint_{\text{surf}} \nabla \times \mathbf{E} \cdot \mathbf{n}\, da = \oint_{\Gamma} \mathbf{E} \cdot d\ell. \tag{1.55}$$

Gradient Theorem

The line integral of the gradient of a scalar function may be evaluated as

$$\int_{\mathbf{a}}^{\mathbf{b}} \nabla f \cdot d\ell = f(\mathbf{b}) - f(\mathbf{a}), \tag{1.56}$$

where the integral depends only on the value of the function at the end points.

1.6 INTEGRAL TRANSFORMS

Laplace Transform

The Laplace transform of the function $f(t)$ is defined as the integral

$$F(s) = \int_0^{\infty} e^{-st} f(t)\, dt. \tag{1.57}$$

This integral is written as $F(s) = L[f(t)]$. The Laplace transform of a sum of functions obeys the linearity property

$$L\left[c_1 f_1(t) + c_2 f_2(t)\right] = c_1 F_1(s) + c_2 F_2(s), \tag{1.58}$$

where c_1 and c_2 are constants. Other properties of the Laplace transform include the following:

Change of scale property

$$L\left[f(at)\right] = \frac{1}{a} F\left(\frac{s}{a}\right). \tag{1.59}$$

Shifting property

$$L\left[e^{at} f(t)\right] = F(s-a). \tag{1.60}$$

Second shifting property

$$L\left[f(t-a)\right]=e^{-as}F(s). \tag{1.61}$$

Transform of a derivative

$$L\left[f'(t)\right]=sF(s)-f(0). \tag{1.62}$$

Transform of a second derivative

$$L\left[f''(t)\right]=s^2F(s)-sf(0)-f'(0). \tag{1.63}$$

Transform of a periodic function

$$L\left[f(t)\right]=\frac{1}{1-e^{-sz}}\int_0^z e^{-st}f(t)\,dt. \tag{1.64}$$

The Laplace transform of a function $f(t)$ may be evaluated in MATLAB using the following syntax:

```
syms a s t
laplace(f)      % returns F(s) given f(t)
% the default variables of laplace are t and s
laplace(1)      % returns 1/s
laplace(t)      % returns 1/s^2
laplace(exp(a*t)) % returns 1/(s-a)
laplace(sin(a*t)) % returns a/(s^2+a^2)
laplace(cos(a*t)) % returns s/(s^2+a^2)
```

Inverse Laplace Transform

The inverse Laplace transform may be evaluated as a contour integral

$$L^{-1}\left[F(s)\right]=\frac{1}{2\pi i}\int_{\gamma-i\infty}^{\gamma+i\infty}e^{st}F(s)\,ds=f(t). \tag{1.65}$$

The inverse Laplace transform obeys the linearity property

$$L^{-1}\left[c_1F_1(s)+c_2F_2(s)\right]=c_1f_1(t)+c_2f_2(t), \tag{1.66}$$

where c_1 and c_2 are constants. Other properties of the inverse Laplace transform $L^{-1}[F(s)] = f(t)$ include the following:

The change of scale property

$$\frac{1}{a}L^{-1}\left[F\left(\frac{s}{a}\right)\right] = f(at). \tag{1.67}$$

Shifting property

$$L^{-1}\left[F(s-a)\right] = e^{at}f(t). \tag{1.68}$$

Second shifting property

$$L^{-1}\left[e^{-as}F(s)\right] = f(t-a). \tag{1.69}$$

Inverse transform of a derivative

$$L^{-1}\left[F'(s)\right] = -tf(t). \tag{1.70}$$

Inverse transform of a second derivative

$$L^{-1}\left[F''(s)\right] = (-t)^2 f(t). \tag{1.71}$$

Inverse transform of an nth derivative

$$L^{-1}\left[F^{(n)}(s)\right] = (-t)^n f(t). \tag{1.72}$$

Convolution property

$$L^{-1}\left[F(s)G(s)\right] = \int_0^t f(u)g(t-u)du. \tag{1.73}$$

The convolution integral is the abbreviated $f*g(t)$, where $f*g = g*f$.

The inverse Laplace transform of a function $f(t)$ may be evaluated in MATLAB using the following syntax:

```
syms a s t
ilaplace(f)            % returns f(t) given F(s)
% the default variables of ilaplace are t and s
ilaplace(1/s)          % returns 1
ilaplace(1/s^2)        % returns t
ilaplace(1/(s-a))      % returns exp(a*t)
ilaplace(a/(s^2+a^2)) % sin(a*t)
ilaplace(s/(s^2+a^2)) % returns cos(a*t)
```

Fourier Transform

The Fourier transform of a function $f(t)$ is defined as the integral

$$F(\omega)=\frac{1}{\sqrt{2\pi}}\int_{-\infty}^{\infty} f(t)e^{-i\omega t}\,dt. \tag{1.74}$$

The Fourier transform of a function $f(t)$ may be evaluated with default variables x and w in MATLAB using the following syntax:

```
fourier(f)      % returns F(w) given f(x)
fourier(f,v)    % returns F(v) given f(x)
fourier(x,u,v)% returns F(v) given f(u)
```

If the Fourier transform is known, the function $f(t)$ may be retrieved from the inverse transform defined and evaluated in MATLAB using the following syntax:

$$f(t)=\frac{1}{\sqrt{2\pi}}\int_{-\infty}^{\infty} F(\omega)e^{i\omega t}\,dt. \tag{1.75}$$

```
ifourier(F)      % returns f(x) given F(w)
ifourier(f,v)    % returns f(x) given F(v)
ifourier(x,u,v)% returns f(u) given F(v)
```

Fast Fourier Transform

The Discrete Fourier Transform (DFT) may be evaluated using the Fast Fourier Transform (FFT) algorithm in MATLAB. The FFT of a time series given by $x = f(t)$ may be evaluated with default variables x and w in MATLAB as

```
y= fft(x) % returns the Fast Fourier transform of time series x
y= fft(x,n) % returns the n-point FFT. If n is greater that the
            % if n is greater than the number of points in x
            % then x is padded with zeros to length n
```

1.7 THE DIRAC DELTA FUNCTION

The Dirac delta function is a non-analytic function defined by

$$\delta(x)=\begin{cases}0, & x\neq 0\\ \infty, & x=0\end{cases}. \tag{1.76}$$

The Dirac delta function may be written in MATLAB as

```
dirac(x)     % returns the Dirac delta function of x
```

The delta function can also be written as a derivative of the Heaviside step function $\theta(x)$

$$\delta(x) = \frac{d\theta(x)}{dx}, \quad \text{where } \theta(x) = \begin{cases} 1, & x > 0 \\ 0, & x \le 0 \end{cases}. \tag{1.77}$$

The derivative of the Heaviside step function and its derivative may be evaluated in MATLAB as

```
heaviside(x)              % returns the Heaviside function of x
diff(heaviside(x),x)      % returns dirac(x)
```

The delta function is even

$$\delta(-x) = \delta(x) \tag{1.78}$$

$$\delta(kx) = \frac{1}{|k|}\delta(x). \tag{1.79}$$

The integral properties of the delta function are given below. The delta function has a unit area so that

$$\int_{-\infty}^{\infty} \delta(x)dx = 1. \tag{1.80}$$

If the delta function is located at an integration limit

$$\int_{-\infty}^{0} \delta(x)dx = \int_{0}^{\infty} \delta(x) = \frac{1}{2} \tag{1.81}$$

or more generally

$$\int_{a}^{b} \delta(x-c)dx = \begin{cases} 1, & c \in (a,b) \\ 0, & c \notin [a,b] \\ \frac{1}{2}, & c \in \{a,b\} \end{cases}. \tag{1.82}$$

Integrals over the dirac delta function may be evaluated in MATLAB using the following syntax:

```
int(dirac(x),-inf, inf)          % returns 1
int(dirac(x), 0, inf)            % returns 1
                                 % result must be multiplied by ½
int(dirac(x-1), 1, inf)          % returns 1
                                 % result must be multiplied by ½
int(dirac(x-1), 2, inf)          % returns 0
```

Integrals of functions multiplied by the delta function are evaluated as

$$\int_{-\infty}^{\infty} f(x)\delta(x)dx = f(0) \tag{1.83}$$

$$\int_{-\infty}^{\infty} f(x)\delta(x-a)dx = f(a) \tag{1.84}$$

$$\int_{-\infty}^{\infty} f(x)\delta'(x-a)dx = -f'(a). \tag{1.85}$$

Integrals over functions multiplied by the dirac delta function may be evaluated in MATLAB using the following syntax:

```
int(sin(x)*dirac(x-a),-inf, inf)        % returns sin(a)
int(besselj(x)*dirac(x-pi), 0, inf)     % returns besselj(pi)
int(dirac(x)*sin(x)/x, 0, inf)          % returns 1
```

In three dimensions, the Dirac delta function can be written as a product of one-dimensional delta functions

$$\delta^3(\mathbf{r}) = \delta(x)\delta(y)\delta(z). \tag{1.86}$$

The three-dimensional delta function has similar integral properties as in one dimension

$$\iiint_{\text{all space}} \delta^3(\mathbf{r})\, dv = 1 \tag{1.87}$$

$$\iiint_{\text{all space}} f(\mathbf{r})\delta^3(\mathbf{r}-\mathbf{r}')\, dv = f(\mathbf{r}'). \tag{1.88}$$

Figure 1.5 shows the displacement vector $\mathbf{r} - \mathbf{r}'$ in three dimensions. The delta function may be expressed in terms of the divergence and the Laplacian

$$\nabla \cdot \frac{\mathbf{r} - \mathbf{r}'}{\left|\mathbf{r} - \mathbf{r}'\right|^3} = 4\pi\delta^3(\mathbf{r} - \mathbf{r}'), \tag{1.89}$$

where the divergence is with respect to the unprimed coordinates. Taking the divergence of the relation

$$\nabla \frac{1}{\left|\mathbf{r} - \mathbf{r}'\right|} = -\frac{\mathbf{r} - \mathbf{r}'}{\left|\mathbf{r} - \mathbf{r}'\right|^3} \tag{1.90}$$

gives

$$\nabla^2 \frac{1}{\left|\mathbf{r} - \mathbf{r}'\right|} = -4\pi\delta^3(\mathbf{r} - \mathbf{r}'). \tag{1.91}$$

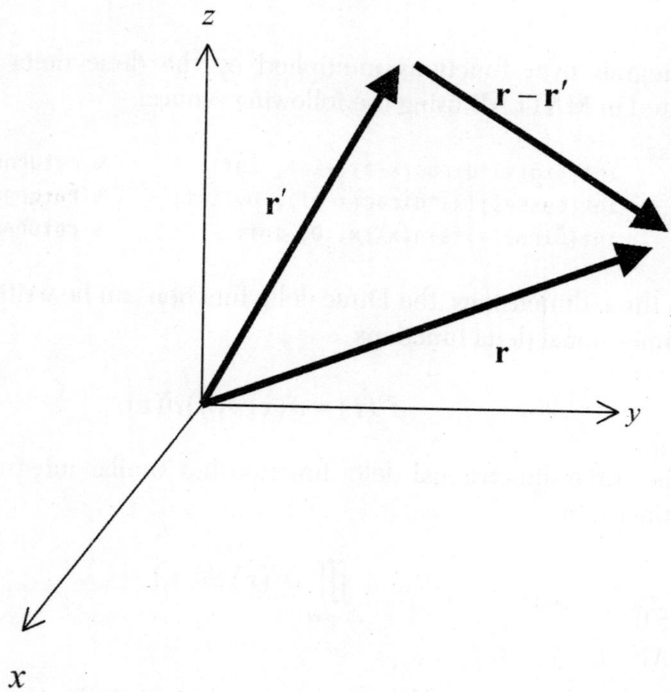

FIGURE 1.5 **Displacement vector r – r′ in three dimensions.**

1.8 SPECIAL FUNCTIONS

Bessel Functions

Bessel's differential equation is given by

$$\frac{d^2y}{dx^2} + \frac{1}{x}\frac{dy}{dx} + \left(1 - \frac{n^2}{x^2}\right)y = 0, \tag{1.92}$$

which has solutions

$$y(x) = c_1 J_n(x) + c_2 Y_n(x), \tag{1.93}$$

where $J_n(x)$ are the Bessel functions and $Y_n(x)$ are Bessel functions of the second kind. The Hankel functions $H_1^{(1)}(x) = J_n(x) + iY_n(x)$ and $H_1^{(2)}(x) = J_n(x) - iY_n(x)$ are also linearly independent solutions to Bessel's differential equation.

The following MATLAB code solves Bessel's differential equation:

```
dsolve('D2y + D1y/x + (1+n^2/x^2)*y=0', 'x')
pretty(ans)          % enhances the output format
```

Bessel and Hankel functions may be evaluated in MATLAB using the following syntax:

```
besselj(x) % Bessel function
bessely(x) % Bessel function of the second kind
besselh(x) % Hankel function
```

Modified Bessel Functions

The differential equation

$$\frac{d^2y}{dx^2} + \frac{1}{x}\frac{dy}{dx} - \left(1 + \frac{n^2}{x^2}\right)y = 0 \tag{1.94}$$

has solutions

$$y(x) = c_1 I_n(x) + c_2 K_n(x), \tag{1.95}$$

where $I_n(x)$ are modified Bessel functions and $K_n(x)$ are modified Bessel functions of the second kind. The modified Bessel functions may be evaluated in MATLAB using the following syntax:

```
besseli(x) % modified Bessel function
besselk(x) % modified Bessel function of the second kind
```

Airy Functions

Airy's differential equation is given by

$$\frac{d^2y}{dx^2} - xy = 0 \tag{1.96}$$

with solutions

$$y(x) = c_1 Ai(x) + c_2 Bi(x), \tag{1.97}$$

where $Ai(x)$ and $Bi(x)$ are Airy functions of the first and second kind, respectively. The following MATLAB code solves Airy's differential equation:

```
dsolve('D2y-x*y=0', 'x')
pretty(ans)
```

Airy functions may also be expressed in MATLAB using the following syntax:

```
airy(x)   % airy function of the first kind
airy(0,x) % same as airy(x)
airy(1,x) % the derivative of airy(x) w.r.t. x
airy(2,x) % the airy function of the second kind
airy(3,x) % the derivative of airy(2,x) w.r.t. x
```

REFERENCES

H. M. Schey. (2005). *Div, Grad, Curl, and All That: An Informal Text on Vector Calculus, 4th edition*, W. W. Norton and Company, New York.

D. J. Griffiths. (1999). *Introduction to Electrodynamics, 3rd edition*, Prentice-Hall.

P. Moon and D. E. Spencer. (2003). *Field Theory Handbook, 2nd edition*, John T. Zubal, Inc.

EXERCISES

1.1 Given the vectors $\mathbf{A} = 2\mathbf{i} + \mathbf{j} - 3\mathbf{k}$, $\mathbf{B} = 3\mathbf{i} - 2\mathbf{j} + \mathbf{k}$ and $\mathbf{C} = \mathbf{i} + 2\mathbf{j} + \mathbf{k}$ use MATLAB to calculate:
(a) $\mathbf{A} - 2\mathbf{B} + \mathbf{C}$
(b) the angle between $\mathbf{A}$ and $\mathbf{C}$
(c) the magnitude $|\mathbf{A} + \mathbf{C}|$
(d) the dot product $\mathbf{A} \cdot (\mathbf{B} + \mathbf{C})$
(e) the cross product $\mathbf{A} \times (\mathbf{B} + \mathbf{C})$
(f) the vector product $\mathbf{A} \cdot (\mathbf{A} \times \mathbf{C})$
(g) the triple cross product $\mathbf{A} \times \mathbf{B} \times \mathbf{C}$.

1.2 Calculate the gradient of the scalar field $S(r,z)$ in cylindrical coordinates (r,z,ϕ) with axial symmetry.

1.3 Calculate the divergence and the curl of the vector field $\mathbf{A} = A_\phi(r,z)\hat{\phi}$ in cylindrical coordinates (r,z,ϕ) with axial symmetry.

1.4 Calculate the gradient of the scalar field $S(\alpha,\beta)$ in bicylindrical coordinates (α,β,z) with planar symmetry.

1.5 Calculate the divergence and the curl of the vector field $\mathbf{A} = A_z(\alpha,\beta)\hat{\mathbf{z}}$ in bicylindrical coordinates (α,β,z) with planar symmetry.

1.6 Calculate the gradient of the scalar field $S(\alpha,\beta)$ in elliptic cylindrical coordinates (α,β,z) with planar symmetry.

1.7 Calculate the divergence and the curl of the vector field $\mathbf{A} = A_z(\alpha,\beta)\hat{\mathbf{z}}$ in elliptic cylindrical coordinates (α,β,z) with planar symmetry.

1.8 Calculate the gradient of the scalar field $S(r,\theta)$ in spherical coordinates (r,θ,ϕ) with axial symmetry.

1.9 Calculate the divergence and the curl of the vector field $\mathbf{A} = A_\phi(r,\theta)\hat{\phi}$ in spherical coordinates (r,θ,ϕ) with axial symmetry.

1.10 Calculate the gradient of the scalar field $S(\xi,\eta)$ in oblate spheroidal coordinates (ξ,η,ϕ) with axial symmetry.

1.11 Calculate the divergence and the curl of the vector field $\mathbf{A} = A_\phi(\xi,\eta)\hat{\phi}$ in oblate spheroidal coordinates (ξ,η,ϕ) with axial symmetry.

1.12 Calculate the gradient of the scalar field $S(\xi,\eta)$ in prolate spheroidal coordinates (ξ,η,ϕ) with axial symmetry.

1.13 Calculate the divergence and the curl of the vector field $\mathbf{A} = A_\phi(\xi,\eta)\hat{\phi}$ in prolate spheroidal coordinates (ξ,η,ϕ) with axial symmetry.

1.14 Calculate the gradient of the scalar field $S(\alpha,\beta)$ in toroidal coordinates (α,β,ϕ) with axial symmetry.

1.15 Calculate the divergence and the curl of the vector field $\mathbf{A} = A_\phi(\alpha,\beta)\hat{\phi}$ in toroidal coordinates (α,β,ϕ) with axial symmetry.

1.16 Calculate the vector Laplacian of the field $\mathbf{A} = A_\phi(r,\phi)\hat{\phi}$ in cylindrical coordinates with axial symmetry.

1.17 Calculate the vector Laplacian of the field $\mathbf{A} = A_z(\alpha,\beta)\hat{\mathbf{z}}$ in bicylindrical coordinates (α,β,z) with planar symmetry.

1.18 Calculate the vector Laplacian of the field $\mathbf{A} = A_z(\alpha,\beta)\hat{\mathbf{z}}$ in elliptic cylindrical coordinates (α,β,z) with planar symmetry.

1.19 Calculate the vector Laplacian of the field $\mathbf{A} = A_\phi(r,\theta)\hat{\phi}$ in spherical coordinates with axial symmetry.

1.20 Calculate the vector Laplacian of the field $\mathbf{A} = A_\phi(\xi,\eta)\hat{\phi}$ in oblate spheroidal coordinates (ξ,η,ϕ) with axial symmetry.

1.21 Calculate the vector Laplacian of the field $\mathbf{A} = A_\phi(\xi,\eta)\hat{\phi}$ in prolate spheroidal coordinates (ξ,η,ϕ) with axial symmetry.

1.22 Calculate the vector Laplacian of the field $\mathbf{A} = A_\phi(\alpha,\beta)\hat{\phi}$ in toroidal coordinates (α,β,ϕ) with axial symmetry.

1.23 Use integration by parts to show

$$x\frac{d}{dx}\delta(x) = -\delta(x).$$

1.24 Evaluate the following integrals involving the Dirac delta function

(a) $\int_{-\infty}^{\infty} \exp(i\pi \sin ax)\delta\left(x - \frac{\pi}{2a}\right)dx$

(b) $\int_{0}^{2\pi} \ln\cosh\left(\frac{x-1}{x+1}\right)\delta(x-\pi)\,dx$

(c) $\int_{-\infty}^{\infty} \sin(x)\delta'(x-a)\,dx$

1.25 Calculate the following Laplace transforms in MATLAB

(a) $L[\sin(\omega t)]$

(b) $L[\sinh(at)]$

(c) $L\left[e^{-\lambda t}\sin(\omega t)\right]$.

1.26 Calculate the following inverse Laplace transforms in MATLAB

(a) $L^{-1}\left[\frac{1}{\sqrt{s-a}+b}\right]$

(b) $L^{-1}\left[\frac{1}{\sqrt{s}(s-a)}\right]$

(c) $L^{-1}\left[\frac{1-e^{-a\sqrt{s}}}{s}\right]$

Note that the above inverse Laplace transforms may be written in terms of the error functions or complementary error functions.

1.27 Calculate the following inverse Laplace transforms in MATLAB

(a) $L^{-1}\left[\dfrac{1}{\sqrt{s^2+a^2}}\right]$

(b) $L^{-1}\left[\dfrac{s^2}{\left(s^2+a^2\right)^{3/2}}\right]$

(c) $L^{-1}\left[\dfrac{e^{-a/s}}{s}\right]$.

Note that the above inverse Laplace transforms may be written in terms of Bessel functions.

1.28 Parabolic cylindrical coordinates describe confocal (sharing the same focal point) parabolas with symmetry about the x-axis. In terms of the Cartesian coordinates

$$x = \frac{1}{2}\left(u^2 - v^2\right)$$
$$y = uv$$
$$z = z.$$

Use MATLAB to plot the projection of parabolic cylindrical surfaces in the x-y plane.

Show that the Laplacian is written as

$$\nabla^2 S = \frac{1}{u^2+v^2}\left(\frac{\partial^2 S}{\partial u^2}+\frac{\partial^2 S}{\partial v^2}\right)+\frac{\partial^2 S}{\partial z^2}.$$

1.29 Paraboloidal coordinate surfaces describe confocal paraboloids of revolution about the z-axis. In terms of the Cartesian coordinates

$$y = uv\sin\phi$$
$$x = uv\sin\phi$$
$$z = \frac{1}{2}\left(u^2 - v^2\right).$$

Use MATLAB to plot the projection of parabolic cylindrical surfaces in the x-z plane.

Show that the Laplacian is written as

$$\nabla^2 S = \frac{1}{u\left(u^2+v^2\right)}\frac{\partial}{\partial u}\left(u\frac{\partial S}{\partial u}\right)+\frac{1}{v\left(u^2+v^2\right)}\frac{\partial}{\partial v}\left(v\frac{\partial S}{\partial v}\right)+\frac{1}{u^2v^2}\frac{\partial^2 S}{\partial \phi^2}.$$

Chapter 2

Solutions to Laplace's Equation

In This Chapter

- Separation of Variables
- Conformal Mapping
- Finite Difference Methods
- Monte Carlo Methods
- Finite Element Method

2.1 Separation of Variables

Cartesian Coordinates

In Cartesian coordinates, Laplace's equation $\nabla^2 S(x,y,z) = 0$ is written as

$$\frac{\partial^2 S}{\partial x^2} + \frac{\partial^2 S}{\partial y^2} + \frac{\partial^2 S}{\partial z^2} = 0, \tag{2.1}$$

where $-\infty < x < \infty$, $-\infty < y < \infty$, $-\infty < z < \infty$. Applying separation of variables we assume a product solution

$$S(x,y,z) = X(x)Y(y)Z(z). \tag{2.2}$$

Substituting this form of S into Laplace's equation and dividing by S gives

$$\frac{1}{X}\frac{\partial^2 X}{\partial x^2} + \frac{1}{Y}\frac{\partial^2 Y}{\partial y^2} + \frac{1}{Z}\frac{\partial^2 Z}{\partial z^2} = 0. \tag{2.3}$$

Each term in the above equation must be equal to a constant if the sum is zero for all x, y, and z, since these variables may vary independently

$$\frac{1}{X}\frac{\partial^2 X}{\partial x^2} = \alpha^2 \quad \frac{1}{Y}\frac{\partial^2 Y}{\partial y^2} = \beta^2 \quad \frac{1}{Z}\frac{\partial^2 Z}{\partial z^2} = -\gamma^2, \tag{2.4}$$

where $\alpha^2 + \beta^2 = \gamma^2$. The solutions to these differential equations may be written as linear combinations of sine, cosine, and exponential functions so that we may construct a product solution

$$S(x,y,z) = \begin{Bmatrix} \sin(\alpha x) \\ \cos(\alpha x) \end{Bmatrix} \begin{Bmatrix} \sin(\beta y) \\ \cos(\beta y) \end{Bmatrix} \begin{Bmatrix} \exp(\gamma z) \\ \exp(-\gamma z) \end{Bmatrix} \tag{2.5}$$

with constants determined by the boundary conditions.

As an example of the application of separation of variables, consider a unit square region with boundary condition $S = 0$ on three sides and $S = 1$ on the other side. Since there is no z dependence $\alpha^2 = -\beta^2$ equation (2.5) become

$$S(x,y) = \{A\sin(\alpha x) + B\cos(\alpha x)\}\{C\sinh(\alpha y) + D\cosh(\alpha y)\}. \tag{2.6}$$

Applying the boundary conditions $S(0,y) = S(1,y) = 0$, where the sine function is zero given $\alpha = n\pi$ and relabeling the constants and writing the solution as a sum

$$S(x,y) = \sum_{n=0}^{\infty}\left(A_n \cosh(n\pi y) + B_n \sinh(n\pi y)\right)\sin(n\pi x). \tag{2.7}$$

The condition $S(x,1) = 0$ gives $B_n = -A_n \cosh(n\pi)/\sinh(n\pi)$ and

$$S(x,y) = \sum_{n=0}^{\infty} A_n\left(\cosh(n\pi y) - \frac{\cosh(n\pi)}{\sinh(n\pi)}\sinh(n\pi y)\right)\sin(n\pi x). \tag{2.8}$$

Finally, A_n is determined by applying the boundary condition at $S(x,1) = 1$

$$1 = \sum_{n=0}^{\infty} A_n \sin(n\pi x). \tag{2.9}$$

Multiplying both sides of this expression by $\sin(n'\pi x)$ and integrating

$$\int_0^1 \sin(n'\pi x)dx = \int_0^1 \sum_{n=0}^{\infty} A_n \sin(n\pi x)\sin(n'\pi x)dx. \tag{2.10}$$

Making use of orthogonality, where only terms with $n = n'$ remain, and using the identity $\sin^2(n\pi) = (1 - \cos 2n\pi)/2$, we have

$$\frac{\left(1 - \cos(n\pi)\right)}{n\pi} = \frac{A_n}{2}, \tag{2.11}$$

where A_n is zero for even n so the sum is over odd n

$$S(x,y) = \frac{4}{\pi}\sum_{n=1,3,5\ldots}^{\infty} \frac{1}{n}\left(\cosh(n\pi y) - \frac{\cosh(n\pi)}{\sinh(n\pi)}\sinh(n\pi y)\right)\sin(n\pi x). \tag{2.12}$$

The following MATLAB program calculates equation (2.12) and plots the solution over the unit square:

```
% set up a 20x20 spatial grid
imax=20;
jmax=20;

% set the upper limit if the sum
nmax=201;

% initialize the S array
S=zeros(imax,jmax);

% spatial loop over the grid
for i=1:imax
    for j=1:jmax

    x=(i-1)/(imax-1);
    y=(j-1)/(jmax-1);

% spatial loop over the grid
for n=1:2:nmax

    G1=4*sin(n*pi*x)/(n*pi);
    G2=cosh(n*pi*y)-sinh(n*pi*y)*(cosh(n*pi)/sinh(n*pi));
    S(i,j)=S(i,j)+ G1*G2;

end
end
end

%create a mesh plot of the solution
mesh(S)
```

In the following coordinate systems we consider Laplace's equation with axial symmetry and no variation in the ϕ-direction, or with planar symmetry and no variation in the z-direction. These results will be useful for comparison to QuickField problems that have either axial or x-y symmetry.

Cylindrical Coordinates

In cylindrical coordinates with axial symmetry, Laplace's equation $\nabla^2 S(r,z) = 0$ is written as

$$\frac{1}{r}\frac{\partial}{\partial r}\left(r\frac{\partial S}{\partial r}\right) + \frac{\partial^2 S}{\partial z^2} = 0, \tag{2.13}$$

where $r \geq 0, -\infty < z < \infty$. Substituting $S(r,z) = R(r)Z(z)$ with separation constant k^2 gives the differential equations

$$\frac{1}{r}\frac{\partial}{\partial r}\left(r\frac{\partial R}{\partial r}\right) = -k^2 \mathrm{R} \qquad \frac{\partial^2 Z}{\partial z^2} = k^2 Z \tag{2.14}$$

so that we may construct our solution

$$S(r,z) = \begin{Bmatrix} J_0(kr) \\ N_0(kr) \end{Bmatrix} \begin{Bmatrix} \exp(kz) \\ \exp(-kz) \end{Bmatrix}, \tag{2.15}$$

where $J_0(kr)$ and $N_0(kr)$ are Bessel functions of zero order. Laplace's equation in cylindrical coordinates without variation in the z-direction $\nabla^2 S(r,\phi) = 0$ is written as

$$\frac{1}{r}\frac{\partial}{\partial r}\left(r\frac{\partial S}{\partial r}\right) + \frac{1}{r^2}\frac{\partial^2 S}{\partial \phi^2} = 0, \tag{2.16}$$

where $r \geq 0, 0 < \phi \leq 2\pi$. Substituting $S(r,\phi) = R(r)\Phi(\phi)$

$$S(r,\phi) = \begin{Bmatrix} r^n \\ r^{-n} \end{Bmatrix} \begin{Bmatrix} \cos(n\phi) \\ \sin(n\phi) \end{Bmatrix} \tag{2.17}$$

with terms that may include $a_0 \ln(r) + b_0\phi + c_0$, where a_0, b_0 and c_0 are constants.

Bicylindrical Coordinates

The Laplacian $\nabla^2 S(\alpha,\beta)$ is written in bicylindrical coordinates

$$\frac{\partial^2 S}{\partial \alpha^2} + \frac{\partial^2 S}{\partial \beta^2} = 0, \tag{2.18}$$

where $-\infty < \alpha < \infty$, $0 \le \beta \le 2\pi$. Substituting $S(\alpha,\beta) = \Lambda(\alpha)\Omega(\beta)$ with separation constant p^2 gives

$$S(\alpha,\beta) = \begin{Bmatrix} \exp(p\alpha) \\ \exp(-p\alpha) \end{Bmatrix} \begin{Bmatrix} \sin(p\beta) \\ \cos(p\beta) \end{Bmatrix}. \tag{2.19}$$

Elliptic Cylindrical Coordinates

The Laplacian $\nabla^2 S(\alpha,\beta) = 0$ is written in elliptic cylindrical coordinates

$$\frac{\partial^2 S}{\partial \alpha^2} + \frac{\partial^2 S}{\partial \beta^2} = 0, \tag{2.20}$$

where $0 \le \alpha$, $0 \le \beta \le 2\pi$. Substituting $S(\alpha,\beta) = \Lambda(\alpha)\Omega(\beta)$ with separation constant p^2 gives

$$S(\alpha,\beta) = \begin{Bmatrix} \exp(p\alpha) \\ \exp(-p\alpha) \end{Bmatrix} \begin{Bmatrix} \sin(p\beta) \\ \cos(p\beta) \end{Bmatrix}. \tag{2.21}$$

Spherical Coordinates

In spherical coordinates with axial symmetry, Laplace's equation $\nabla^2 S(r,\theta) = 0$ is written as

$$\frac{\partial}{\partial r}\left(r^2 \frac{\partial S}{\partial r}\right) + \frac{1}{\sin\theta}\frac{\partial}{\partial \theta}\left(\sin\theta \frac{\partial S}{\partial \theta}\right) = 0. \tag{2.22}$$

Substituting $S(r,\theta) = R(r)\Theta(\theta)$ with separation constant $\ell(\ell+1)$ we obtain

$$S(r,\theta) = \begin{Bmatrix} r^{\ell} \\ r^{-(\ell+1)} \end{Bmatrix} \begin{Bmatrix} P_\ell(\cos\theta) \\ Q_\ell(\cos\theta) \end{Bmatrix}, \tag{2.23}$$

where P_ℓ and Q_ℓ are Legendre functions of the first and second kind, respectively.

As an example of a spherically symmetric solution to (2.22) consider a spherical surface of radius R with boundary conditions $S(R,\theta) = V_0$ and $S(r\to\infty,\theta) = 0$. Since there is no θ dependence we take $\ell = 0$ so that (2.23) becomes

$$S(r) = A + \frac{B}{r}. \tag{2.24}$$

The boundary condition at infinity gives $A = 0$. The boundary condition at $r = R$ gives $B = RV_0$ and we have

$$S(r) = \frac{V_0 R}{r}. \tag{2.25}$$

Note that the same result may be obtained by setting the radial part of (2.22) equal to zero and integrating twice.

Prolate Spheroidal Coordinates

Laplace's equation in prolate spheroidal coordinates with axial symmetry $\nabla^2 S(\xi,\eta) = 0$ is written as

$$\frac{1}{\sinh\xi}\frac{\partial}{\partial\xi}\left(\sinh\xi\frac{\partial S}{\partial\xi}\right) + \frac{1}{\sin\eta}\frac{\partial}{\partial\eta}\left(\sin\eta\frac{\partial S}{\partial\eta}\right) = 0, \tag{2.26}$$

where $\xi \geq 0$, $0 \leq \eta \leq \pi$. Substituting $S(\xi,\eta) = \Upsilon(\xi)\Theta(\eta)$ with separation constant $\ell(\ell + 1)$ gives

$$S(\xi,\eta) = \begin{Bmatrix} P_\ell(\cosh\xi) \\ Q_\ell(\cosh\xi) \end{Bmatrix} \begin{Bmatrix} P_\ell(\cos\eta) \\ Q_\ell(\cos\eta) \end{Bmatrix}. \tag{2.27}$$

We now consider a football shaped prolate spheroidal surface $\xi = \xi_0$ with potential V_0. There is no η dependence so that (2.27) becomes

$$S(\xi) = AP_0(\cosh\xi) + BQ_0(\cosh\xi), \tag{2.28}$$

where $P_0(x) = 1$. Applying the boundary conditions $S(\xi\to\infty) = 0$ and $S(\xi\to\xi_0) = V_0$ we have

$$S(\xi) = V_0\frac{Q_0(\cosh\xi)}{Q_0(\cosh\xi_0)}. \tag{2.29}$$

This is equivalent to

$$S(\xi) = V_0 \frac{\ln \tanh\left(\dfrac{\xi}{2}\right)}{\ln \tanh\left(\dfrac{\xi_0}{2}\right)}, \tag{2.30}$$

which may also be obtained by setting the first term of (2.26) equal to zero and integrating twice. Fortified by this simple result it might be tempting to find the potential between two opposing electrodes in the shape of hyperboloids of revolution by setting the second term in (2.26) equal to zero and integrating twice. However, this procedure would not yield a solution over the region between the electrodes, where ξ is varying. Instead we must use (2.27)

$$S(\xi,\eta) = \sum_{\ell=0}^{\infty} P_\ell(\cosh\xi)\left\{A_\ell P_\ell(\cos\eta) + B_\ell Q_\ell(\cos\eta)\right\}, \tag{2.31}$$

where the $Q_\ell(\cosh\ \xi)$ are divergent at $\xi = 0$ and are discarded. The constants A_ℓ and B_ℓ are determined by boundary conditions at $\eta = \eta_0$, making use of the orthogonality of $P_\ell(\cosh\ \eta)$ and $Q_\ell(\cosh\ \eta)$.

Oblate Spheroidal Coordinates

Laplace's equation in oblate spheroidal coordinates with axial symmetry $\nabla^2 S(\xi,\eta) = 0$ is written as

$$\frac{1}{\cosh\xi}\frac{\partial}{\partial\xi}\left(\cosh\xi\frac{\partial S}{\partial\xi}\right) + \frac{1}{\cos\eta}\frac{\partial}{\partial\eta}\left(\cos\eta\frac{\partial S}{\partial\eta}\right) = 0, \tag{2.32}$$

where $\xi \geq 0$, $0 \leq \eta \leq \pi$. Substituting $S(\xi,\eta) = \Upsilon(\xi)\Theta(\eta)$ with separation constant $\ell(\ell+1)$ gives

$$S(\xi,\eta) = \begin{Bmatrix} P_\ell(i\sinh\xi) \\ Q_\ell(i\sinh\xi) \end{Bmatrix} \begin{Bmatrix} P_\ell(\cos\eta) \\ Q_\ell(\cos\eta) \end{Bmatrix}. \tag{2.33}$$

Toroidal Coordinates

The Laplacian $\nabla^2 S(\alpha,\beta)$ is written in toroidal coordinates as

$$\frac{\partial}{\partial\alpha}\left(\frac{\sinh\alpha}{\cosh\alpha - \cos\beta}\frac{\partial S}{\partial\alpha}\right) + \frac{\partial}{\partial\beta}\left(\frac{\sinh\alpha}{\cosh\alpha - \cos\beta}\frac{\partial S}{\partial\beta}\right) = 0, \tag{2.34}$$

where $\alpha \geq 0, -\pi \leq \beta < \pi$. Substituting $S(\alpha,\beta) = \Upsilon(\alpha)\Theta(\beta)$ with separation constant υ^2 gives

$$S(\alpha,\beta) = \begin{Bmatrix} P_{\upsilon-1/2}(\cosh\alpha) \\ Q_{\upsilon-1/2}(\cosh\alpha) \end{Bmatrix} \begin{Bmatrix} \cos\nu\beta \\ \sin\nu\beta \end{Bmatrix}. \tag{2.35}$$

2.2 CONFORMAL MAPPING

The conformal mapping method makes use of a property of functions of a complex variable that enables a solution to Laplace's equation in two dimensions. Given a complex variable $z = x + iy$, we may express a function of the variable $f(z)$ as a sum of real and imaginary parts

$$f(z) = u(x,y) + iv(x,y), \tag{2.36}$$

where $u = \mathrm{Re}(f)$ and $v = \mathrm{Im}(f)$. The transformation $w = f(z)$ is conformal over a region if $f'(z) \neq 0$. If $f(z)$ is analytic over a region of the complex plane then u and v satisfy the Cauchy-Riemann equations

$$\begin{aligned} \frac{\partial u}{\partial x} &= \frac{\partial v}{\partial y} \\ \frac{\partial v}{\partial x} &= -\frac{\partial u}{\partial y}. \end{aligned} \tag{2.37}$$

If follows that both u and v satisfy Laplace's equation

$$\frac{\partial^2}{\partial x^2}\begin{pmatrix} u \\ v \end{pmatrix} + \frac{\partial^2}{\partial y^2}\begin{pmatrix} u \\ v \end{pmatrix} = 0. \tag{2.38}$$

Furthermore, contours of constant u are perpendicular to contours of constant v. This feature gives us the added bonus of finding both equipotentials and electric field lines in electrostatic problems. If we can find a function that satisfies both Laplace's equation and the specified boundary conditions, the uniqueness theorem tells us that we have found the only solution to Laplace's equation over the region with specific boundary conditions.

If a scalar function $S(x,y)$ satisfies Laplace's equation in the x-y plane then

$$\frac{\partial^2 S}{\partial x^2} + \frac{\partial^2 S}{\partial y^2} = 0. \tag{2.39}$$

Color images for many of the figures may be found on the CD-ROM.

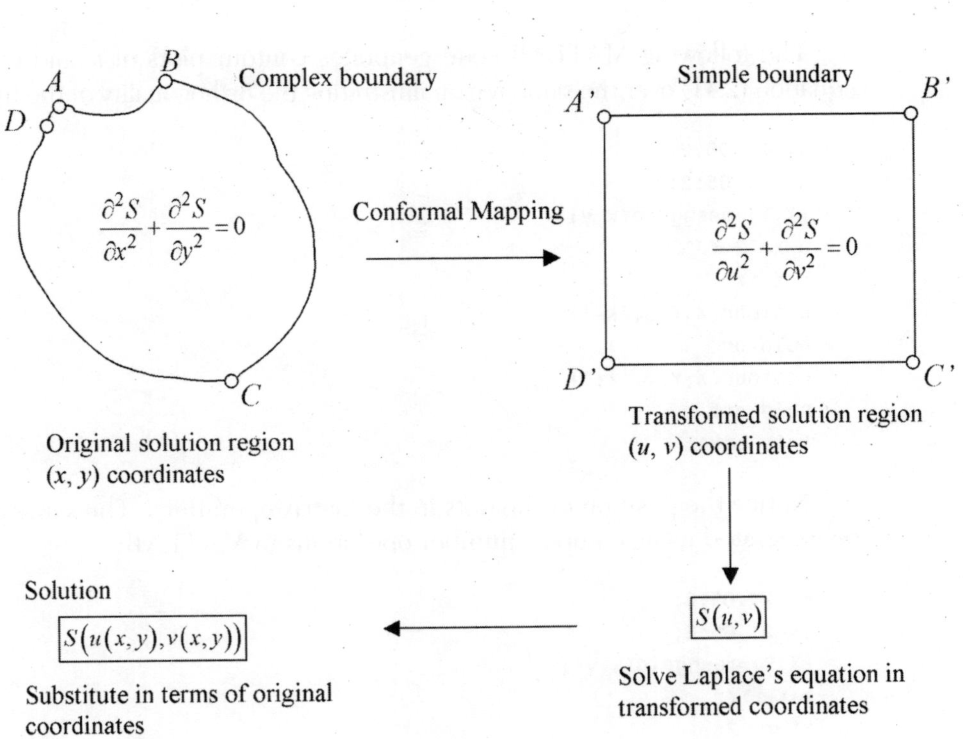

FIGURE 2.1 **Procedure for transforming boundaries using conformal mapping for the solution of Laplace's equation.**

A transform that maps a region of the z-plane to a region in the w-plane preserves the form of Laplace's equations written in terms of the new coordinates u and v

$$\frac{\partial^2 S}{\partial u^2}+\frac{\partial^2 S}{\partial v^2}=0. \tag{2.40}$$

In this way we may transform the original complicated boundary into a simple one as shown schematically in Figure 2.1. Once $S(u,v)$ is determined from (2.40), then $S(x,y)$is found by substituting $S(u(x,y),\, v(x,y))$.

As an example, consider the transformation $w=f(z)=z^2$. This transformation maps the first quadrant of the z-plane to the upper half of the w-plane. In terms of the original z-coordinates, the w components may be written as

$$\begin{aligned} u &= \mathrm{Re}(w) = x^2 - y^2 \\ v &= \mathrm{Im}(w) = 2xy, \end{aligned} \tag{2.41}$$

where each of the functions satisfies Laplace's equation.

The following MATLAB code generates contour plots of u and v given by equation (2.41) over the same region illustrating the orthogonality of the functions:

```
x=-2:.05:2;
y=-2:.05:2;
[X,Y]=meshgrid(x,y);
u=X.^2-Y.^2;
v=X.*Y;
contour(X,Y,u,'k-')
hold on
contour(X,Y,v,'k:')
hold off
axis equal
```

Notice the position of the dots in the matrix operations. The same plots can be generated using complex number operations in MATLAB:

```
x=-2:.05:2;
y=-2:.05:2;
[X,Y]=meshgrid(x,y);
z=X+i*Y;
z1=z.^2;
u=real(z1);
v=imag(z1);
contour(X,Y,u,'k-')
hold on
contour(X,Y,v,'k:')
hold off
axis equal
```

The output of the above code is shown in Figure 2.2. Another example of a complex potential satisfying Laplace's equation is that of a dipole described by $S(z) = \mu/z$. The following code generates a dipole field with moment $\mu = 1$:

```
x=-2:.01:2;
y=-2:.01:2;
[X,Y]=meshgrid(x,y);
z=X+i*Y;
z1=z.\1; %note the left-hand division here
u=real(z1);
v=imag(z1);
contour(X,Y,u,'k-')
hold on
contour(X,Y,v,'k:')
hold off
axis equal
```

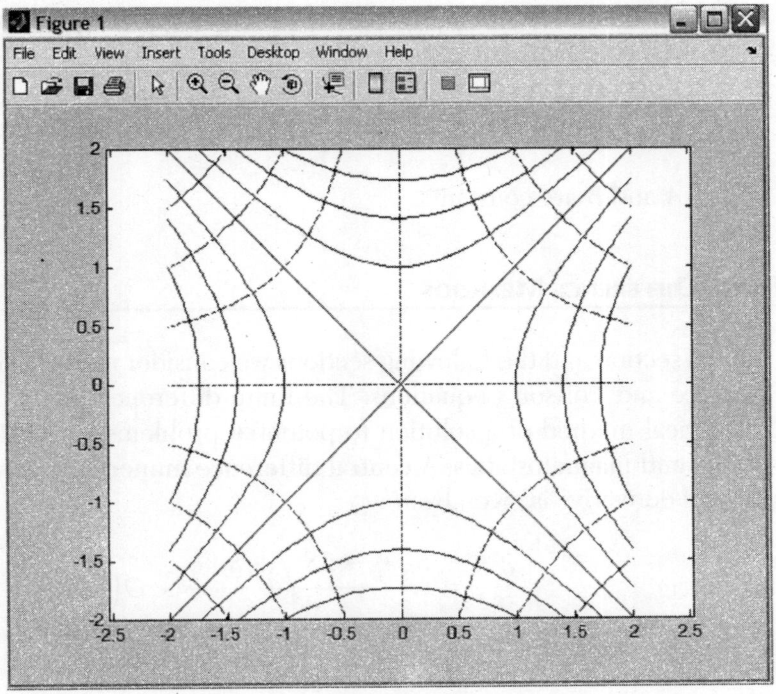

FIGURE 2.2 **Contour map of the real and imaginary parts of the complex potential z^2.**

Complex regions in the w- or z-plane may often be transformed to the half-plane above the x- or u-axis or on to a unit circle. Appendix 3 gives a table of conformal mapping on the half-space and the unit circle. Poisson's integral formulas may then be used to obtain the solution to Laplace's equation. If a scalar field is specified everywhere along the x-axis $S(x,0) = G(x')$ then

$$S(x,y) = \frac{1}{\pi}\int_{-\infty}^{\infty}\frac{yG(x')dx'}{y^2 + (x - x')^2}. \tag{2.42}$$

Poisson's formula for a unit circle gives the scalar field inside a circular region with boundary condition $S(1,\theta) = F(\theta')$

$$S(r,\theta) = \frac{1}{2\pi}\int_{0}^{2\pi}\frac{(1 - r^2)F(\theta')d\theta'}{1-2r\cos(\theta - \theta') + r^2}. \tag{2.43}$$

The Schwarz-Christoffel transformation

$$\frac{dw}{dz} = A(z - x_1)^{\alpha_1/\pi-1}(z - x_2)^{\alpha_2/\pi-1} \cdots (z - x_n)^{\alpha_n/\pi-1} \tag{2.44}$$

maps the boundaries of polygons with interior angles α_j in the w-plane to the x-axis of the z-plane. Integrating this expression gives

$$w = A\int (z - x_1)^{\alpha_1/\pi - 1} (z - x_2)^{\alpha_2/\pi - 1} \cdots (z - x_n)^{\alpha_n/\pi - 1} dz + B, \quad (2.45)$$

where A and B are constants.

2.3 FINITE DIFFERENCE METHODS

In this section and the following sections we consider numerical solutions to the Laplace and Poisson's equations. The finite difference method is the simplest numerical method of a solution to potential problems encountered in electrostatics and magnetostatics. A central difference numerical approximation to the second derivative is given by

$$\frac{\partial^2 S}{\partial x^2} = \frac{S_{i+1,j} - 2S_{i,j} + S_{i-1,j}}{\Delta x^2} + O(\Delta x^2) \quad (2.46)$$

and

$$\frac{\partial^2 S}{\partial y^2} = \frac{S_{i,j+1} - 2S_{i,j} + S_{i,j-1}}{\Delta y^2} + O(\Delta y^2), \quad (2.47)$$

where (i, j) are indices corresponding the x- and y-axes, respectively. If we let $\Delta x = \Delta y = \Delta$, we can write the discrete form of the Laplacian in two dimensions

$$\frac{\partial^2 S}{\partial x^2} + \frac{\partial^2 S}{\partial y^2} = \frac{S_{i+1,j} + S_{i-1,j} + S_{i,j+1} + S_{i,j-1} - 4S_{i,j}}{\Delta^2}. \quad (2.48)$$

Laplace's equation gives us $S_{i,j}$ as the average of nearest neighbors

$$S_{i,j} = \frac{S_{i+1,j} + S_{i-1,j} + S_{i,j+1} + S_{i,j-1}}{4}. \quad (2.49)$$

Boundary conditions are typically applied to the sides of the solution region as well as on inner boundaries. The Dirichlet condition assigns a constant value of S. The Neumann condition specifies the normal derivative $\partial S/\partial n$. For the vanishing normal derivative (or zero normal derivative) the finite difference equation (2.49) is modified according to

$$\text{left edge: } S_{1,j} = \frac{2S_{2,j} + S_{1,j+1} + S_{1,j-1}}{4} \quad (2.50)$$

$$\text{right edge: } S_{i\max,j} = \frac{2S_{i\max-1,j} + S_{i\max,j+1} + S_{i\max,j-1}}{4} \tag{2.51}$$

$$\text{top edge: } S_{i,j\max} = \frac{S_{i+1,j\max} + S_{i-1,j\max} + 2S_{i,j\max-1}}{4} \tag{2.52}$$

$$\text{bottom edge: } S_{i,1} = \frac{S_{i+1,1} + S_{i-1,1} + 2S_{i,2}}{4}. \tag{2.53}$$

Corner grid points are assigned the average of the nearest two grid points. For example, the lower left corner $S_{1,1} = (S_{1,2} + S_{2,1})/2$.

Figure 2.3 shows the finite difference computational molecule for the calculation of $S_{i,j}$. As a simple example, consider the finite difference approximation to the scalar field at the center of the square region shown in Figure 2.4. Using this computational molecule, the scalar field is correctly calculated at the center as $(1 + 2 + 3 + 4)/4 = 5/2$. A more complex example is shown in Figure 2.5, where an L-shaped region is discretized on a rectangular finite difference grid. In solving the finite difference equations, the values of $S_{i,j}$ are specified on the boundaries and initialized to zero on interior points. Equation (2.49), or an iterative procedure such as the successive-overrelaxation

$$S_{i,j}^{k+1} = S_{i,j}^{k} + \frac{\omega}{4}\left(S_{i+1,j}^{k} + S_{i-1,j}^{k} + S_{i,j+1}^{k} + S_{i,j-1}^{k} - 4S_{i,j}^{k}\right), \tag{2.54}$$

can be applied for greater accuracy and stability, where the optimum parameter ω is between 1 and 2.

The following MATLAB code solves Laplace's equation for the scalar field S over a square region with $S = 0$ on three sides and $S = 1$ on the remaining side implementing the finite difference method.

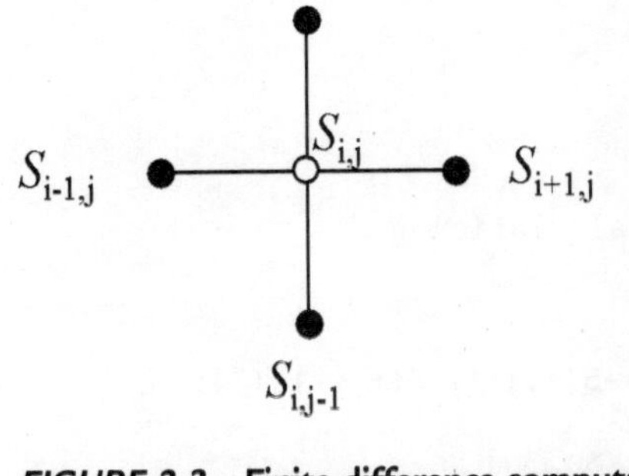

FIGURE 2.3 Finite difference computational molecules in 2D Cartesian coordinates.

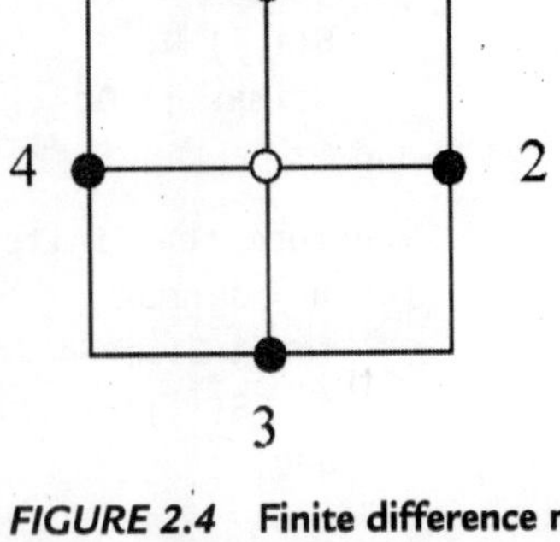

FIGURE 2.4 Finite difference mesh of a square boundary.

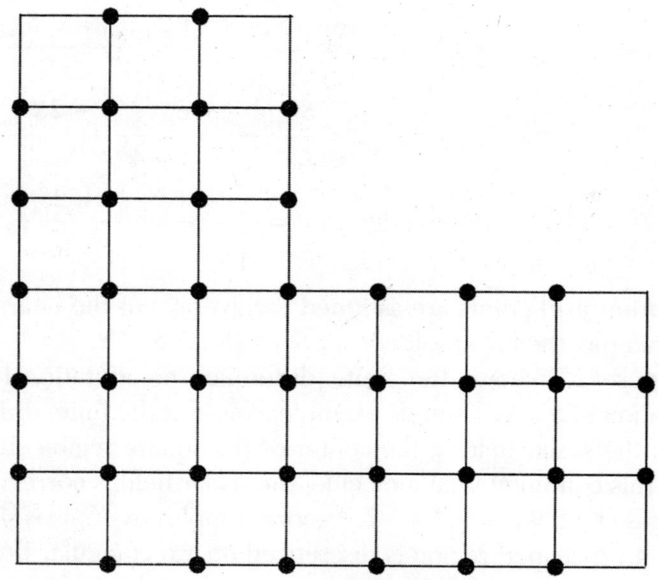

FIGURE 2.5 **Finite difference discretization of an L-shaped region.**

```
%set the number of grid points and iteration number
imax = 20;
jmax = 20;
nmax=1000;

%initialize the grid
S=zeros(imax,jmax);

%specify the boundary conditions on the top and bottom sides
for i = 1:imax
    S(i,1)=0;
    S(i,jmax)=1;
end

%specify the boundary conditions on the right and left sides
for j=2:jmax-1
    S(1,j)=0;
    S(imax,j)=0;
end

%perform the finite difference calculation
for n = 1:nmax
    for i=2:imx-1 for j=2:jmax-1
        S(i,j)=(S(i+1,j)+S(i-1,j)+S(i,j+1)+S(i,j-1))/4;
    end
    end
end
```

```
%create a surface plot of the scalar field
mesh(S)
```

Figure 2.6 shows the output of this procedure. The maximum number of iterations should be large enough so that the desired accuracy is obtained. This may be achieved by performing the finite difference loop until the maximum variation of any grid value between iterations of the outer n loop is below a specified threshold.

The finite difference solution to Poisson's equation $\nabla^2 S = -\rho$ may also be obtained, where (2.49) is modified accordingly

$$S_{i,j} = \frac{S_{i+1,j} + S_{i-1,j} + S_{i,j+1} + S_{i,j-1} + \Delta^2 \rho_{i,j}}{4}. \tag{2.55}$$

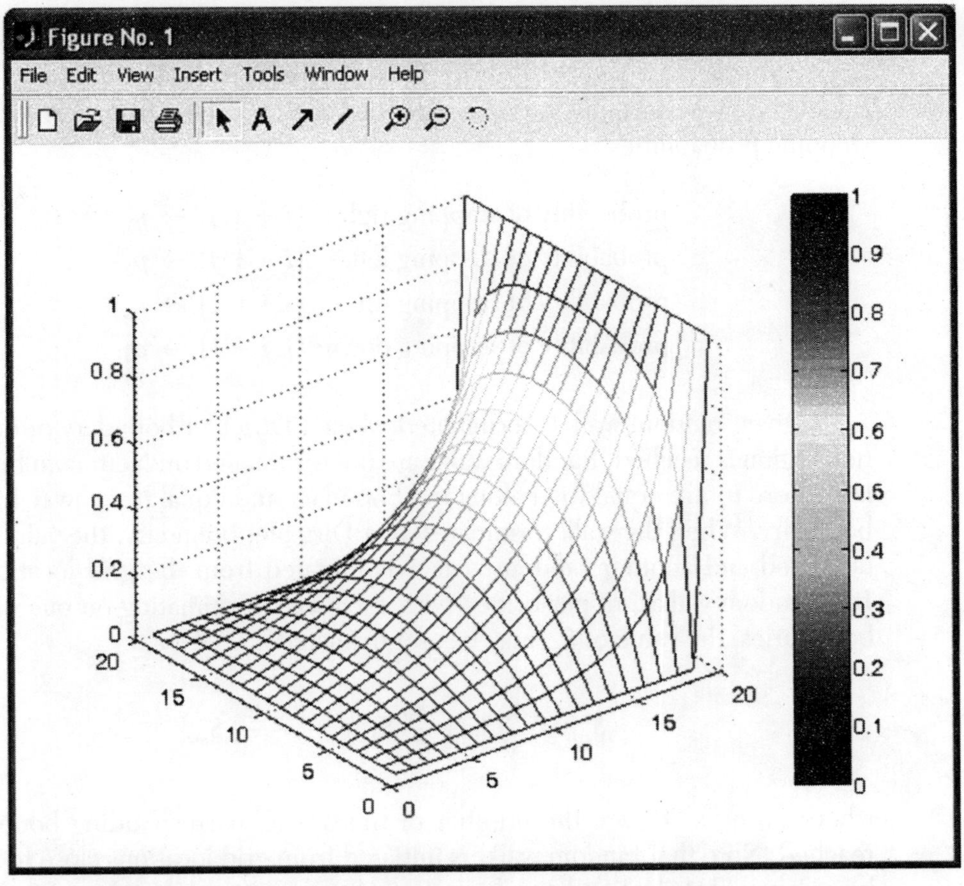

FIGURE 2.6 **Finite difference calculation of the scalar field over a square region with Dirichlet boundary conditions on each of the four sides.**

The successive over relaxation scheme can also be applied in solving Poisson's equation

$$S_{i,j}^{k+1} = S_{i,j}^{k} + \frac{\omega}{4}\left(S_{i+1,j}^{k} + S_{i-1,j}^{k} + S_{i,j+1}^{k} + S_{i,j-1}^{k} - 4S_{i,j}^{k} + \Delta^2 \rho_{i,j}\right). \quad (2.56)$$

2.4 MONTE CARLO METHOD

The Monte Carlo method can be used to calculate the scalar field satisfying Laplace's equation over a solution region with specified boundary conditions. The finite difference approximation to $S_{i,j}$ may be written as

$$p_1 S_{i+1,j} + p_2 S_{i-1,j} + p_3 S_{i,j+1} + p_4 S_{i,j-1}, \quad (2.57)$$

where the coefficients p_n are all ¼. Instead of solving the finite difference equations directly, an approximation to the scalar field at the location (i,j) may be calculated by performing multiple random walks beginning from (i,j) with equal stepping probabilities

$$\begin{aligned} &\text{probability of stepping right: } (i+1,j) \to p_1 \\ &\text{probability of stepping left: } (i-1,j) \to p_2 \\ &\text{probability of stepping up: } (i,j+1) \to p_3 \\ &\text{probability of stepping down: } (i,j-1) \to p_4. \end{aligned} \quad (2.58)$$

A given random walk is terminated when a Dirichlet boundary of the solution region is reached. Random walks are not terminated on Neumann boundaries, where p_n are equal to ¼ along the boundary and equal to ½ away from the boundary. When the walk terminates on a Dirichlet boundary, the value of S is tabulated and another random walker is initiated from the grid location (i,j). If N random walks are performed with each walk terminating on one of the m boundaries, the value of $S_{i,j}$ may be approximated

$$S_{i,j} = \frac{n_1}{N} S_1 + \frac{n_2}{N} S_2 + \cdots + \frac{n_m}{N} S_m, \quad (2.59)$$

where n_1, n_2 … n_m are the number of times each corresponding boundary is reached. Note that random walkers initiated from grid locations close to a given boundary will strike that boundary more often resulting in a value of S closer to the value of the nearby boundary.

The Monte Carlo method may be applied to Poisson's equation, where $S_{i,j}$ is now

$$p_1 S_{i+1,j} + p_2 S_{i-1,j} + p_3 S_{i,j+1} + p_4 S_{i,j-1} + \frac{\Delta^2}{4}\rho_{i,j}. \tag{2.60}$$

The stepping probabilities are the same as before except the value $\Delta^2\rho_{i,j}/4$ must be accumulated at each step for every random walk

$$S_{i,j} = \frac{n_1}{N}S_1 + \frac{n_2}{N}S_2 + \cdots + \frac{n_m}{N}S_m + \frac{1}{N}\sum_{\text{all } i,j \text{ visited}} \frac{\Delta^2}{4}\rho_{i,j}. \tag{2.61}$$

As an example, if $\rho_{i,j}$ is nonzero except at one grid location (50, 50) and if that site is visited 66 times during the course of 100 random walks, then

$$\frac{1}{N}\sum_{\text{all } i,j \text{ visited}} \frac{\Delta^2}{4}\rho_{i,j} = \frac{1}{100} \times 66 \times \frac{\Delta^2}{4}\rho_{50,50}. \tag{2.62}$$

The Monte Carlo method is often compared to a drunken walk, or a *'tour-de-wino,'* that randomly courses through the grid. In solving Poisson's equation, our random walker also picks up charged *'mud'* on his shoes as he treks through the grid.

The Monte Carlo method may be adapted for other coordinate systems. In axial symmetric cylindrical coordinates $\nabla^2 S(r,z)$ is discretized as

$$\frac{1}{r}\frac{\partial S}{\partial r} + \frac{\partial^2 S}{\partial r^2} + \frac{\partial^2 S}{\partial z^2}$$
$$\to \frac{1}{i\Delta r}\frac{S_{i,j} - S_{i-1,j}}{\Delta r} + \frac{S_{i+1,j} - 2S_{i,j} + S_{i-1,j}}{\Delta r^2} + \frac{S_{i,j+1} - 2S_{i,j} + S_{i,j-1}}{\Delta z^2}. \tag{2.63}$$

Setting the above expression equal to zero and solving for $S_{i,j}$ gives

$$\underbrace{\frac{i-1}{2i + 2i\left(\frac{\Delta r}{\Delta z}\right)^2 - 1}}_{p_1} S_{i-1,j} + \underbrace{\frac{i}{2i + 2i\left(\frac{\Delta r}{\Delta z}\right)^2 - 1}}_{p_2} S_{i+1,j}$$
$$+ \underbrace{\frac{i}{2i + 2i\left(\frac{\Delta z}{\Delta r}\right)^2 - \left(\frac{\Delta z}{\Delta r}\right)^2}}_{p_3} S_{i,j-1} + \underbrace{\frac{i}{2i + 2i\left(\frac{\Delta z}{\Delta r}\right)^2 - \left(\frac{\Delta z}{\Delta r}\right)^2}}_{p_4} S_{i,j+1} \tag{2.64}$$

The stepping probabilities are no longer equal as in the Cartesian case. However, we can see $p_1 + p_2 + p_3 + p_4 = 1$, as required.

A disadvantage of the Monte Carlo method is its greater computation time compared to other numerical techniques. One very significant advantage of the technique is that it allows the calculation of S at one or more points of interest without having to calculate the scalar field over the entire solution region. Other advantages include ease in coding, simple extension to three dimensions, and adaptability to other coordinate systems including those developed in this chapter.

The following MATLAB code calculates the scalar field at the center of a square solution region with boundary conditions S = 1, 2, 3, 4 on the bottom, right, top, and left boundaries, respectively.

```
imax=101;  % number of x grid divisions
jmax=101;  % number of y grid divisions

nmax=10000;  % number of random walks

%boundary conditions
bottom_bc=1;
right_bc=2;
top_bc=3;
left_bc=4;

S=0;        %initialize the scalar potential

for n=1:nmax

% begin the random walk at the center of the grid
i=51;
j=51;

% perform random walk until a boundary is reached
while (i~=1)&(j~=1)&(i~=imax)&(j~=jmax)
    r=rand;

    if r<0.25
        i=i+1;                          %step right
    elseif (r>=0.25)&(r<0.5)
        i=i-1;                          %step left
    elseif (r>=0.5)&(r<0.75)
        j=j+1;                          %step up
    else
        j=j-1;                          %step down
end
end
```

```
% random walker reaches a side
if (j==1)&(i~=1)&(i~=imax)
    S=S+bottom_bc;
elseif (j==jmax)&(i~=1)&(i~=imax)
    S=S+top_bc;
elseif (i==1)&(j~=1)&(j~=jmax)
    S=S+left_bc;
elseif (i==imax)&(j~=1)&(j~=jmax)
    S=S+right_bc;
end
end
S=S/nmax    %divide by the total number of random walks
```

Here, the value of S at the center of the square region should equal the average of the four walls or $(1 + 2 + 3 + 4)/4 = 2.5$. Three executions of this program output 2.4957, 2.4963, and 2.5027. When working with Monte Carlo estimates, it is customary to include error bars or standard deviations when reporting results. Standard deviations may also be reported at a few points in a larger solution region.

Note that it is not necessary to assign conditions to the corners in this procedure because the walker will never reach a corner without first terminating on a boundary. It will be possible, however, to reach corners, where Neumann boundary conditions terminate. Periodic boundary conditions may also be applied, where random walkers step from one side to the opposite side of the grid.

The Monte Carlo method may be used to calculate the scalar potential in electrostatics, the magnetic scalar potential, and components of the magnetic vector potential in magnetostatics, as well as the temperature distribution in heat transfer problems. In electrostatics with 2D x-y symmetry, if a walker is located on the boundary between two dielectric media ε_1 and ε_2, the probability of stepping into ε_1 is $\varepsilon_1/2(\varepsilon_1 + \varepsilon_2)$. The probability of stepping into ε_2 is $\varepsilon_2/2(\varepsilon_1 + \varepsilon_2)$ and the probability of stepping tangential to the surface is ¼ in either direction. In 2D x-y magnetostatics, if the particle is located at the boundary between two permeable media μ_1 and μ_2, then the probability of stepping into μ_1 is $\mu_1/(\mu_1 + \mu_2)$, the probability of stepping into μ_2 is $\mu_2/2(\mu_1 + \mu_2)$, and the probability is ¼ for stepping tangential to the surface in either direction. These stepping probabilities may be worked out between different media in other coordinate systems from the finite difference coefficients. (See also problems at the end of Chapter 4 and Chapter 5.)

Other Monte Carlo techniques may be used to solve problems in electrostatics, magnetostatics, and heat transfer. An example is the Exodus method, where a large number of walkers begin at a grid point and diffuse toward the boundaries. No random number generator is used in the Exodus method. If 4000 walkers begin at one grid location then each neighboring grid point will have 1000 walkers after the first step. The walkers continue to disperse until some specified fraction (say 99.99%) reach a boundary. The scalar field at the originating grid point is then calculated according to (2.59).

2.5 FINITE ELEMENT METHOD

The Finite Element Method (FEM) is a more powerful method for calculating the fields in regions with complex shapes and inhomogeneous material properties. In this section an outline of the FEM is given and considers the scalar field to be the electric potential U whose negative gradient gives the electric field.

Whereas the finite difference method solves for the scalar field at each grid location, the FEM gives a continuous potential distribution over a spatial region. First, the solution region is divided into one-, two-, or three-dimensional finite elements that may have various shapes depending on the number of nodes per element. QuickField uses triangular finite elements with three nodes per element as shown in Figure 2.7. In two-dimensional Cartesian coordinates, the electric scalar potential U_e is approximated in the element

$$U_e(x,y) = a + bx + cy. \tag{2.65}$$

A uniform electric field in the element is thus

$$\mathbf{E}_e = -\nabla U_e(x,y) = -(b\hat{\mathbf{i}} + c\hat{\mathbf{j}}). \tag{2.66}$$

In cylindrical coordinates with axial symmetry, the potential inside each triangular finite element is approximated by

$$U_e(r,z) = a + br + cz \tag{2.67}$$

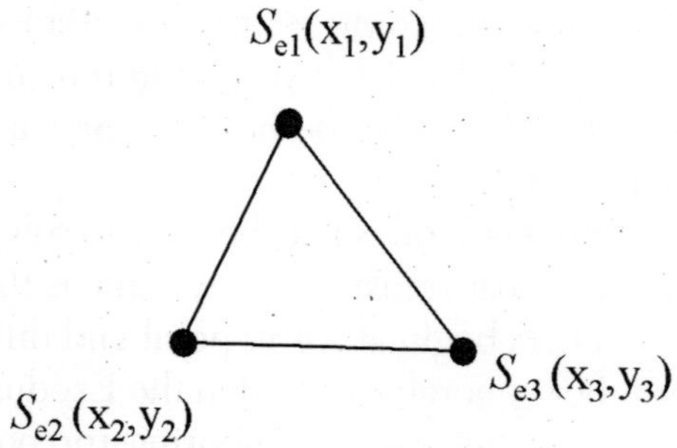

FIGURE 2.7 **Triangular, three-node finite element.**

corresponding to a uniform electric field in the element

$$\mathbf{E}_e = -\nabla U_e(r,z) = -(b\hat{\mathbf{r}} + c\hat{\mathbf{z}}). \tag{2.68}$$

The coefficients (a,b,c) in Figure 2.7 may be obtained by solving (2.65) evaluated at all three nodes

$$\begin{pmatrix} U_{e1} \\ U_{e2} \\ U_{e3} \end{pmatrix} = \begin{pmatrix} 1 & x_1 & y_1 \\ 1 & x_2 & y_2 \\ 1 & x_3 & y_3 \end{pmatrix} \begin{pmatrix} a \\ b \\ c \end{pmatrix}. \tag{2.69}$$

Inverting this matrix gives the coefficients

$$\begin{pmatrix} a \\ b \\ c \end{pmatrix} = \begin{pmatrix} 1 & x_1 & y_1 \\ 1 & x_2 & y_2 \\ 1 & x_3 & y_3 \end{pmatrix}^{-1} \begin{pmatrix} U_{e1} \\ U_{e2} \\ U_{e3} \end{pmatrix}. \tag{2.70}$$

We may then construct coordinate dependent shape functions $\alpha_i(x,y)$ so that the potential at points inside the element is given by

$$U_e(x,y) = \sum_{i=1}^{3} \alpha_i(x,y) U_{ei}. \tag{2.71}$$

To solve for the potential over a solution region such as illustrated in Figure 2.8, it is necessary to find the potential of nodes that are not specified by boundary conditions. This is accomplished by the construction of the energy functional for each element

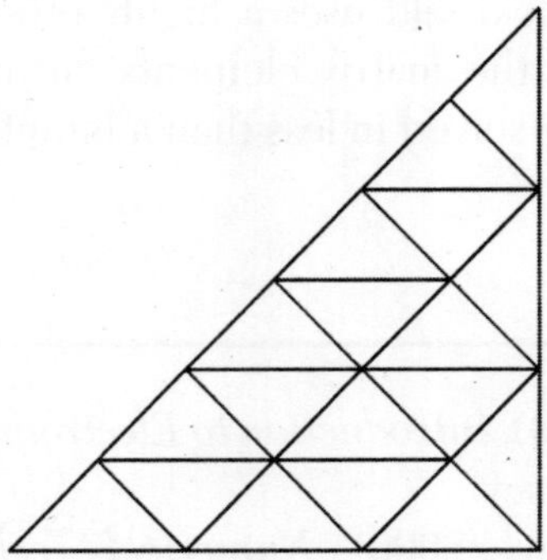

FIGURE 2.8 **Triangular solution region divided into 18 triangular finite elements.**

$$W_e = \frac{\varepsilon}{2}\begin{pmatrix} U_{e1} & U_{e2} & U_{e3} \end{pmatrix}\begin{bmatrix} C_{11} & C_{12} & C_{13} \\ C_{21} & C_{22} & C_{23} \\ C_{31} & C_{32} & C_{33} \end{bmatrix}\begin{pmatrix} U_{e1} \\ U_{e2} \\ U_{e3} \end{pmatrix}. \tag{2.72}$$

The total energy contained in the solution region is obtained by assembling the contribution from each triangular element

$$W_{\text{total}} = \frac{\varepsilon}{2}\begin{pmatrix} U_1 & U_2 & \cdots & U_n \end{pmatrix}\begin{bmatrix} C_{11} & C_{12} & & C_{1n} \\ C_{21} & C_{22} & & \vdots \\ & & \ddots & \\ C_{n1} & \cdots & & C_{nn} \end{bmatrix}\begin{pmatrix} U_1 \\ U_2 \\ \vdots \\ U_n \end{pmatrix}, \tag{2.73}$$

where n is the total number of nodes. Equation (2.73) may be written in matrix notation

$$W_{\text{total}} = \frac{\varepsilon}{2}\mathbf{U}^{-1}\mathbf{C}\mathbf{U}, \tag{2.74}$$

where $\mathbf{C}$ is the global coefficient matrix. It turns out that the potential satisfying Laplace's or Poisson's equation also minimizes the total energy contained in the solution region. The minimal energy requirement is implemented by setting the partial derivative of the total energy with respect to each node potential equal to zero

$$\frac{\partial W_{\text{total}}}{\partial U_i} = 0. \tag{2.75}$$

This results in a set of n algebraic equations to be solved for each potential value. Techniques for solving equation (2.75) include the band matrix and iteration methods. QuickField uses a highly efficient algorithm that does not require the storage of the matrix elements during the solution. Problems on most computers can be solved in less than a minute.

REFERENCES

David J. Griffiths. (1999). *Introduction to Electrodynamics 3rd edition,* Prentice-Hall, Inc., New Jersey.

Mathew N. O. Sadiku. (2000). *Numerical Techniques in Electromagnetics 2nd edition,* CRC Press. (Note that the first edition of this text features FORTRAN code with C++ in the second edition.)

Dean G. Duffy. (2003). *Advanced Engineering Mathematics with MATLAB 2nd edition*, Chapman & Hall/CRC Press.

Stanley J. Farlow. (1993). *Partial Differential Equations for Scientists and Engineers*, Dover Publications, Inc., New York.

Murray R. Spiegel. (1964). *Schaum's Outline of Complex Variables*, McGraw-Hill.

EXERCISES

2.1 Solve Laplace's equation for the scalar field S over a square region of unit length with $S(0,y) = S(1,y) = 0$, $S(x,0) = 1$, and $\partial S/\partial n = 0$ on the boundary $(x,1)$.

2.2 Solve Laplace's equation for the scalar field S between two cylindrical surfaces of radii R_1 and R_2 with respective boundary values $S(R_1) = 1$ and $S(R_1) = 2$.

2.3 Show that Poisson's equation $\nabla^2 S(r,z) = f(r,z)$ has the form

$$\frac{1}{r}\frac{\partial}{\partial r}\left(r\frac{\partial S}{\partial r}\right) + \frac{\partial^2 S}{\partial z^2} = f(r,z)$$

in cylindrical coordinates with axial symmetry.

2.4 Show that Poisson's equation $\nabla^2 S(\alpha, \beta) = f(\alpha, \beta)$ has the form

$$\frac{\partial^2 S}{\partial \alpha^2} + \frac{\partial^2 S}{\partial \beta^2} = \frac{a^2}{(\cosh \alpha - \cos \beta)^2} f(\alpha, \beta)$$

in bicylindrical coordinates with axial symmetry.

2.5 Show that Poisson's equation $\nabla^2 S(\alpha, \beta) = f(\alpha, \beta)$ has the form

$$\frac{\partial^2 S}{\partial \alpha^2} + \frac{\partial^2 S}{\partial \beta^2} = a^2\left(\sinh^2 \alpha + \sin^2 \beta\right) f(\alpha, \beta)$$

in elliptic cylindrical coordinates with axial symmetry.

2.6 Show that Poisson's equation $\nabla^2 S(r,\theta) = f(r,\theta)$ has the form

$$\frac{1}{r^2}\frac{\partial}{\partial r}\left(r^2\frac{\partial S}{\partial r}\right) + \frac{1}{r^2 \sin\theta}\frac{\partial}{\partial \theta}\left(\sin\theta\frac{\partial S}{\partial \theta}\right) = f(r,\theta)$$

in spherical coordinates with axial symmetry.

2.7 Show that Poisson's equation $\nabla^2 S(\xi,\eta) = f(\xi,\eta)$ has the form

$$\frac{1}{\cosh\xi}\frac{\partial}{\partial \xi}\left(\cosh\xi\frac{\partial S}{\partial \xi}\right) + \frac{1}{\cos\eta}\frac{\partial}{\partial \eta}\left(\cos\eta\frac{\partial S}{\partial \eta}\right)$$
$$= a^2\left(\sinh^2 \xi + \sin^2 \eta\right) f(\xi,\eta)$$

in oblate spheroidal coordinates with axial symmetry.

2.8 Show that Poisson's equation $\nabla^2 S(\xi,\eta) = f(\xi,\eta)$ has the form

$$\frac{1}{\sinh\xi}\frac{\partial}{\partial\xi}\left(\sinh\xi\frac{\partial S}{\partial\xi}\right) + \frac{1}{\sin\eta}\frac{\partial}{\partial\eta}\left(\sin\eta\frac{\partial S}{\partial\eta}\right)$$
$$= a^2\left(\sinh^2\xi + \sin^2\eta\right) f(\xi,\eta)$$

in prolate spheroidal coordinates with axial symmetry.

2.9 Show that Poisson's equation $\nabla^2 S(\alpha,\beta) = f(\alpha,\beta)$ has the form

$$\frac{\partial}{\partial\alpha}\left(\frac{\sinh\alpha}{\cosh\alpha - \cos\beta}\frac{\partial S}{\partial\alpha}\right) + \frac{\partial}{\partial\beta}\left(\frac{\sinh\alpha}{\cosh\alpha - \cos\beta}\frac{\partial S}{\partial\beta}\right)$$
$$= \frac{a^2\sinh\alpha}{(\cosh\alpha - \cos\beta)^3} f(\alpha,\beta)$$

in toroidal coordinates with axial symmetry.

2.10 An oblate spheroidal surface is maintained at a constant potential V_0. Write an expression for the scalar field outside the spheroidal surface.

2.11 Use MATLAB to create contour plots of the real and imaginary parts of the complex potentials corresponding to a

(a) Uniform flow

$$\Omega(z) = V_0 e^{-i\delta} z$$

(b) Source at $z = a$

$$\Omega(z) = k\ln(z - a)$$

(c) Sink at $z = a$

$$\Omega(z) = -k\ln(z - a)$$

(d) Superposition of source and sink

$$\Omega(z) = -k\ln\frac{z + a}{z - a}.$$

2.12 The boundary of the first quadrant of the Cartesian coordinate system is maintained at $S(x,0) = 1$ for $x \geq 0$ and $S(0,y) = 0$ for $y \geq 0$. Calculate $S(x,y)$ on the first quadrant by conformal mapping to the upper half-plane using Poisson's integral formula.

2.13 Write the Monte Carlo stepping probabilities for the solution of Laplace's equation in
(a) cylindrical coordinates with axial symmetry
(b) bicylindrical coordinates with planar symmetry
(c) elliptic cylindrical coordinates with planar symmetry
(d) spherical coordinates with axial symmetry
(e) oblate spheroidal coordinates with axial symmetry
(f) prolate spheroidal coordinates with axial symmetry
(g) toroidal coordinates with axial symmetry.

2.14 Write the Monte Carlo stepping probabilities in the above coordinate systems in three dimensions without axial or planar symmetry.

2.15 Write a MATLAB program for the Monte Carlo solution of Laplace's equation in three-dimensional Cartesian Coordinates. Calculate the scalar field at the center of a cube with five faces at $S = 0$ and the other face at $S = 1$.

2.16 Write a MATLAB program for the Monte Carlo solution of Laplace's equation in three-dimensional cylindrical coordinates. Calculate the scalar field at the center of a cylinder with end caps at $S = 0$ and the cylindrical surface at $S(R,\phi) = \cos(\phi)$.

Chapter 3 A Walk Through QuickField

In This Chapter

- What is QuickField?
- Basic Organization of QuickField
- Specifying Problem Types and Properties
- Model Construction
- Problem Solving and Post Processing
- Electrostatics Simulation
- Magnetostatics Simulation

3.1 What is QuickField?

QuickField is a Finite Element Method CAD modeling software that supports a variety of electromagnetic, thermal, and stress analysis problems that have either two-dimensional *x-y* symmetry or axial symmetry. Problem types supported include Electrostatics, Magnetostatics, Transient Magnetics, AC Magnetics, DC Conduction, AC Conduction, Transient and Steady State Thermal Analysis, Stress Analysis, and Electrical Circuits. Multiphysics analysis capabilities include various couplings between Thermal, Stress, Electrostatic, Magnetostatic, Time-harmonic Magnetic, AC and DC Conduction. Table 3.1 gives an overview of analysis types supported by QuickField.

TABLE 3.1

Analysis & Symmetry	Material Properties	Sources	Boundary Conditions	Post Processing Results
Electrostatic				
· X-Y Symmetry · Axial Symmetry	· Isotropic Permittivity · Orthotropic Permittivity · Air $\varepsilon = \varepsilon_0$	· Voltages · Point Charges · Surface Charges · Volume Charges	· Dirichlet Condition (Specified Voltages) · Neumann Condition (Specified Charge Density) · Constant Potential with Specified Total Charge	· Voltages · Electric Fields · Field Gradients · Flux Density · Surface Charges · Capacitances · Electric Forces · Torques · Electric Energy
Magnetostatic				
· X-Y Symmetry · Axial Symmetry	· Isotropic Permeability · Orthotropic Permeability · Air $\mu = \mu_0$ · **B-H** curves · Ferromagnets & Superconductors	· Line Currents · Current Density · Uniform Field · Permanent Magnets	· Dirichlet Condition (Vector Potential **A**) · Neumann Condition (Tangential **B** Field) · Superconductor (Zero Normal **B**)	· Vector Potential **A** · Magnetic Fields (Flux Density **B**) · Field Intensity **H** · Magnetic Forces · Torques · Magnetic Energy · Flux Linkages · Inductances (Self & Mutual)
AC Magnetic				
· X-Y Symmetry · Axial Symmetry	· Isotropic Permeability · Orthotropic Permeability · Air $\mu = \mu_0$ · Conductivity	· Line Currents (Time Harmonic) · Current Density (Time Harmonic) · Uniform AC Magnetic Field · AC Voltage	· Dirichlet Condition (Vector Potential **A**) · Neumann Condition (Tangential **B** Field) · Superconductor (Zero Normal **B**)	· Vector Potential **A** · Magnetic Fields (Flux Density **B**) · Field Intensity **H** · Maxwell and Lorentz Forces (Peak & Time Avg.) · Maxwell and Lorentz Torques (Peak & Time Avg.) · Magnetic Energy · Flux Linkages

(continued)

TABLE 3.1 (*Continued*).

Analysis & Symmetry	Material Properties	Sources	Boundary Conditions	Post Processing Results
				· Inductance (Self & Mutual) · Eddy Current · Joule Heat · Impedances
Transient Magnetic				
· X-Y Symmetry · Axial Symmetry	· Isotropic Permeability · Orthotropic Permeability · Air $\mu = \mu_0$ · **B-H** curves-Ferromagnets & Super-conductors	· Line Currents (Time-Dependent) · Current Density (Time-Dependent) · Uniform Field · Permanent Magnets	· Dirichlet Condition (Vector Potential **A**) · Neumann Condition (Tangential **B** Field) · Superconductor (Zero Normal **B**)	· Vector Potential **A** · Magnetic Fields (Flux Density **B**) · Field Intensity **H** · Magnetic Forces · Magnetic Torques · Magnetic Energy · Flux Linkages · Inductances (Self & Mutual) · Eddy Currents
DC Conduction				
· X-Y Symmetry · Axial Symmetry	· Isotropic Resistivity · Orthotropic Resistivity	· Voltage · Current Density	· Dirichlet Condition (Specified Voltage) · Normal Derivative (Surface Current Density) · Constant Potential (Given Constraint)	· Voltages · Current Density · Electric Fields · Total Currents · Power Loss
AC Conduction				
· X-Y Symmetry · Axial Symmetry	· Isotropic Resistivity · Orthotropic Resistivity · Isotropic Permittivity	· AC Voltages · Current Density (Time Harmonic)	· Dirichlet Condition (Specified Voltages) · Normal Derivative (Surface Current Densities)	· Voltages · Current Density · Electric Fields · Total Currents· · Power Loss

(*continued*)

TABLE 3.1 (*Continued*).

Analysis & Symmetry	Material Properties	Sources	Boundary Conditions	Post Processing Results
	· Orthotropic Permittivity		· Constant Potential (Given Constraint)	· Voltages · Electric forces (Peek & Time Avg.) · Electric Torques (Peak & Time Avg.)
Thermal				
· X-Y Symmetry · Axial Symmetry	· Thermal Conductivity (Isotropic for Temperature Dependent) · Thermal Conductivity (Orthotropic Temperature Independent) · Specific Heat (Temperature Dependent)	· Volume Heat Densities (Constant and Temperature Dependent) · Convective Sources · Radiative Sources · Joule Heat Sources (Imported from DC or AC conduction or AC or transient Magnetic Analysis	· Specified Temperatures · Boundary Heat Flows · Convection · Radiation · Specified Constraints for Constant Temperature Boundaries	· Temperatures · Thermal Gradients · Heat Flux Density · Total Heat Loss or Gain Over Subregion of Model · Time Dependence of Quantities (Transient Analysis)
Stress				
· X-Y Symmetry · Axial Symmetry	· Isotropic Materials · Orthotropic Materials	· Body Forces · Pressure · Thermal Strain · Imported Electric or Magnetic Force from Electric or Magnetic Analysis	· Prescribed Displacements · Elastic Spring Supports	· Displacement · Stress Components · Principle Stress · von Mises Stress · Tresca · Mohr-Coulomb · Drucker-Prager · Hill Criteria
Circuit Analysifs (AC Magnetics and Transient Magnetics)	· Resistors · Capacitors · Inductors · Conductivity & Permittivity of Connected Model Blocks	· AC Voltage and Current · DC Voltage and Current · Transient Voltage and Current	· Electrical Wires	· Voltage · Current · Frequency Response

3.2 BASIC ORGANIZATION OF QUICKFIELD

The basic steps in setting up, solving, and analyzing problem results in QuickField are outlined in Figure 3.1. These steps include specifying the problem type, designing the geometrical model, and assigning material properties and boundary conditions. Finally, the problem is solved and the results are analyzed. Color images for many of the figures may be found on the CD-ROM.

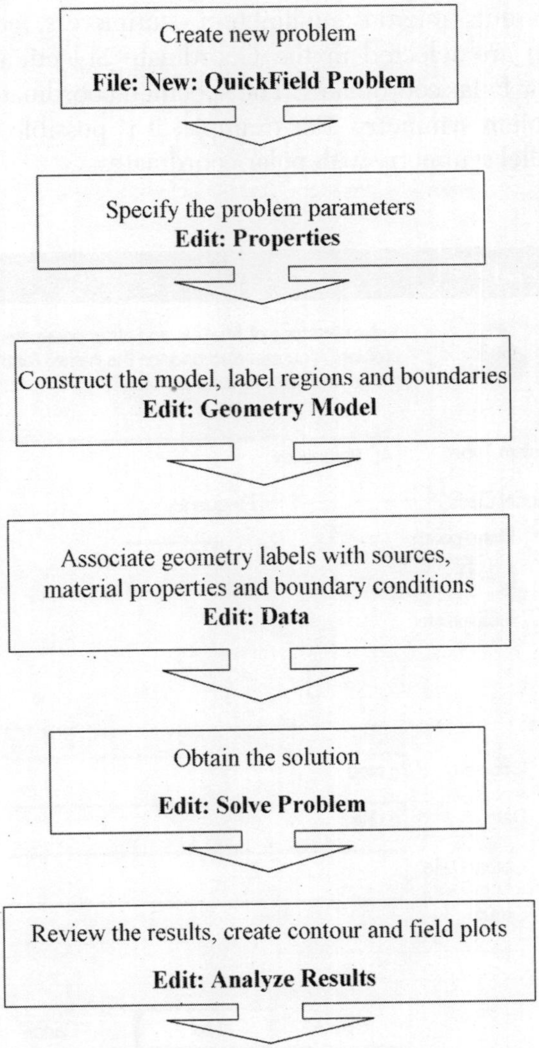

FIGURE 3.1 **Outline of steps for creating and solving a problem in QuickField.**

3.3 SPECIFYING PROBLEM TYPES AND PROPERTIES

Once the Problem Type (Electrostatics, Magnetostatics, AC Magnetics, etc.) has been chosen (File → New → QuickField Problem → OK) the filename and destination folder are chosen, and the model class is selected as either Plane-Parallel (with x-y symmetry) or Axisymmetric (with r-z symmetry). The Precision is chosen as Draft, Normal, or High. Preexisting model files, Data, or Library Data may also be selected. The frequency is specified for AC Magnetic and AC Conduction Problems. Circuit files may be selected for Transient or AC Magnetic problem types as shown in Figure 3.2.

Length units (microns, millimeters, centimeters, meters, kilometers, inches, feet, miles) are selected in the Coordinate Selection Box along with either Cartesian or Polar coordinates. The specified coordinate system is independent of the problem symmetry. For example, it is possible to have a problem with Plane-Parallel symmetry with polar coordinates.

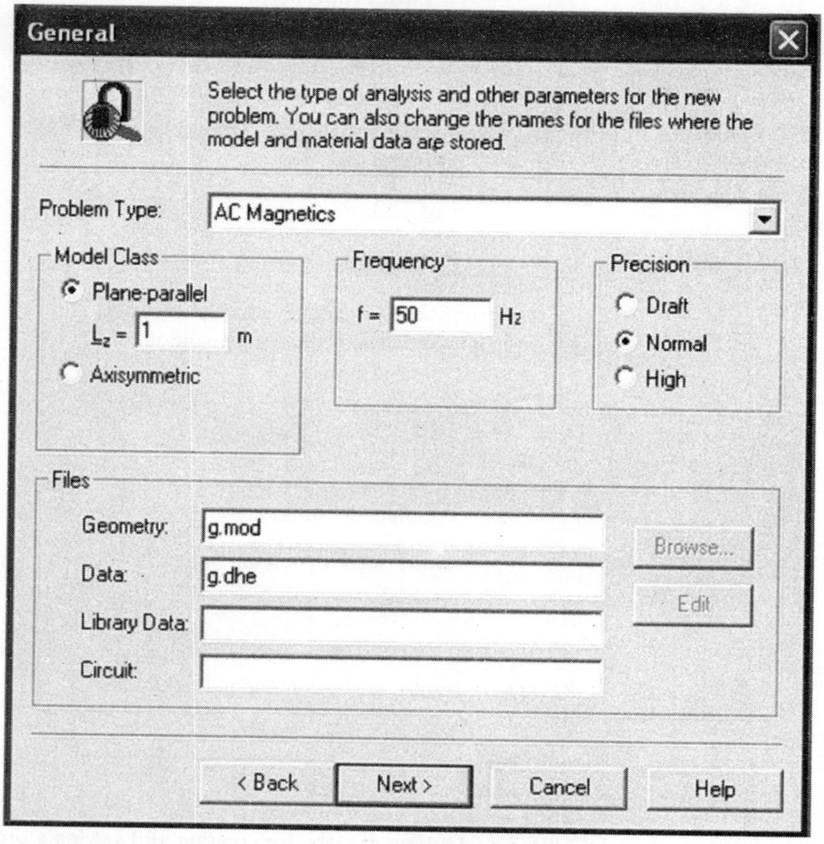

FIGURE 3.2 **Window for specifying problem properties in QuickField.**

3.4 MODEL CONSTRUCTION

To begin model construction, select Edit Geometry Model from the Edit menu or click the Geometry icon in the left-hand tree of the QuickField screen. Click OK in response to the create model file prompt.

Geometric models are built by connecting vertices with edges to form the outer boundary of the solution region with inner subregions or blocks. Geometric objects such as squares, circles, and ellipses may be added from the Insert Shape button on the Model toolbar shown in Figure 3.3.

Edges, blocks, and vertices can be labeled by first double clicking on the desired geometrical object and then entering the label in the Selection Properties window. Material properties and boundary conditions associated with each label can be entered by clicking on the corresponding label icon on the left of the screen and entering the desired values.

Selected blocks can be meshed by selecting the Build Mesh icon on the Model toolbar. For the Student Version of QuickField, the total number of nodes is limited to 250, with no node limit in the Professional Version. The mesh density over a region may be adjusted manually by setting the spacing in the vertex Selection Properties window. When working with the Student Version of QuickField, it is often necessary to increase the mesh spacing in order to keep the total number of nodes under 250. It is best to increase the spacing over portions of the solution region, where a less accurate solution is desired. Figure 3.4 shows the manual node spacing resulting in a mesh with eight triangular elements.

Parts of the geometric model may be translated as shown in Figure 3.5 or duplicated and translated as shown in Figure 3.6. This feature enables rapid construction of models with repetitive structures.

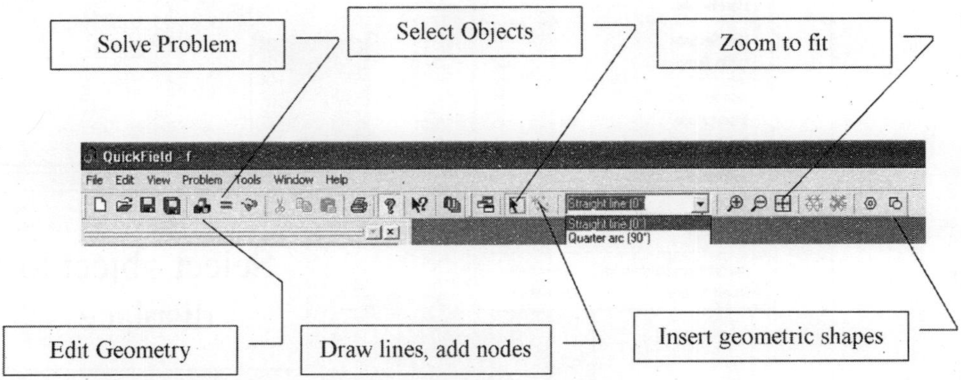

FIGURE 3.3 **QuickField General toolbar (left icons) and Model toolbar (right icons) showing select operations.**

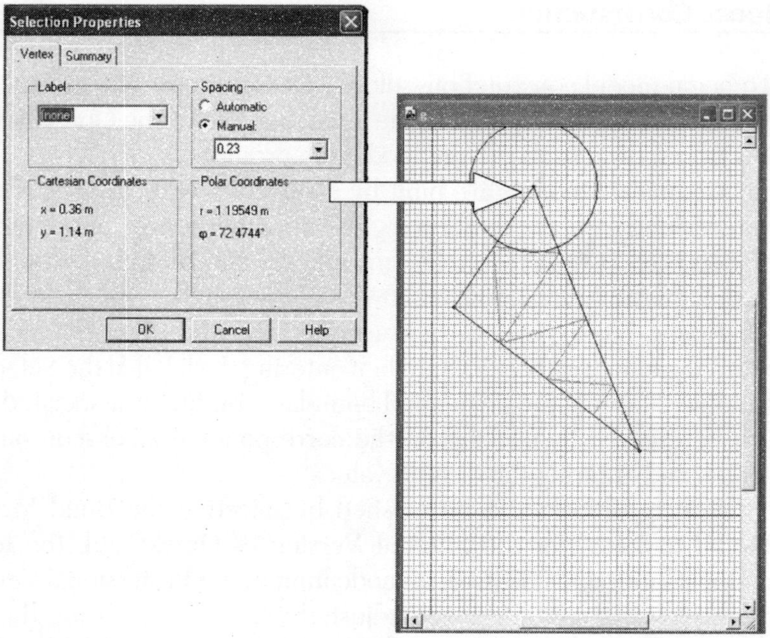

FIGURE 3.4 **Illustration of manual node spacing resulting in a mesh with eight triangular elements.**

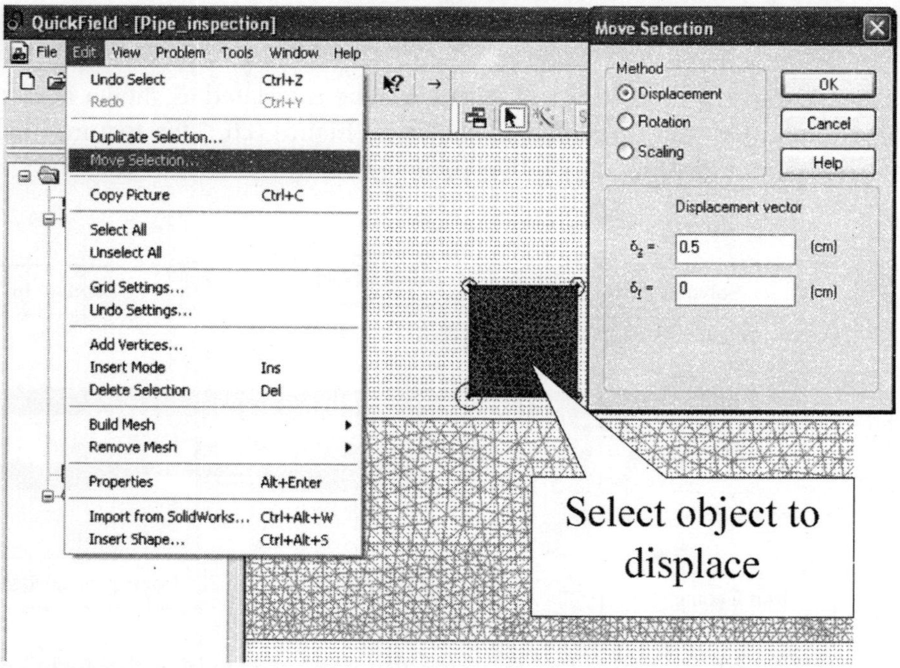

FIGURE 3.5 **Displacing part of a geometric model.**

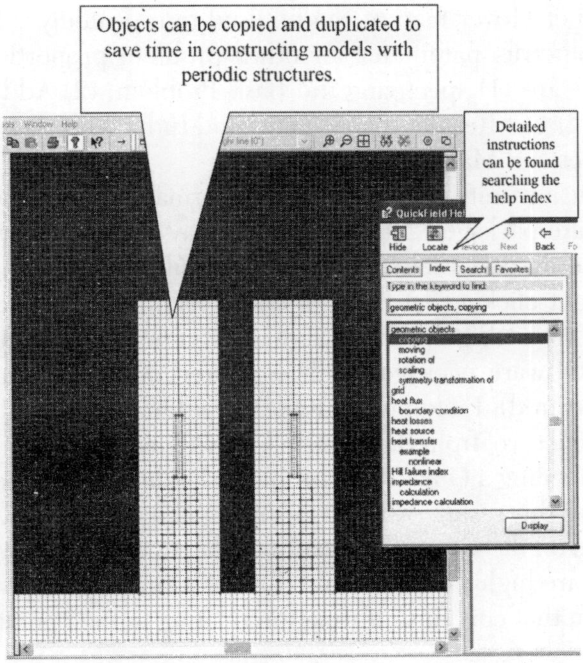

FIGURE 3.6 **Duplicating periodic elements in a geometric model.**

3.5 PROBLEM SOLVING AND POST PROCESSING

Problems are solved by choosing Solve Problem from the Problem pull-down menu or by clicking the Solve (equal sign) icon on the General toolbar shown in Figure 3.3. Results are viewed in the Post Processing window that is displayed after the problem is solved (answering 'Yes' to view the results). Contour and vector field plots of the calculated scalar and vector fields can be made. Graphical options are available for adjusting vector arrow lengths, color of intensity maps, and density of equipotentials lines, etc. (by right-clicking on the plot and selecting 'Field Picture…').

Field quantities can be plotted along selected open or closed contours using the Add to Contour icon on the Post Processor toolbar. Contours can be specified along model edges simply by clicking on the corresponding edge. Contours may also be drawn free-style or specified by entering the beginning and end-point contour coordinates (selecting 'Add Lines…' in the Contour pull-down menu). Right clicking the field plot and selecting 'XY-plot' from the pop-up menu generates the contour plot. Results may be viewed in tabular form by selecting 'Table' from the same pop-up menu. Integral Values are also viewed on the left of the screen by selecting 'Integral Values…' from this pop-up window or by clicking the Integral Calculator icon on the Post Processor toolbar.

The Label Mover feature can be used to repeatedly solve a problem while varying geometric, parameter, or other problem properties. The basic Label Mover steps are (1) specifying the Base Problem, (2) Add Values, (3) Record Steps, and (4) Get Results. Capacitance and Inductance Wizards are available in Electrostatics and Magnetics modules.

Numerical results displayed in table format may be copied and pasted into other software packages. Numerical data may also be exported as a data file by choosing the 'Export Field' from the File pull-down menu. Field plots can be saved by right clicking on the plot and selecting 'Copy Picture' or by making screenshots by hitting the 'Print Scrn' button on the keyboard and then pasting into other software packages such as PowerPoint (cropping the entire screen picture as desired). Field plots are best saved in different file formats by selecting 'Export Picture' from the File pull-down menu.

When opening a QuickField problem that has already been solved, it is possible to view the results without resolving the problem by clicking the 'View Results' icon on the General toolbar (eyeglasses-shaped icon). All the features in QuickField are highly intuitive and there is a running help window on the right of the screen that contains information relating to the operation at hand.

3.6 ELECTROSTATICS SIMULATION

Figures 3.7 through 3.18 show the steps involved in creating a problem in Electrostatics. This simulation is that of a line charge with linear charge density modeled in x-y symmetry. The solution will differ from that of line charge in free space because of the influence of the edges of the solution region that act as a grounded conducting box surrounding the line charge.

FIGURE 3.7 **Opening a new problem.**

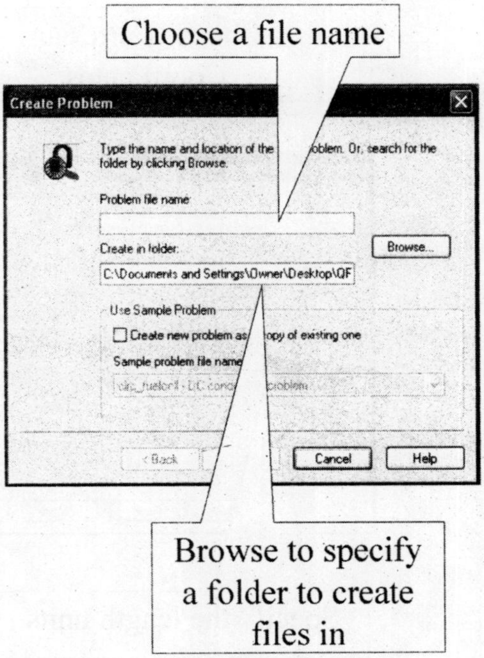

FIGURE 3.8 **Specifying the problem filename and destination folder.**

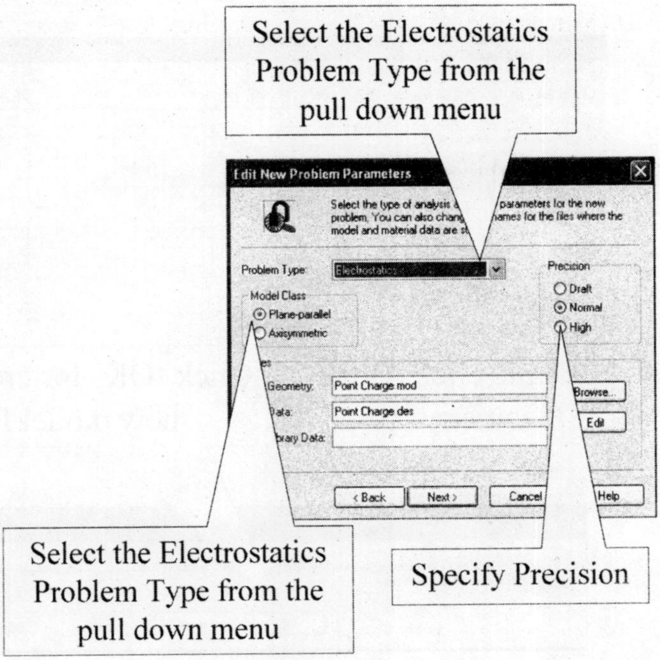

FIGURE 3.9 **Selecting problem type and symmetry.**

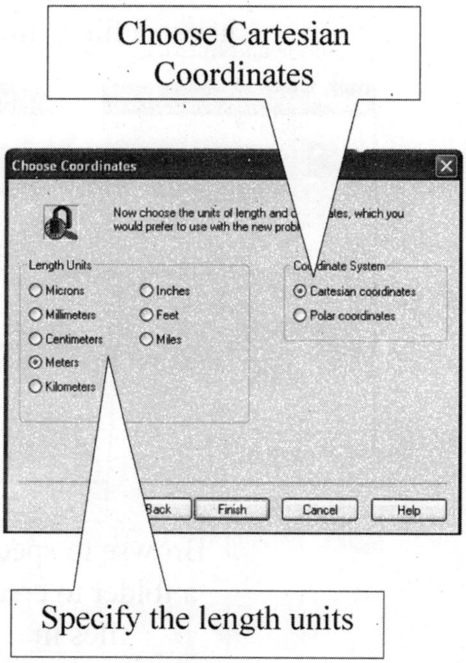

FIGURE 3.10 **Specifying length units and coordinate system.**

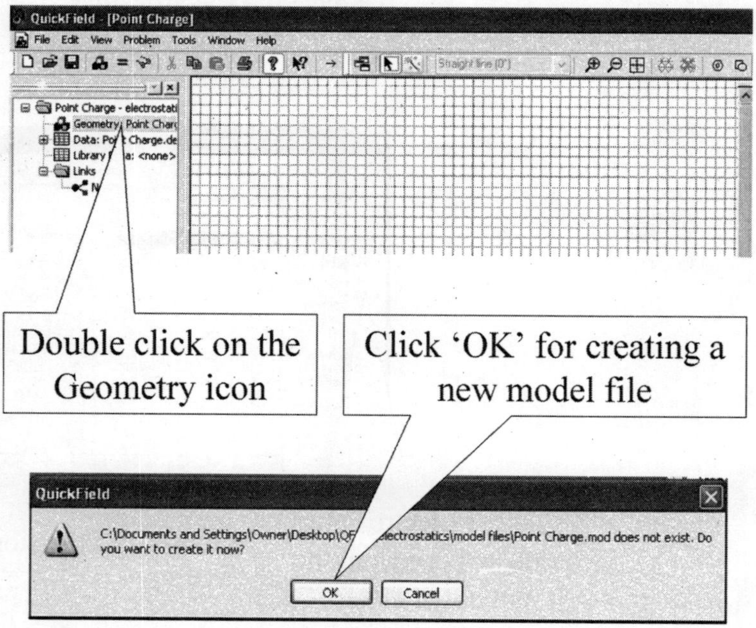

FIGURE 3.11 **Creating a new model file.**

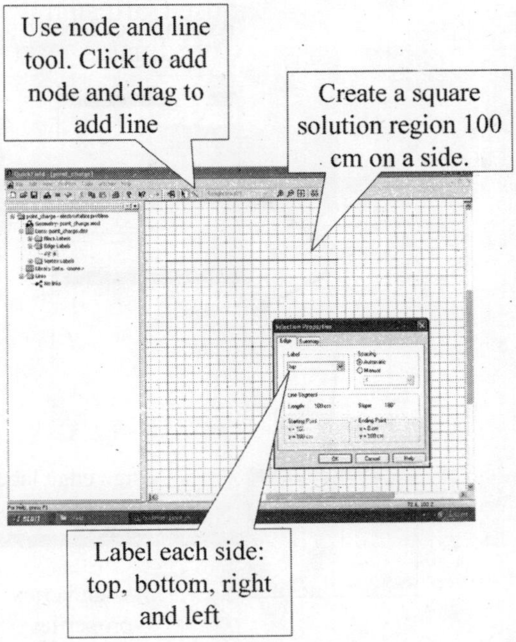

FIGURE 3.12 **Creating a square solution region.**

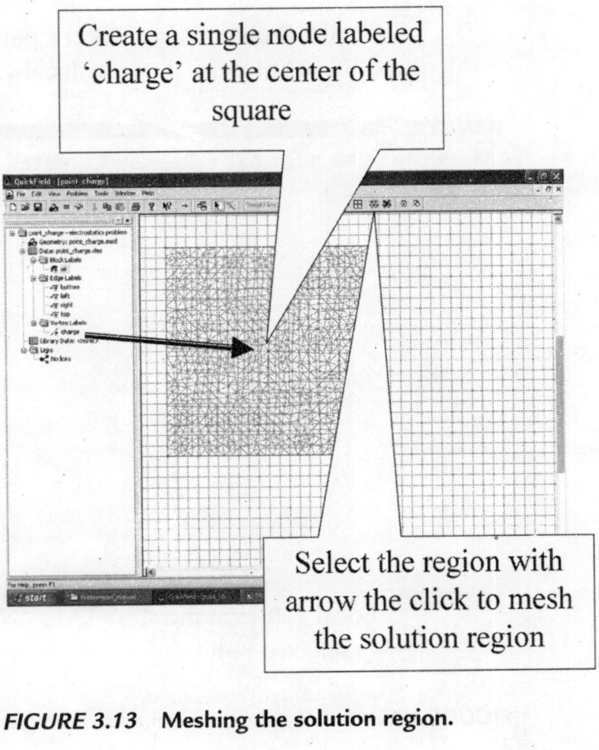

FIGURE 3.13 **Meshing the solution region.**

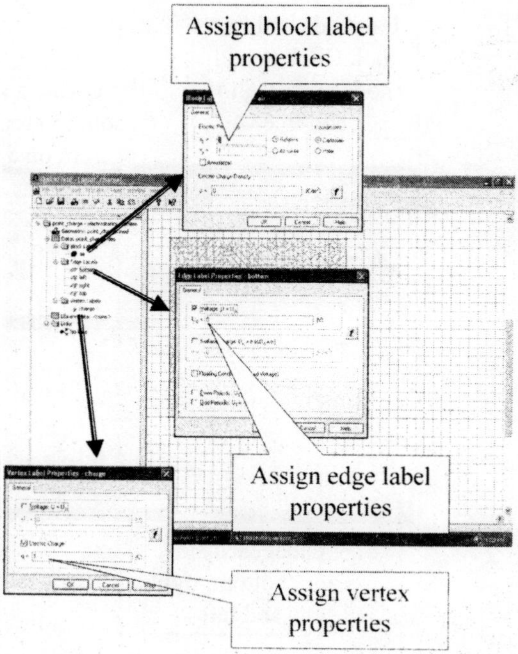

FIGURE 3.14 **Assigning block, edge, and vertex properties.**

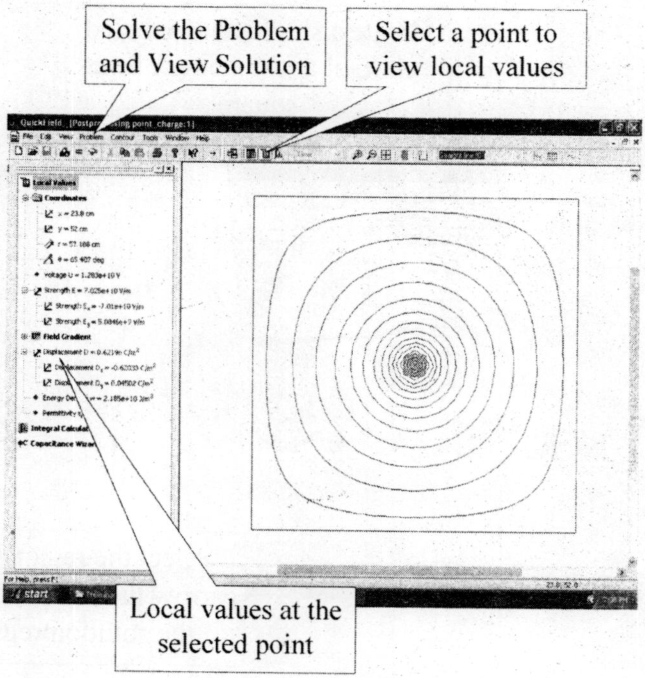

FIGURE 3.15 **Solving the problem and viewing the solution.**

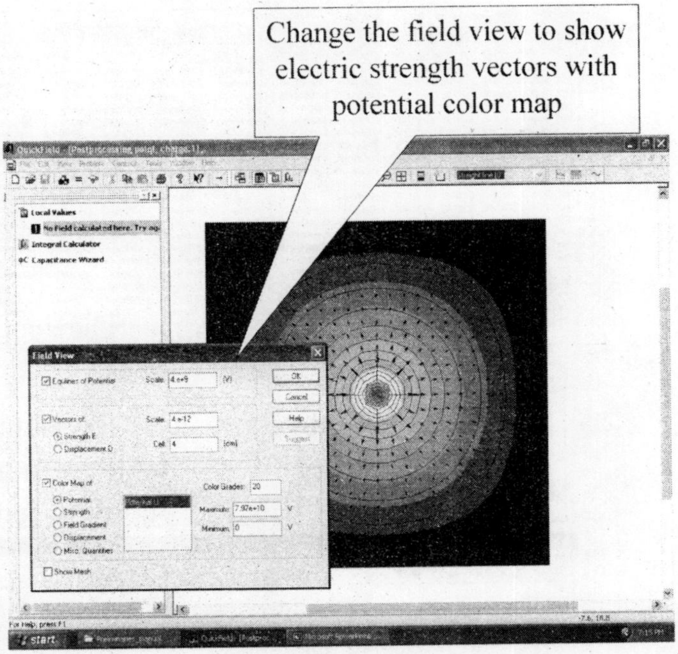

FIGURE 3.16 **Viewing a color map and vector plot.**

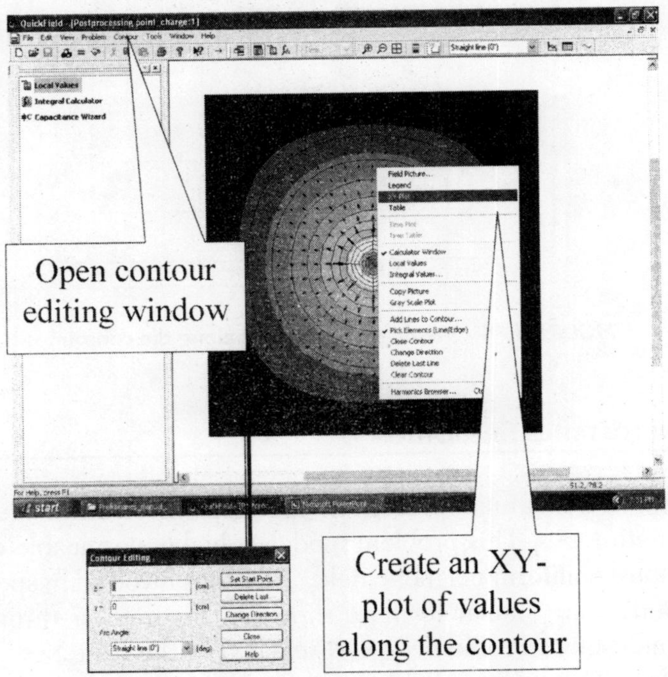

FIGURE 3.17 **Creating a contour.**

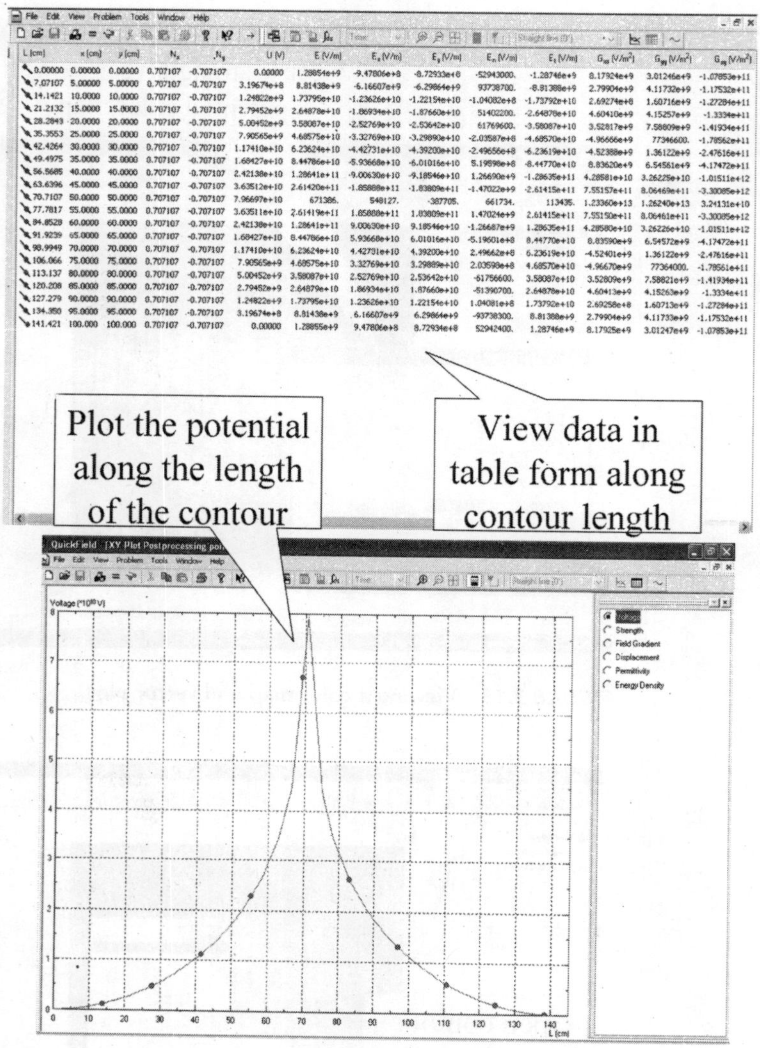

FIGURE 3.18 **Viewing the potential along the contour.**

3.7 MAGNETOSTATICS SIMULATION

Figures 3.19 through 3.28 show the steps involved in creating a problem in Magnetostatics. This problem models a highly permeable elliptic cylinder in a previously uniform magnetic field. The steps involved in specifying the problem properties are similar to those shown in Figures 3.7 through 3.11 except the Magnetostatics Problem Type is selected in the Edit New Problem Parameter window. The uniform field is created by applying boundary conditions to the sides of the model as illustrated in Figure 3.22.

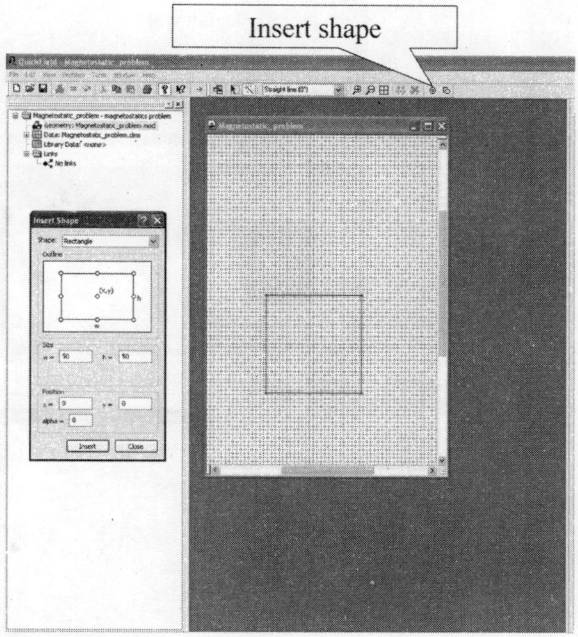

FIGURE 3.19 **Inserting a square shape that will serve as the boundary of the solution region.**

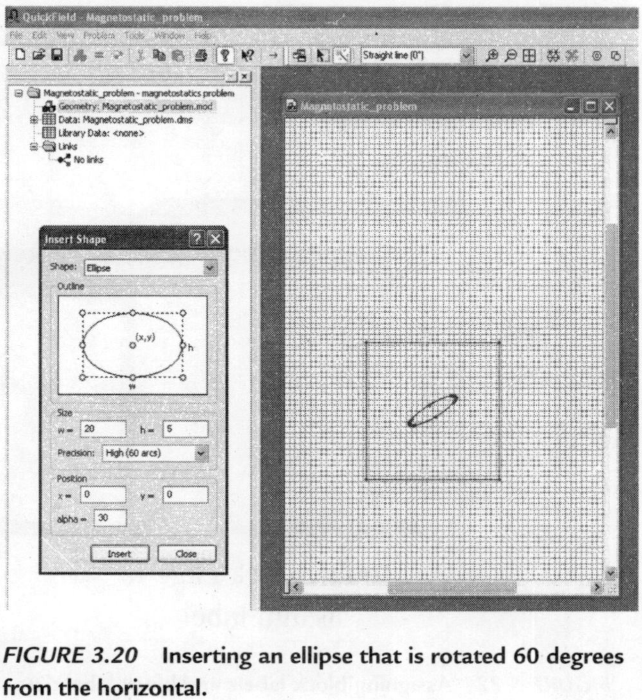

FIGURE 3.20 **Inserting an ellipse that is rotated 60 degrees from the horizontal.**

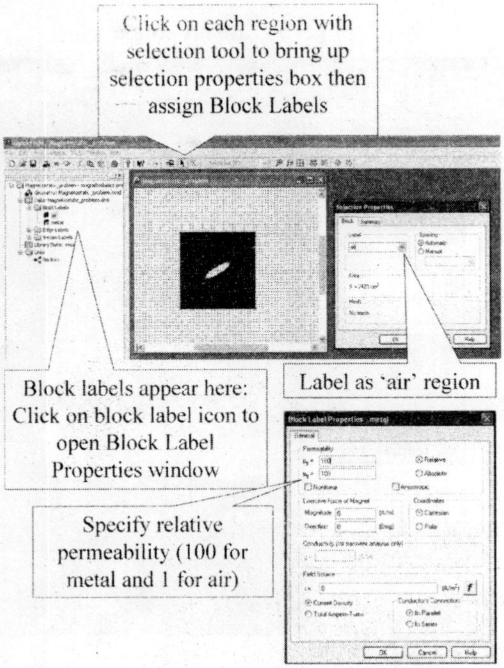

FIGURE 3.21 **Assigning block labels and block label data.**

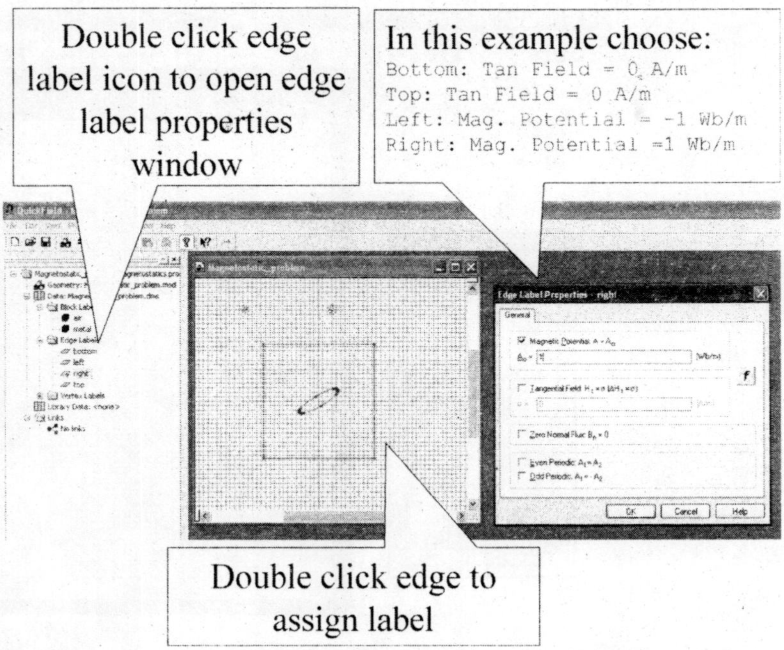

FIGURE 3.22 **Assigning block labels and block label data.**

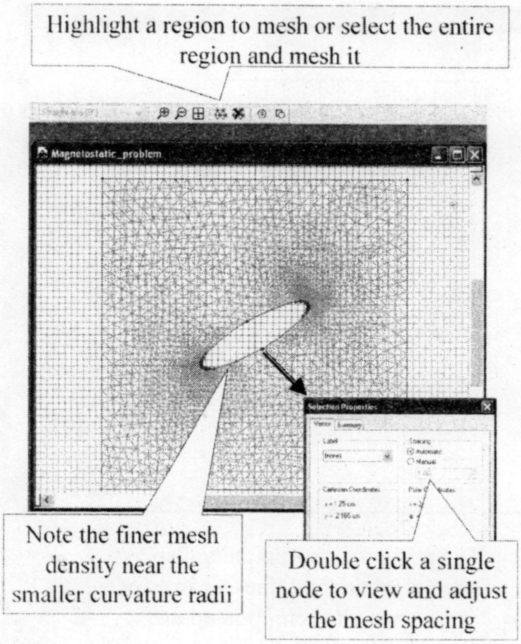

FIGURE 3.23 **Meshing the air region outside the permeable elliptic cylinder. The cylindrical region is subsequently selected and meshed in the same way.**

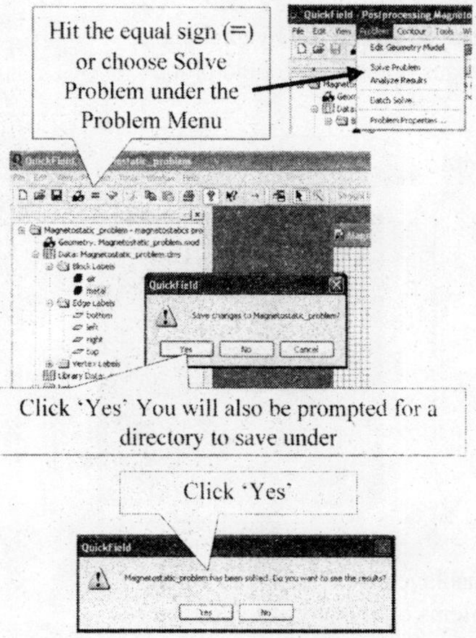

FIGURE 3.24 **Solving the problem and answering 'Yes' to the subsequent prompts.**

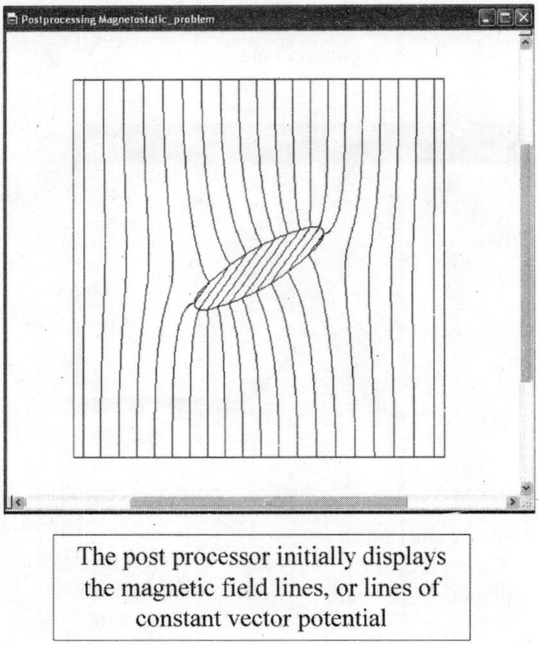

FIGURE 3.25 **Initial output of the post processor showing lines of equal magnetic potential.**

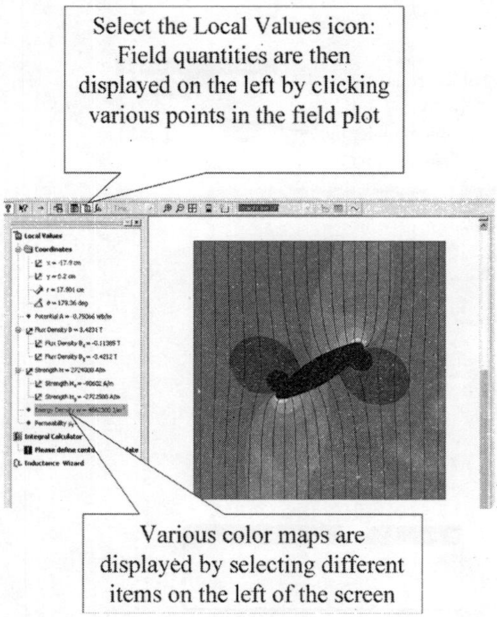

FIGURE 3.26 **Displaying local field values at specific points in the solution region.**

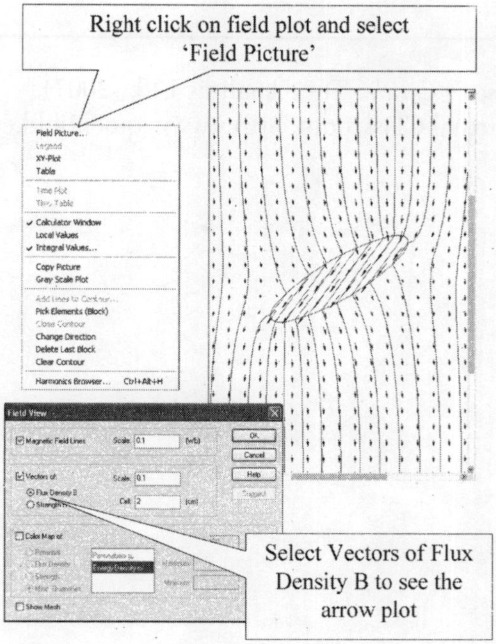

FIGURE 3.27 **Displaying a vector field plotted with the magnetic field lines.**

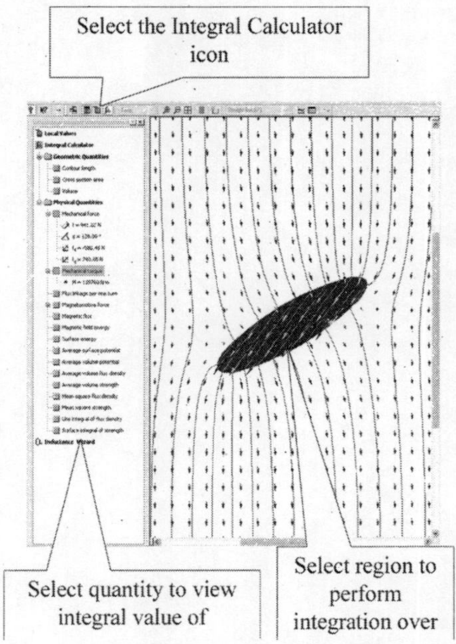

FIGURE 3.28 **Calculating integral quantities over the elliptic cylinder region.**

REFERENCES

QuickField User's Guide: Tera Analysis Ltd. (2007).

QuickField Virtual Classroom. http://www.quickfield.com/demo/index.htm.

EXERCISES

3.1 Construct a triangular region similar to Figure E-3.1 in Electrostatics by connecting three nodes with edges. Assign a relative permittivity to the region. Adjust the spacing on the three corner vertices so that the total number of nodes in the meshed region is as close to 250 as possible. Label and assign boundary conditions to the three sides of the triangle. Solve the problem and investigate field plots, practice making contour plots and calculating integral values.

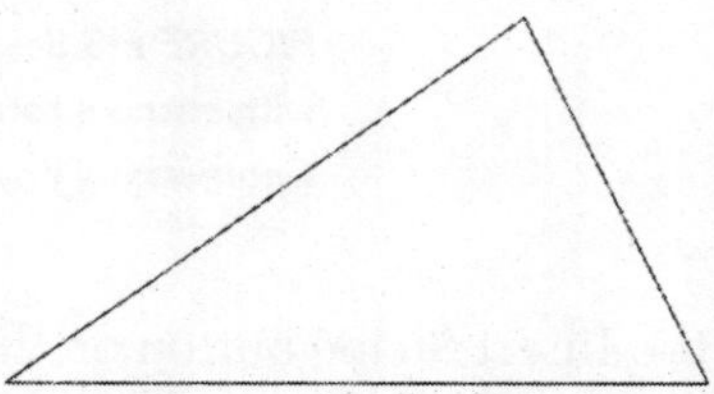

FIGURE E-3.1 **Triangular region. Geometry: *x-y* symmetry.**

3.2 Design a hollow permeable pipe with a missing wedge as shown in Figure E-3.2. Create the model in Magnetostatics with *x-y* symmetry. Assign different current densities to the flat edges and the zero vector potential boundary condition on the curved edges. Solve the problem and investigate field plots, practice making contour plots and calculating integral values.

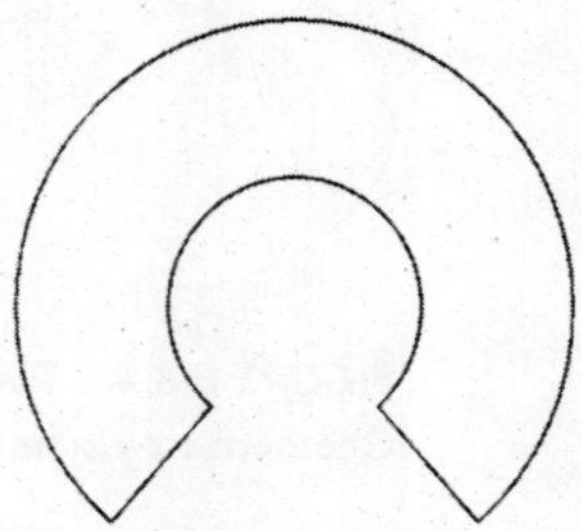

FIGURE E-3.2 **Hollow pipe with wedge missing. Geometry: *x-y* symmetry.**

3.3 Create a rectangular dielectric region with corners formed by circular arcs in Electrostatics with *x-y* symmetry. Assign a relative permittivity to the part and assign alternating positive and negative charge densities to the circular boundaries. Assign the flat boundaries to be zero volts.

Solve the problem and investigate field plots, practice making contour plots and calculating integral values.

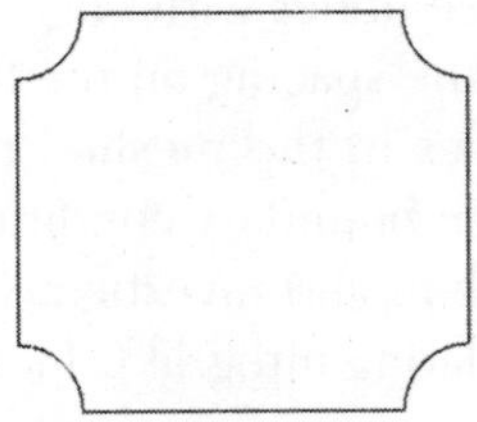

FIGURE E-3.3 **Rectangular region with corners formed by circular arcs. Geometry: *x-y* symmetry.**

3.4 Use the Insert Shape button on the Model toolbar to model a living cell as two coaxial ellipses with a 50-micron semi-major axis of the outer ellipse and roughly A 5-nm separation between the outer and inner ellipse. Create the model in the AC Current module with x-y symmetry. For node-limited versions of Quickfield create a thicker membrane such as shown in Figure E-3.4. Assign permittivity and conductivity values to each region (see Appendix 4 for typical biological parameter values). Place vertices at opposite ends of the cell assigned at different potentials (simulating microelectrodes). Plot the resulting current distribution in the cell for different electrode positions.

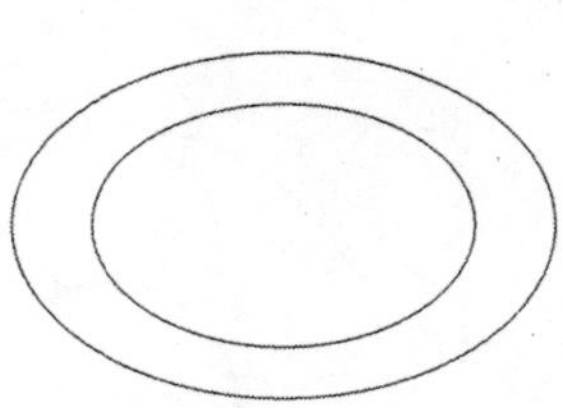

FIGURE E-3.4 **Two coaxial ellipses. Geometry: *x-y* symmetry.**

3.5 Repeat exercise 3.4 investigating the perturbation of injected currents by internal organelles such as the nucleus and endoplasmic reticulum inside the cell. Note that the organelle membranes are also roughly 5 nm in thickness.

3.6 Repeat exercise 3.4 for a spherical living cell. Refer to Chapter 10 for more simulations of living cells under electrical excitation.

Chapter **4** ELECTROSTATICS

In This Chapter

- Coulomb's Law and the Superposition Principle
- Electric Flux and Gauss' Law
- Electric Potential
- QuickField Electrostatics Calculations
- Calculation of Capacitance
- Motion of Charged Particles in Static Electric Fields

4.1 COULOMB'S LAW AND THE SUPERPOSITION PRINCIPLE

The force between two point charges separated by a distance r is proportional to the product of the charges and inversely proportional to the square of their separation. The force on charge q_1 due to q_2 in Figure 4.1 is given by

$$\mathbf{F}_{12} = \frac{1}{4\pi\varepsilon_0}\frac{q_1 q_2}{r^2}\hat{\mathbf{r}}, \tag{4.1}$$

where $\hat{\mathbf{r}}$ is a unit vector that points from q_1 toward q_2. The force is attractive or repulsive depending on whether the charges have opposite or like signs, respectively. The force on q_1 is equal in magnitude and opposite in direction to the force on q_2 so that $\mathbf{F}_{12} = -\mathbf{F}_{21}$ according to Newton's third law. The unit vector $\hat{\mathbf{r}}$ may be written as

$$\hat{\mathbf{r}} = \frac{(\mathbf{r} - \mathbf{r}')}{|\mathbf{r} - \mathbf{r}'|}, \tag{4.2}$$

where q_1 is located at $\mathbf{r}$ and q_2 is located at $\mathbf{r}'$ so that $\mathbf{F}_{12}$ is

$$\mathbf{F}_{12} = \frac{q_1 q_2}{4\pi\varepsilon_0} \frac{(\mathbf{r} - \mathbf{r}')}{|\mathbf{r} - \mathbf{r}'|^3}. \tag{4.3}$$

The electric field contribution at $\mathbf{r}$ due to charge q_2 at $\mathbf{r}'$ is given by $\mathbf{F}_{12}/q_1$ or

$$\mathbf{E}(\mathbf{r}) = \frac{q_2}{4\pi\varepsilon_0} \frac{(\mathbf{r} - \mathbf{r}')}{|\mathbf{r} - \mathbf{r}'|^3}. \tag{4.4}$$

ON THE CD

Color images for many of the figures may be found on the CD-ROM.

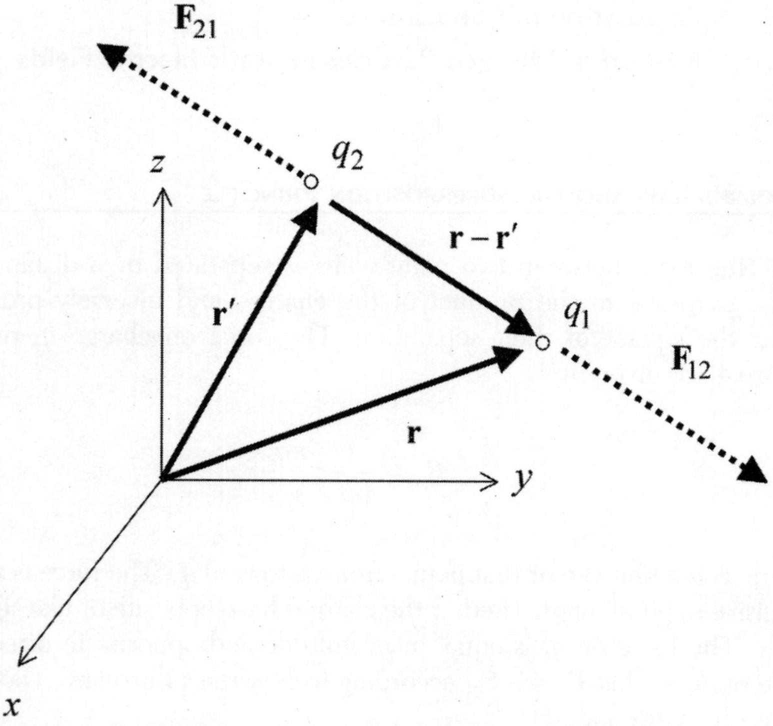

FIGURE 4.1 **Equal and opposite forces between two point charges separated by r–r′.**

In general, we can write that the force acting on a point charge q in an electric field $\mathbf{E}$ is given by $\mathbf{F} = q\mathbf{E}$. For a collection of point charges q_i located at $\mathbf{r}_i$ the superposition principle gives the electric field at the location $\mathbf{r}$

$$\mathbf{E}(\mathbf{r}) = \frac{1}{4\pi\varepsilon_0}\sum_i \frac{q_i(\mathbf{r}-\mathbf{r}_i)}{|\mathbf{r}-\mathbf{r}_i|^3}. \tag{4.5}$$

For a continuous charge distribution $\rho(\mathbf{r}')$, the electric field at $\mathbf{r}$ is given by

$$\mathbf{E}(\mathbf{r}) = \frac{1}{4\pi\varepsilon_0}\iiint_{\text{vol}} \frac{\rho(\mathbf{r}')(\mathbf{r}-\mathbf{r}')}{|\mathbf{r}-\mathbf{r}'|^3}dv', \tag{4.6}$$

where the integral is evaluated over the primed coordinates of the charge distribution.

4.2 ELECTRIC FLUX AND GAUSS' LAW

Electric Flux

The infinitesimal electric flux $d\Phi$ through a surface element da is the product of the perpendicular electric field $\mathbf{E} \cdot \hat{\mathbf{n}}$ times da

$$d\Phi = \mathbf{E}\cdot d\mathbf{a} = \mathbf{E}\cdot\hat{\mathbf{n}}da, \tag{4.7}$$

where $\hat{\mathbf{n}}$ is the unit vector perpendicular to da. The total flux Φ is obtained by integrating $\mathbf{E}\cdot d\mathbf{a}$ over the entire surface

$$\Phi = \iint_{\text{surf}} \mathbf{E}\cdot d\mathbf{a}. \tag{4.8}$$

The units of Φ are that of the electric field times area or $\text{Nm}^2/\text{C} = \text{V}\cdot\text{m}$.

Gauss' Law

Figure 4.2 shows two Gaussian surfaces with one surface surrounding a single point charge and the second surface surrounding no charge. Gauss' law states that the net flux through any closed surface is proportional to the charge enclosed so that the integral over surface 1 is

$$\oiint_{\text{surf}} \mathbf{E}\cdot d\mathbf{a} = \frac{q_{\text{enclosed}}}{\varepsilon_0}. \tag{4.9}$$

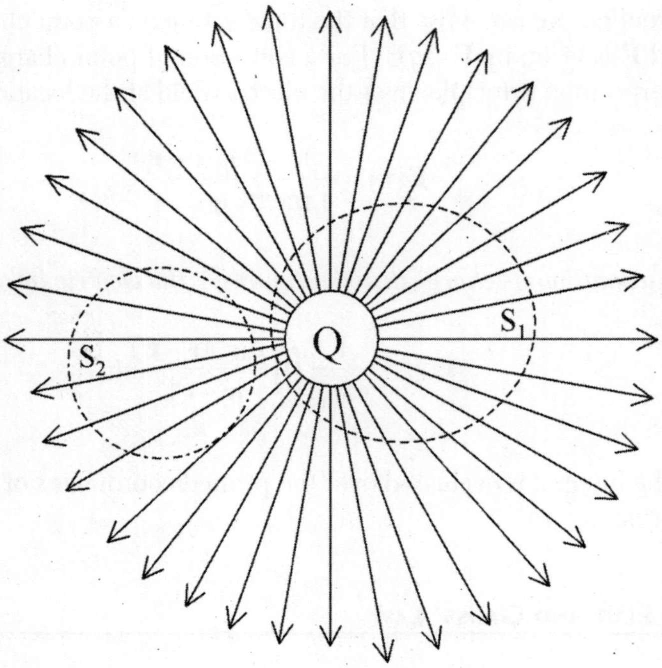

FIGURE 4.2 **The net flux through Gaussian surface S_1 is equal to Q/ε_0. The net flux through surface S_2 is zero.**

The mathematical boundary of the enclosing Gaussian surface need not coincide with the boundary of any physical surface. The electric flux through any given closed surface is zero if that surface surrounds no net charge. For example, the total flux will be zero through any arbitrary surface enclosing two equal point charges of opposite sign.

Calculation of the Electric Field from Gauss' Law

Gauss' law may be used to calculate the electric field resulting from sufficiently symmetric charge distributions with a suitable choice of surrounding Gaussian surface. The Gaussian surface should have the same symmetry as the enclosed charge distribution so that the electric field lines will be perpendicular to that surface. As an example, the electric field lines will be normal to a spherical Gaussian surface surrounding a point charge q located at the origin $\mathbf{r}' = 0$ so that

$$\oiint_{\text{surf}} \mathbf{E} \cdot d\mathbf{a} = E \oiint_{\text{surf}} da = E_r \cdot 4\pi r^2. \tag{4.10}$$

This must be equal to q/ε_0 giving the radial component of electric field

$$E_r = \frac{1}{4\pi\varepsilon_0}\frac{q}{r^2}, \tag{4.11}$$

where $\mathbf{E} = E_r\hat{\mathbf{r}}$.

Divergence of the Electric Field

Gauss' law may be used to calculate the divergence of the electric field, where the total charge is given by a volume integral over the charge density so that equation (4.9) becomes

$$\oiint_{\text{surf}} \mathbf{E}\cdot d\mathbf{a} = \frac{1}{\varepsilon_0}\iiint_{\text{vol}} \rho(\mathbf{r}')dv'. \tag{4.12}$$

Gauss' divergence theorem transforms the surface integral of the normal electric field into a volume integral of the electric field's divergence

$$\iiint_{\text{vol}} \nabla\cdot\mathbf{E}\,dv' = \frac{1}{\varepsilon_0}\iiint_{\text{vol}} \rho(\mathbf{r}')dv'. \tag{4.13}$$

The integrands must be equal since both integrals are evaluated over the same arbitrary volume. The divergence of electric field at every point in space is numerically equal to the charge density divided by the permittivity or

$$\nabla\cdot\mathbf{E} = \frac{\rho}{\varepsilon_0} \tag{4.14}$$

so we have that $\nabla\cdot\mathbf{E} = 0$ in charge free regions. Also, the spatial distribution of ρ may be calculated given $\mathbf{E}$. For example, $\mathbf{E}(\mathbf{r})$ is given by equation (4.4) for a single point charge located at $\mathbf{r}'$ so that

$$\varepsilon_0\nabla\cdot\mathbf{E}(\mathbf{r}) = \frac{q}{4\pi}\nabla\cdot\frac{\mathbf{r}-\mathbf{r}'}{|\mathbf{r}-\mathbf{r}'|^3} = q\delta(\mathbf{r}-\mathbf{r}'), \tag{4.15}$$

which is a delta function charge distribution. Equation (4.14) represents three equations and three unknowns (E_x, E_y, E_z) in Cartesian coordinates, but may be used to calculate the electric field in situations, where the charge density varies in one dimension, say $\rho(x)$. Equation (4.14) may also be solved in spherical or cylindrical coordinates, where the charge density depends only on the radial coordinate. For more complex charge distributions, it is more convenient to work with the scalar electric potential.

4.3 ELECTRIC POTENTIAL

The electrostatic potential at a distance r from a single point charge located at the origin $\mathbf{r}' = 0$ of a spherical coordinate system is given by

$$U(r) = \frac{1}{4\pi\varepsilon_0}\frac{q}{r}, \tag{4.16}$$

where the zero of potential is located at $r = \infty$. The electric field is

$$\mathbf{E} = -\nabla U(\mathbf{r}). \tag{4.17}$$

In spherical coordinates the radial part of the gradient is

$$\nabla = \hat{\mathbf{r}}\frac{\partial}{\partial r} \tag{4.18}$$

so that

$$\mathbf{E}(r) = \frac{1}{4\pi\varepsilon_0}\frac{q}{r^2}\hat{\mathbf{r}}. \tag{4.19}$$

For a collection of point charges we have the superposition

$$U(\mathbf{r}) = \frac{1}{4\pi\varepsilon_0}\sum_i \frac{q_i}{|\mathbf{r} - \mathbf{r}_i|}. \tag{4.20}$$

The electric potential for a continuous charge distribution is given by

$$U(\mathbf{r}) = \frac{1}{4\pi\varepsilon_0}\iiint_{\text{vol}} \frac{\rho(\mathbf{r}')}{|\mathbf{r} - \mathbf{r}'|}dv'. \tag{4.21}$$

In calculating the electric field, the ∇ operator may be brought inside the integral because ∇ acts on the unprimed field coordinates while the integral is taken over the primed source coordinates

$$\nabla U(\mathbf{r}) = \frac{1}{4\pi\varepsilon_0}\iiint_{\text{vol}} \nabla\frac{\rho(\mathbf{r}')}{|\mathbf{r} - \mathbf{r}'|}dv'. \tag{4.22}$$

Using the identity

$$\nabla \frac{1}{|\mathbf{r} - \mathbf{r}'|} = -\frac{(\mathbf{r} - \mathbf{r}')}{|\mathbf{r} - \mathbf{r}'|^3} \tag{4.23}$$

we have

$$\mathbf{E}(\mathbf{r}) = \frac{1}{4\pi\varepsilon_0} \iiint_{\text{vol}} \frac{\rho(\mathbf{r}')(\mathbf{r} - \mathbf{r}')}{|\mathbf{r} - \mathbf{r}'|^3} dv'. \tag{4.24}$$

For electrostatics we have that $\mathbf{E} = -\nabla \boldsymbol{U}$ and

$$-\nabla \times \mathbf{E} = \nabla \times \nabla U = 0 \tag{4.25}$$

so that electrostatic fields are curl free.

Delta Function Charge Distribution

The charge distribution of a collection of point charges may be written as the sum of delta functions locating the ith point charge multiplied by q_i

$$\rho(\mathbf{r}') = \sum_i q_i \delta(\mathbf{r}' - \mathbf{r}_i). \tag{4.26}$$

The potential $V(\mathbf{r})$ is then given by an integral over the charge distribution

$$U(\mathbf{r}) = \frac{1}{4\pi\varepsilon_0} \sum_i \iiint_{\text{vol}} \frac{q\delta(\mathbf{r}' - \mathbf{r}_i)}{|\mathbf{r} - \mathbf{r}'|} dv'. \tag{4.27}$$

Using the integral property of the delta function

$$U(\mathbf{r}) = \frac{1}{4\pi\varepsilon_0} \sum_i \frac{q_i}{|\mathbf{r} - \mathbf{r}_i|}. \tag{4.28}$$

4.4 QUICKFIELD ELECTROSTATIC CALCULATION

QuickField allows for the solution of the electric potential U in situations with x-y or axial symmetry. The dielectric permittivity may be inhomogeneous or anisotropic. For spatially varying ε, the permittivity tensor, we have

$$\nabla \cdot \varepsilon\mathbf{E} = \rho. \tag{4.29}$$

where $\mathbf{E} = -\nabla U$ so that

$$\nabla \cdot (\varepsilon \nabla U) = -\rho. \tag{4.30}$$

For zero charge distribution and constant ε the electric potential satisfies Laplace's equation

$$\nabla^2 U = 0. \tag{4.31}$$

In x-y symmetry the permittivity $\varepsilon = (\varepsilon_x, \varepsilon_y)$ is anisotropic if $\varepsilon_x \neq \varepsilon_y$ and

$$\frac{\partial}{\partial x}\left(\varepsilon_x \frac{\partial U}{\partial x}\right) + \frac{\partial}{\partial y}\left(\varepsilon_y \frac{\partial U}{\partial y}\right) = -\rho. \tag{4.32}$$

In axial symmetry the permittivity $\varepsilon = (\varepsilon_r, \varepsilon_z)$ is anisotropic if $\varepsilon_r \neq \varepsilon_z$ and

$$\frac{1}{r}\frac{\partial}{\partial r}\left(\varepsilon_r r \frac{\partial U}{\partial r}\right) + \frac{\partial}{\partial z}\left(\varepsilon_z \frac{\partial U}{\partial z}\right) = -\rho. \tag{4.33}$$

Once the electric potential is known the field components may be calculated according to

$$E_x = -\frac{\partial U}{\partial x} \quad E_y = -\frac{\partial U}{\partial y} \tag{4.34}$$

for x-y symmetry and

$$E_z = -\frac{\partial U}{\partial z} \quad E_r = -\frac{\partial U}{\partial r} \tag{4.35}$$

for problems with axial symmetry.

Boundary Conditions

The window for editing edge label data for electrostatics problems in QuickField is shown in Figure 4.3. To open the edge label window, click on the corresponding edge label in the tree diagram on the left of the screen.

Dirichlet Boundary Conditions

Assign specific potential values on exterior boundaries or edges within the solution region. This boundary condition is typically applied to the outward boundary of the solution region. This condition must be applied to at least one point in the solution region.

Neumann Boundary Conditions

Specify the normal electric displacement on either exterior boundaries as $D_n = \sigma$ or interior boundaries as $D_n^+ - D_n^+ = \sigma$. The discontinuity in the electric displacement is numerically equal to the surface charge density. Both Neumann and Dirichlet boundary conditions may depend on coordinates, where a specific formula is entered instead of a numerical value.

Periodic Boundary Conditions

Can be specified on two edges of the model that are not necessarily parallel. This type of boundary condition is used to reduce model sizes, where only one element of a periodic structure needs to be modeled. It is not necessary to specify the potential or its derivative on a periodic boundary. The potential is given by $U_1 = U_2$ on even periodic boundaries and $U_1 = -U_2$ on odd periodic boundaries.

Constant Potential Boundary Conditions

Are used for isolated conductors whose surfaces are at constant but unknown potential values. Conductors carrying a total charge may be assigned this boundary condition with the total charge specified as a single node on the conductor.

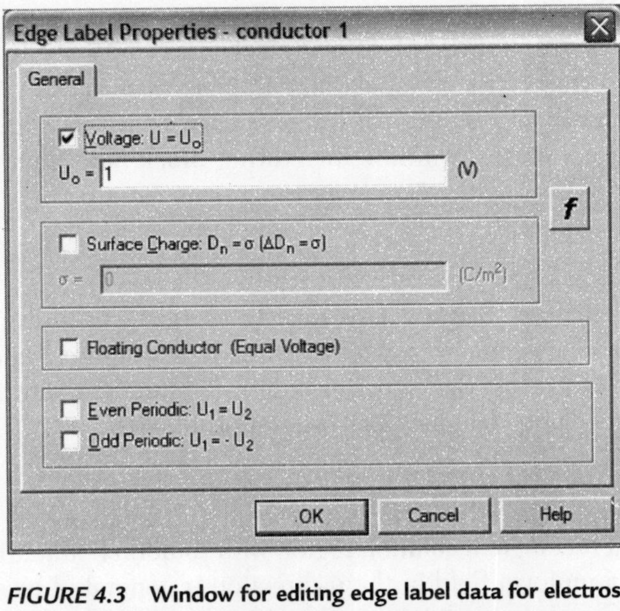

FIGURE 4.3 **Window for editing edge label data for electrostatics problems. To open the edge label window, click on the corresponding edge label in the tree diagram on the left of the screen.**

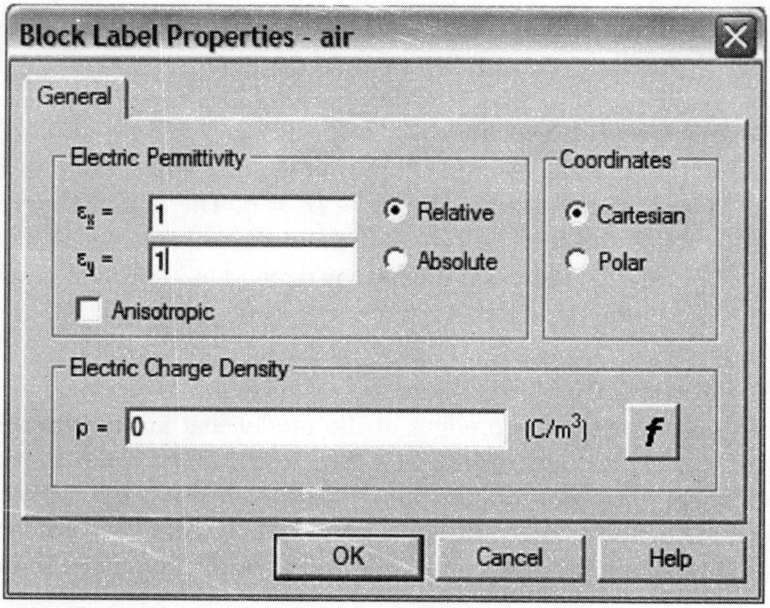

FIGURE 4.4 **Window for editing block label data for electrostatics problems. To open the block label window, click on the corresponding block label in the tree diagram on the left of the screen.**

Material Properties

Block labels are associated with dielectric permittivity that may be isotropic or anisotropic. Block labels may also be associated with electric charge density entered as a numerical value in C/m^3 or as a function of coordinates. Figure 4.4 shows the window for editing block label data for electrostatics problems.

Vertex Conditions

Local potential or charge values may be assigned to individual vertices. The potential or charge values may be specified by a numerical value or as a function of coordinates using the equation editor. Figure 4.5 shows the window for editing vertex label data for electrostatics problems.

Uniform Electric Field and Dipole Sources

Assuming a rectangular solution region with solution boundaries at $x = (0, a)$ and $y = (0, b)$, a uniform field in the y-direction is generated by assigning separate Dirichlet boundary conditions on the electric potential to the edges $y = (0, a)$ so that $U(x = 0) \neq U(x = a)$ and assigning a Neumann boundary condition, where the

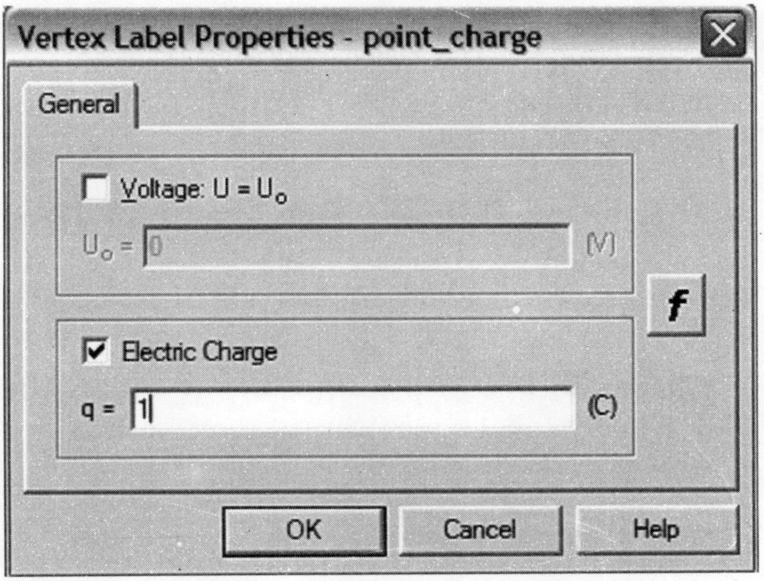

FIGURE 4.5 **Window for editing vertex label data for electrostatics problems: To open the vertex label window, click on the corresponding vertex label in the tree diagram on the left of the screen.**

normal electric field $E_{\perp} = 0$ on the edges $y = (0, b)$. The magnitude of the electric field will then be

$$E = \frac{U(x = a) - U(x = 0)}{a} \tag{4.36}$$

and will point in the direction of decreasing electric potential. Alternatively, a uniform electric field in the x-direction is generated by assigning Dirichlet conditions to the edges $x = (0, b)$ and Neumann conditions to $y = (0, b)$.

In r-z axisymmetry, a uniform field along the z-axis is generated by assigning a Dirichlet boundary condition to the edge $r = a$ and Neumann boundary conditions to $z = (0, b)$. For an accurate determination of the flux distribution near a point, the edges of the solution region should be removed a sufficient distance so as not to influence the field distribution in the region of interest.

Integral Quantities

The total charge is obtained from Gauss' law

$$q = \oint\!\!\oint_{\text{surf}} \mathbf{D} \cdot \mathbf{n} da, \tag{4.37}$$

where $\mathbf{D} = \varepsilon\mathbf{E}$ and $\nabla \cdot \mathbf{D} = \rho_{\text{free}}$, where ρ_{free} is the contribution to the charge density due to free charges and not bound polarization charges.

The electric force acting on the body may be obtained by integration of Maxwell's stress tensor

$$\mathbf{F} = \frac{1}{2}\oiint_{\text{surf}}\left(\mathbf{E}(\mathbf{n}\cdot\mathbf{D}) + \mathbf{D}(\mathbf{n}\cdot\mathbf{E}) - \mathbf{n}(\mathbf{E}\cdot\mathbf{D})\right)da. \tag{4.38}$$

The torque is given by the formula $\mathbf{T} = \mathbf{r}\times\mathbf{F}$

$$\mathbf{T} = \frac{1}{2}\oiint_{\text{surf}}\left((\mathbf{r}\times\mathbf{E})(\mathbf{n}\cdot\mathbf{D}) + (\mathbf{r}\times\mathbf{D})(\mathbf{n}\cdot\mathbf{E}) - (\mathbf{r}\times\mathbf{n})(\mathbf{E}\cdot\mathbf{D})\right)da. \tag{4.39}$$

The total energy W contained in the volume V

$$W = \frac{1}{2}\iiint_{\text{vol}}(\mathbf{E}\cdot\mathbf{D})dv, \tag{4.40}$$

where the electric energy density is given by $u_E = (\mathbf{E}\cdot\mathbf{D})/2$. The total energy stored in the fields is also related to the capacitance by the equivalent forms $W = CU^2/2 \times Q^2/2C$.

Point Charge Inside a Conical Cavity in a Dielectric

A point charge inside a dielectric conical cavity may be modeled in axial symmetry with the point charge located on the z-axis about which the conical surface is symmetric. Figure 4.6 shows the potential distribution inside the conical hole. The finite element mesh is shown in both the air and dielectric regions. A contour plot of induced surface charge density in C/m^2 on the surface of the conical cavity is displayed in Figure 4.7. The surface charge is maximal on the conical surface closer to the point charge in this figure.

Charged Needle Threading a Circular Hole in a Conducting Sheet

A charged needle threading a circular hole in a grounded conducting sheet may be modeled in axial symmetry with the charged needle located on the z-axis about which the hole is symmetric. Figure 4.8 shows the equipotentials and electric field vectors in the region surrounding the needle and the hole. The equipotentials lines are tangential to the charged needle and the conducting surface while the electric field is perpendicular to these surfaces. The needle has a length of 10.0 cm, a total charge of 1.0 C, and is centered in the hole with a radius of 4.0 cm. The floating conductor (equal voltage) boundary condition is assigned to the needle with the total charge specified on a single node at the

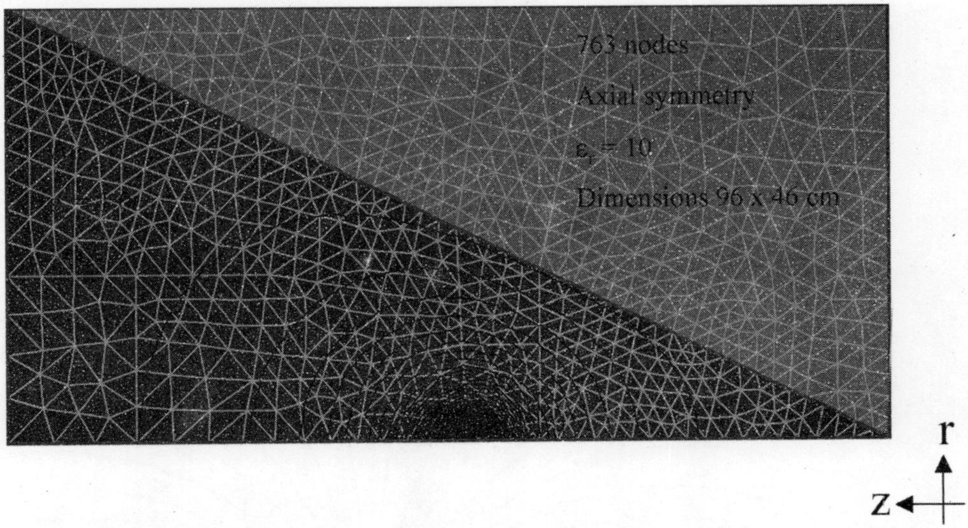

FIGURE 4.6 **Potential distribution resulting from a point charge positioned along the axis of a conical cavity inside a dielectric medium. The finite element mesh is shown in both the air and dielectric regions.**

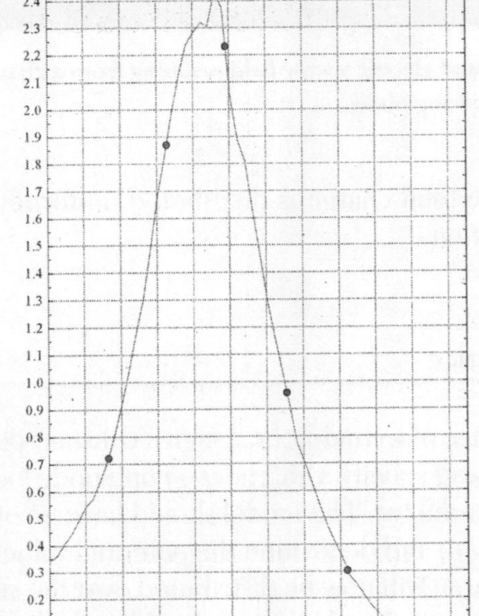

FIGURE 4.7 **Contour plot of induced surface charge density in C/m² on the surface of the conical cavity modeled in Figure 4.6.**

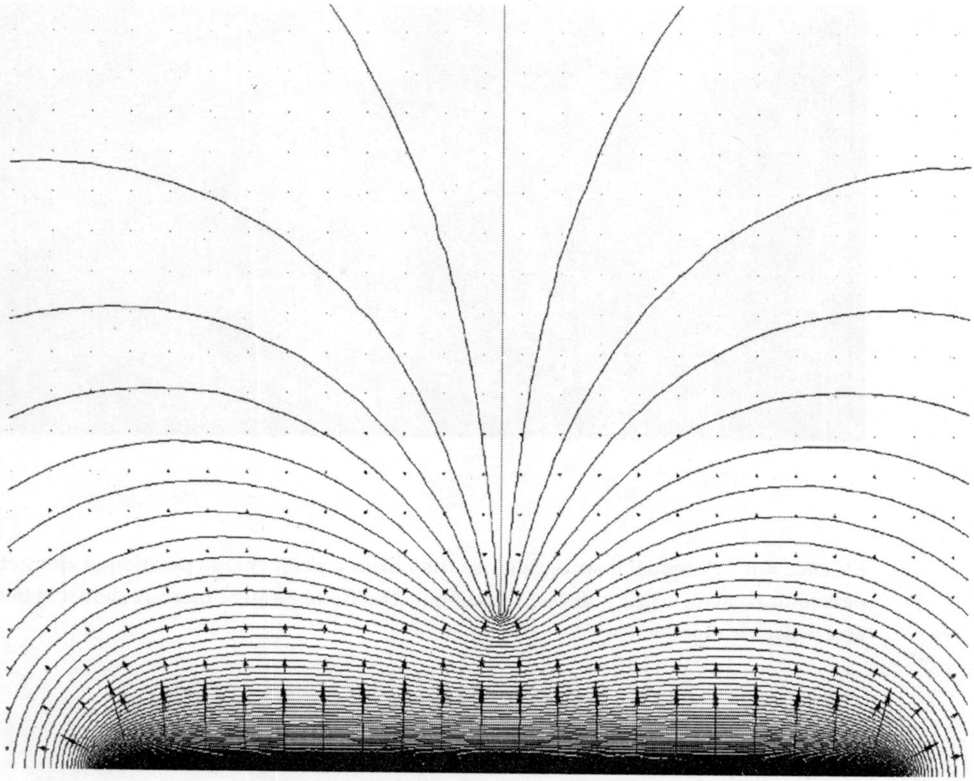

FIGURE 4.8 **Equipotentials and electric vector field resulting from a charged needle inside a circular hole through a conducing plane.**

center of the needle. The total charge is distributed uniformly over the needle with this boundary condition.

4.5 CALCULATION OF CAPACITANCE

To calculate the capacitance of a conductor, assign a constant potential boundary condition to a conductor and specify a charge q on one node located somewhere on the boundary of the conductor. The model should have no other field sources specified. Solve the problem and determine the potential value U on the surface of the conductor. The charge will now be distributed over the surface of the conductor. The capacitance C can then be determined by dividing the total charge on the conductor divided by the surface potential U

$$C = \frac{q}{U}. \tag{4.41}$$

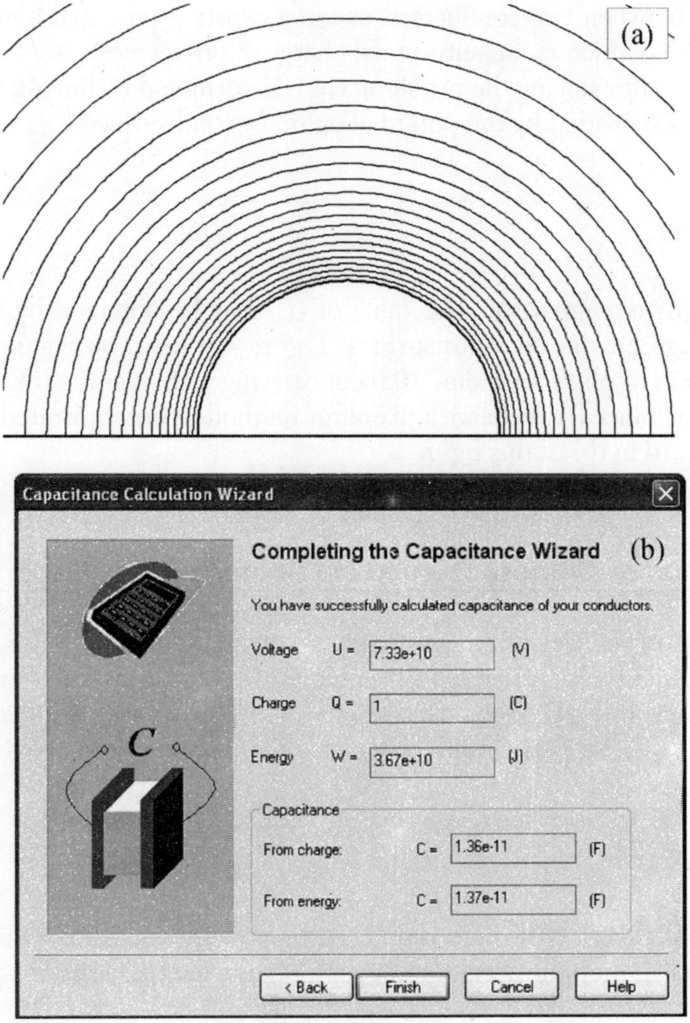

FIGURE 4.9 **(a) Equipotentials resulting from a sphere of radius 10.0 cm carrying a charge of 1.0 C. (b) Capacitance Wizard showing calculation of capacitance from charge and energy methods.**

The capacitance may also be obtained from the total energy W by

$$C = \frac{2W}{U^2} \quad \text{or} \quad C = \frac{q^2}{2W}, \tag{4.42}$$

where W is obtained by performing an integral $\mathbf{E}\cdot\mathbf{D}/2$ over all space. The Capacitance Wizard may be used in QuickField to calculate the capacitance from either the potential or the total field energy. To calculate the mutual capacitance

C_{12} between two conductors, assign a constant potential boundary condition to both conductors. Specify a total charge q_1 on one node of either conducting surface. After solving the problem, C_{12} is determined by the total charge on the first surface divided by the potential on the second surface U_2

$$C_{12} = \frac{q_1}{U_2}. \tag{4.43}$$

As a comparison, the value of C_{21} may be obtained by instead placing the charge on the opposite surface. Figure 4.9 (a) shows equipotentials resulting due to a sphere of radius 10.0 cm carrying a charge of 1.0 C. The calculation of capacitance from charge and energy methods is demonstrated in the Capacitance Wizard in (b) of this figure.

4.6 MOTION OF CHARGED PARTICLES IN STATIC ELECTRIC FIELDS

The trajectories of charged particles in static electric fields may be calculated in QuickField. A particle with charge q in an external electric field experiences a force $\mathbf{F} = q\mathbf{E}$. Newton's second law $\mathbf{F} = m\mathbf{a}$ gives the resulting acceleration $\mathbf{a} = q\mathbf{E}/m$ for particle speeds much less than the speed of light

$$\frac{d^2\mathbf{r}}{dt^2} = \frac{q}{m}\mathbf{E}. \tag{4.44}$$

If the electric field only has components in the x-y plane $\mathbf{E} = E_x(x,y)\mathbf{i} + E_y(x,y)\mathbf{j}$, then the acceleration components may be written as

$$\begin{Bmatrix} \ddot{x} \\ \ddot{y} \\ \ddot{z} \end{Bmatrix} = \begin{Bmatrix} \frac{q}{m}E_x \\ \frac{q}{m}E_y \\ 0 \end{Bmatrix}. \tag{4.45}$$

Particle trajectories may also be simulated in axial symmetry, where there is no velocity component in the ϕ-direction. The electric field distribution for a given charge distribution is first calculated in QuickField. Equations (4.45) are then solved using a Runge-Kutta-Merson method with specified initial position (x_0, y_0, z_0) and velocity components (v_{x0}, v_{y0}, v_{z0}) specified in the

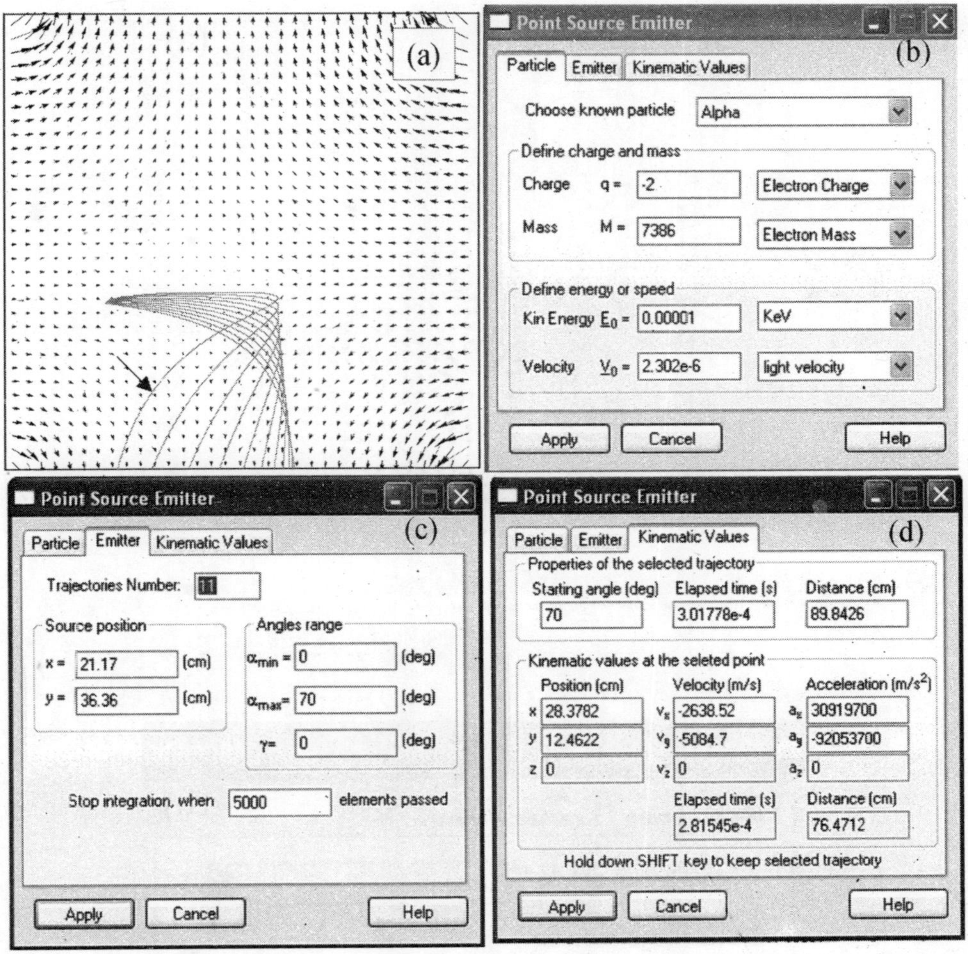

FIGURE 4.10 **(a) Particle trajectories in a non-uniform electric field. Windows for specifying; (b) particle charge, mass, and initial velocity; (c) number of trajectories, source position, and initial angle; (d) display of kinematical values of the upper left trajectory indicated by the arrow.**

Post Processing window. The length of the particles trajectory is obtained by integrating

$$\frac{d\ell}{dt} = \sqrt{\dot{x}^2 + \dot{y}^2 + \dot{z}^2} \tag{4.46}$$

along the path of the particle. Figure 4.10 (a) shows several particle trajectories in a non-uniform electric field. The Particle Source Emitter window is also shown, where values of the particles charge, mass, and initial velocity are

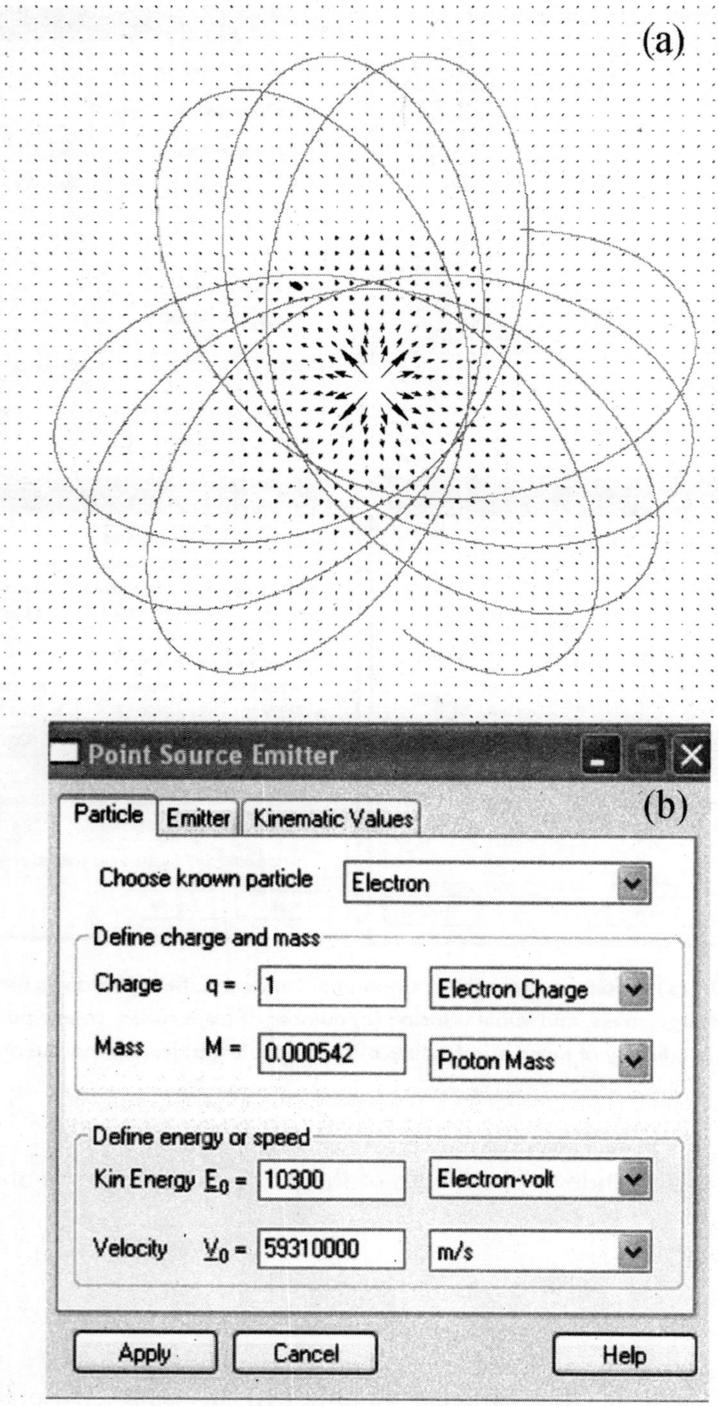

FIGURE 4.11 **(a) Single particle trajectory in an electric field produced by a line charge; (b) window for specifying particle charge, mass, and initial velocity of the particle.**

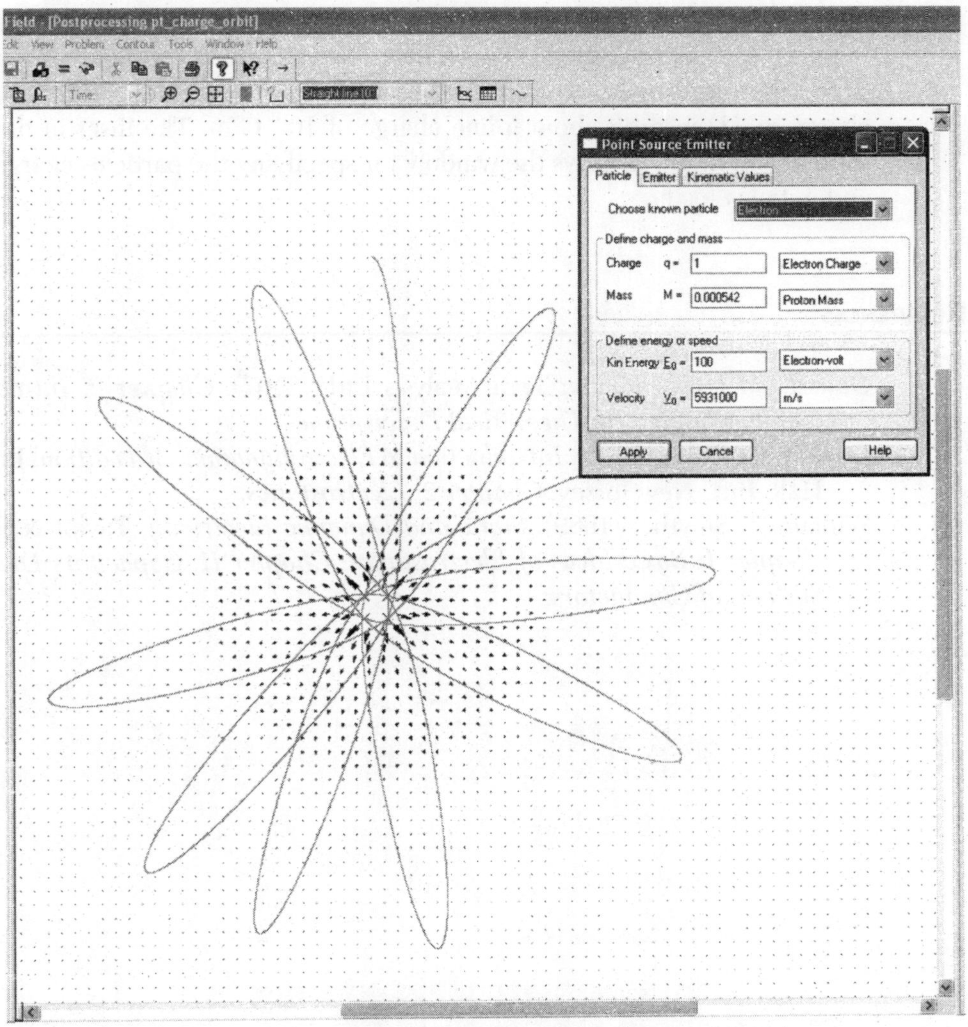

FIGURE 4.12 **Particle with proton mass and charge of –1 (antiproton) in orbit about a line charge of 10^{-6} C/m. Upper right: Window for specifying the particles charge, mass, and initial velocity.**

specified in (b) as well as the number of trajectories with a given source position and projection angle in (c) of this figure. Figure 4.10 (d) displays the calculated kinematical values of a single trajectory indicated by an arrow in (a). Numerical results may be viewed for each trajectory in this window.

Figure 4.11 (a) shows the trajectory of a single particle in the electric field produced by a line charge in *x-y* symmetry. The trajectory of the particle resembles that of a precessing ellipse with one focal point centered on the line charge. In the present case, however, the central force varies as $1/r$ unlike the

Kepler problem, where the force varies as $1/r^2$. The particles charge, mass, and initial velocity are specified in the window shown in (b) of this figure. Figure 4.12 shows the more tightly bound orbit of an antiproton (with a proton mass and charge of –1) in orbit about a line charge of 10^{-6} C/m. The inset in the upper right of this figure displays the window for specifying the particles charge, mass, and initial velocity.

REFERENCES

QuickField User's Guide: Tera Analysis Ltd. (2007). *Chapter 2: Introductory Guide; Chapter 10: Theoretical Description.*

David J. Griffiths. (1999). *Introduction to Electrodynamics 3rd edition*, Prentice-Hall, Inc., New Jersey. *Chapter 2: Electrostatics.*

William R. Smythe. (1989). *Static and Dynamic Electricity*, Taylor & Francis. *Chapter I: Basic Ideas of Electrostatics; Chapter II: Capacitors Dielectrics Systems of Conductors.*

EXERCISES

4.1 The electric field gradient $\mathbf{G} = \nabla\mathbf{E}$ in x-y symmetry has components

$$G_{xx} = \frac{\partial E_x}{\partial x}, \quad G_{yy} = \frac{\partial E_y}{\partial y}, \quad G_{xy} = \frac{1}{2}\left(\frac{\partial E_x}{\partial x} + \frac{\partial E_y}{\partial y}\right).$$

Calculate the electric field gradient at the center of four line charges forming a square with side a.

4.2 The electric field gradient $\mathbf{G} = \nabla\mathbf{E}$ in r-z symmetry has components

$$G_{zz} = \frac{\partial E_z}{\partial z}, \quad G_{rr} = \frac{\partial E_r}{\partial r}, \quad G_{zr} = \frac{1}{2}\left(\frac{\partial E_z}{\partial r} + \frac{\partial E_r}{\partial z}\right).$$

Calculate the electric field gradient along the axis of a thin charged needle.

4.3 Model a single point charge with charge Q located on the z-axis. Compare Gauss' law evaluating the surface integral of $\mathbf{E} \cdot \mathbf{n}$ over the cylindrical and spherical Gaussian surfaces shown in Figure E-4.3.

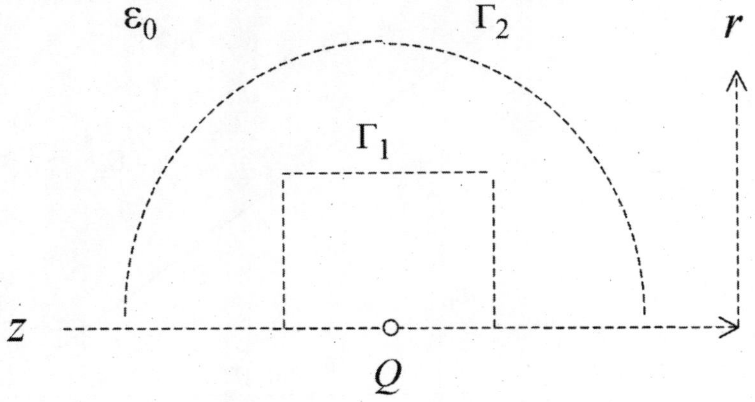

FIGURE E-4.3 **Single point charge surrounded by cylindrical and spherical Gaussian surfaces. Geometry: axial symmetry.**

4.4 Model a line charge with charge per unit length λ distributed in the z-direction, perpendicular to the page. Compare Gauss' law evaluating the surface integral of $\mathbf{E} \cdot \mathbf{n}$ over the rectangular and cylindrical surfaces shown in Figure E-4.4.

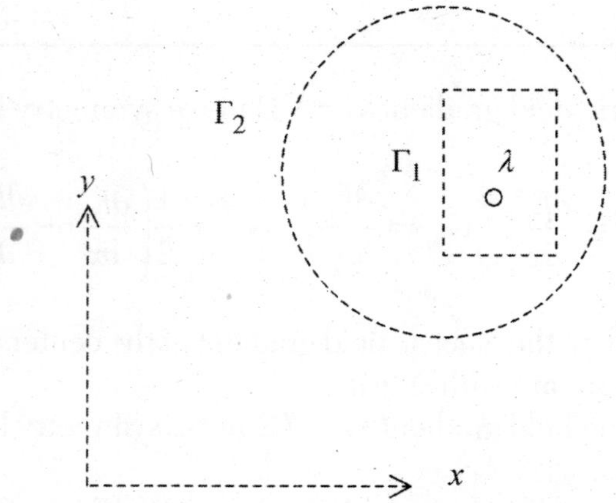

FIGURE E-4.4 **Charge per unit length λ surrounded by rectangular and cylindrical Gaussian surfaces. Geometry: axial symmetry.**

4.5 Model a point charge on the axis of the dielectric cone shown in Figure E-4.5. Plot the electric field distribution and make a contour plot of the induced surface charge on the inner surface of the dielectric cone. Calculate the force acting on the point charge.

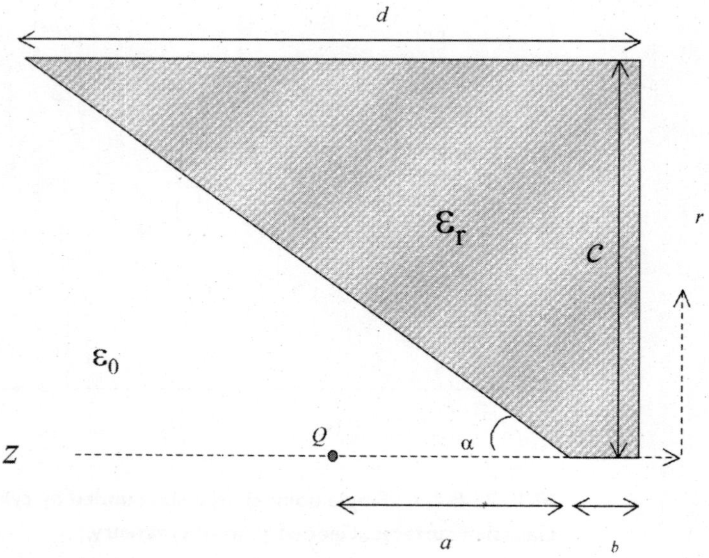

FIGURE E-4.5 **Point charge on the axis of a dielectric cone. Geometry: axial symmetry.**

4.6 Calculate the mutual capacitance of the two concentric conducting toroids shown in Figure E-4.6. Calculate the electric field in between the toroids held at opposite potentials and compare to the field obtained by solving Laplace's equation in toroidal coordinates.

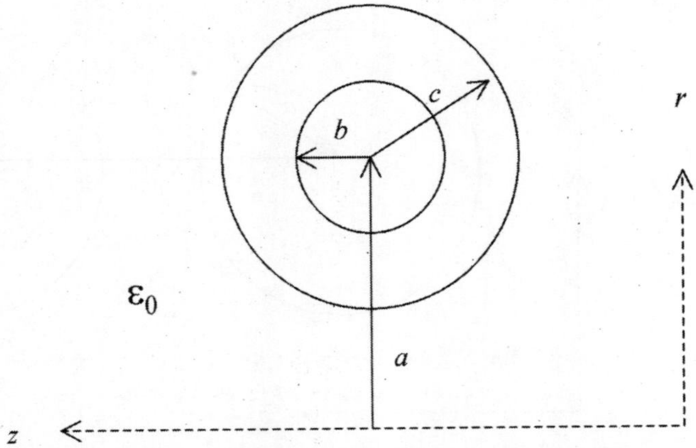

FIGURE E-4.6 **Concentric toroidal capacitor. Geometry: axial symmetry.**

4.7 Calculate the capacitance between the two conducting spheres of radius a with a center-to-center separation of d > $2a$ shown in Figure E-4.7. Plot the equipotentials, electric field vectors in the region between the spheres held at opposite potential. Compare your results to the solution obtained by solving Laplace's equation in bispherical coordinates.

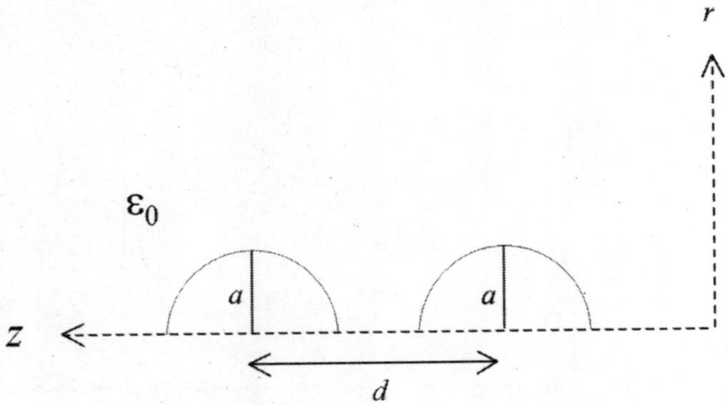

FIGURE E-4.7 **Two conducting spheres of radius *a* with center-to-center separation d > 2*a*. Geometry: axial symmetry.**

4.8 Calculate the capacitance between the two conducting cylinders of radius a with a center-to-center separation of d > $2a$ shown in Figure E-4.8. Calculate the equipotentials and electric field vectors in the region between the cylinders held at opposite potential. Compare your results to the solution obtained by solving Laplace's equation in bicylindrical coordinates.

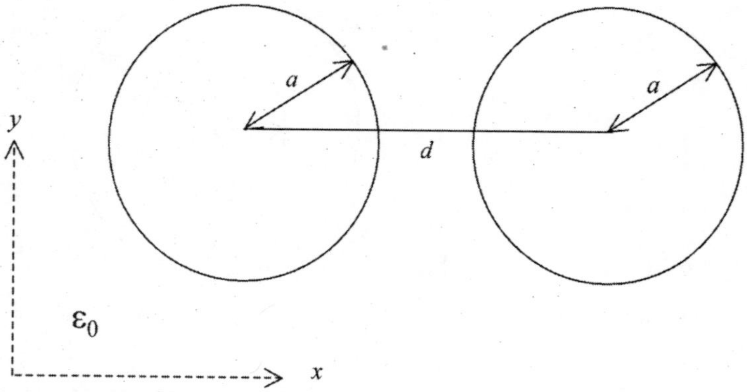

FIGURE E-4.8 **Two cylinders of radius *a* with center-to-center separation *d* > 2*a*. Geometry: axial symmetry.**

4.9 Calculate the mutual capacitance between the conducting sphere of radius a and the conducting plane shown in Figure E-4.9.

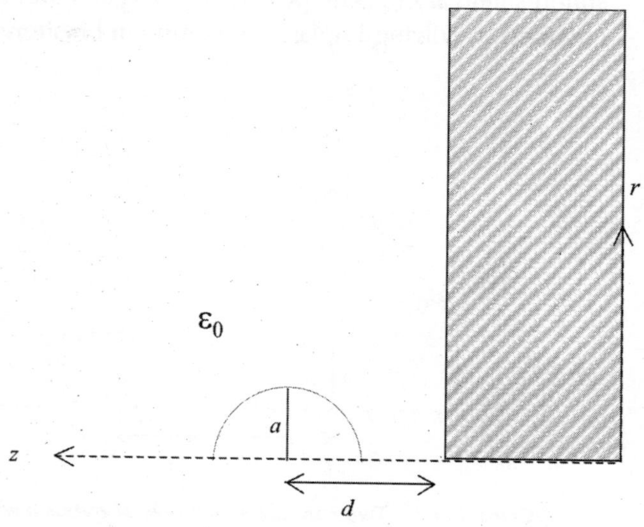

FIGURE E-4.9 **Sphere of radius *a* above a conducting plane. Geometry: axial symmetry.**

4.10 Plot the equipotentials and electric fields resulting from a charged needle passing through the center of a hole in the conducting plane shown in Figure E-4.10. Make a contour plot of the surface charge induced on the conducting plane. Use Label Mover to calculate the force on the needle translated at various positions along the z-axis.

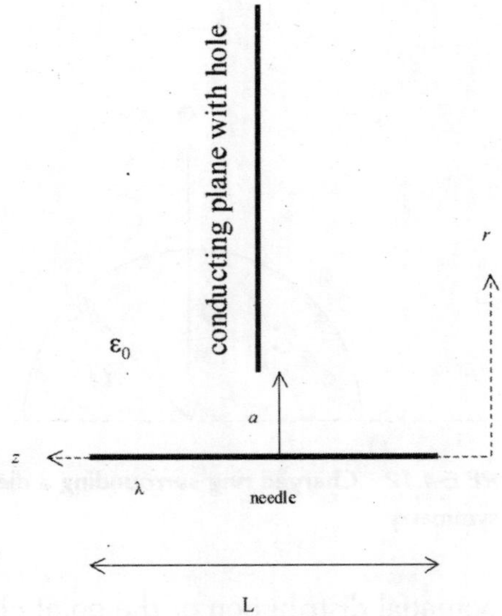

FIGURE E-4.10 **Charged needle threading a hole in a conducting sheet. Geometry: axial symmetry.**

4.11 Plot the potential and electric field distribution of the line charge parallel to a dielectric cylinder shown in Figure E-4.11. Calculate the force acting on the line charge and the dielectric cylinder.

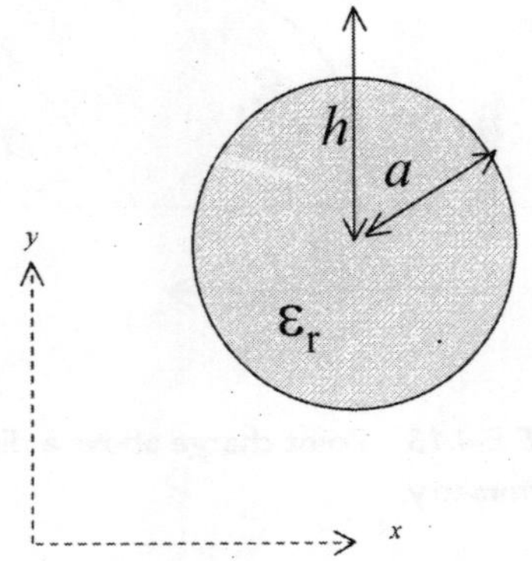

FIGURE E-4.11 **Point charge on the axis of a dielectric cone. Geometry: axial symmetry.**

4.12 Calculate the potential distribution of the charged ring surrounding the dielectric sphere in Figure E-4.12. Make a contour plot of the electric field along the z-axis passing through the sphere. Calculate the induced charge density on the surface of the sphere.

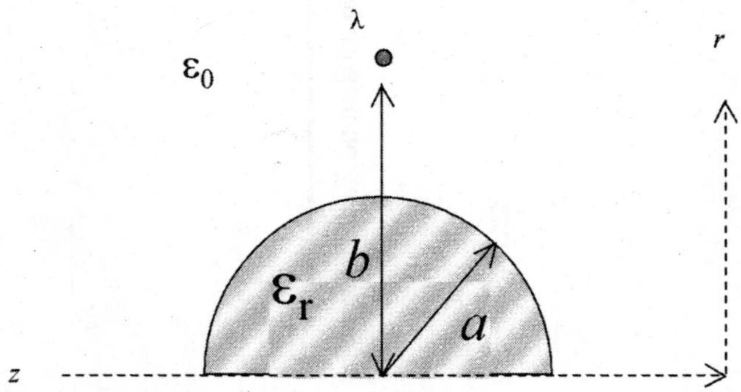

FIGURE E-4.12 **Charged ring surrounding a dielectric sphere. Geometry: axial symmetry.**

4.13 Plot the potential distribution of the point charge above the dielectric sphere shown in Figure E-4.13. Make a contour plot of the induced surface charge distribution on the outer surface of the sphere. Calculate the force acting on the sphere.

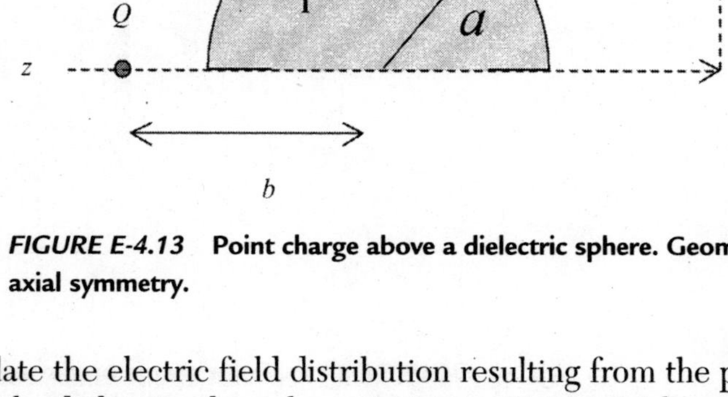

FIGURE E-4.13 **Point charge above a dielectric sphere. Geometry: axial symmetry.**

4.14 Calculate the electric field distribution resulting from the point charge above the dielectric plane shown in Figure E-4.14. Make a contour plot of the induced surface charge on the surface of the dielectric plane. Calculate the force acting on the point charge.

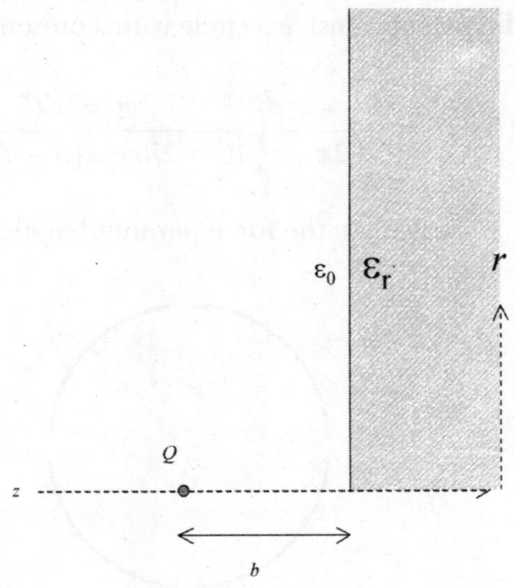

FIGURE E-4.14 **Point charge above a dielectric plane. Geometry: axial symmetry.**

4.15 Plot the electric field and potential distribution near the ring held at potential V_0 shown in Figure E-4.15. Make a contour plot of the z-component electric field along the axis of the ring.

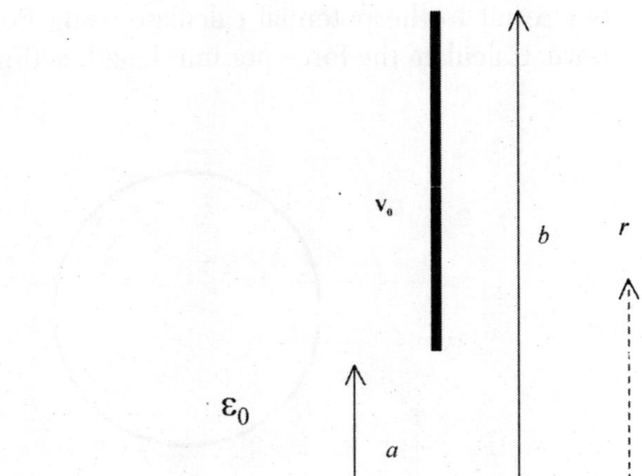

FIGURE E-4.15 **Wide ring held at a constant potential. Geometry: axial symmetry.**

4.16 Calculate the potential and electric field distribution of the four cylindrical quarter shells with alternating potential shown in Figure E-4.16. Compare your result to the potential calculate using Poisson's integral formula giving the potential inside a circle with a prescribed outer boundary condition

$$U(r,\theta) = \frac{R^2 - r^2}{2\pi}\int_0^{2\pi} \frac{g(\theta')d\theta'}{R^2 - 2Rr\cos(\theta - \theta') + r^2}, \quad \text{where } U(R,\theta) = g(\theta).$$

Calculate the force per unit length acting on each quarter shell.

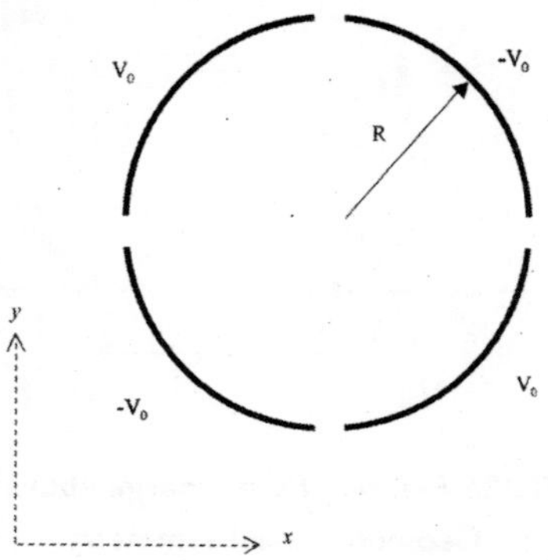

FIGURE E-4.16 **Cylindrical shell with alternating potentials on four cylindrical quarter shells. Geometry: x-y symmetry.**

4.17 Plot the potential and electric field distribution of the two cylindrical half shells with opposite potential shown in Figure E-4.17. Compare your result to the potential calculate using Poisson's integral formula above. Calculate the force per unit length acting on each half shell.

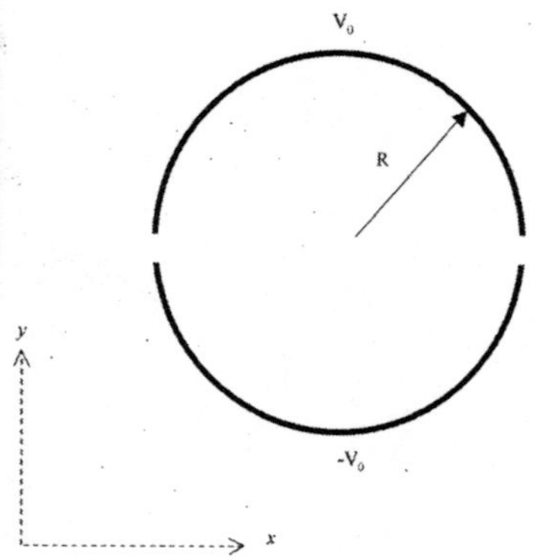

FIGURE E-4.17 **Two cylindrical half shells with opposite potential. Geometry: *x-y* symmetry.**

4.18 Calculate the potential and electric field distribution of the two hemispherical shells with opposite potential shown in Figure E-4.18. Calculate the force acting on the spherical half shells. Calculate the mutual capacitance of two perfect electrical conductors with the same geometry.

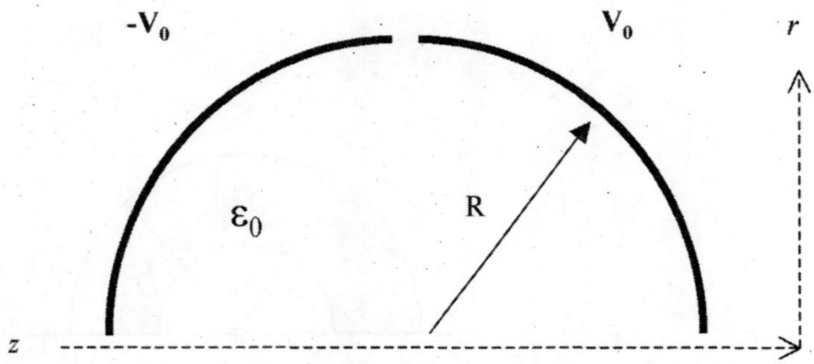

FIGURE E-4.18 **Two hemispherical shells held at opposite potential. Geometry: axial symmetry.**

4.19 Plot the potential distribution of a line charge inside the hollow grounded right triangular conduit shown in Figure E-4.19. Calculate the total charge induced on the interior surface of the conduit.

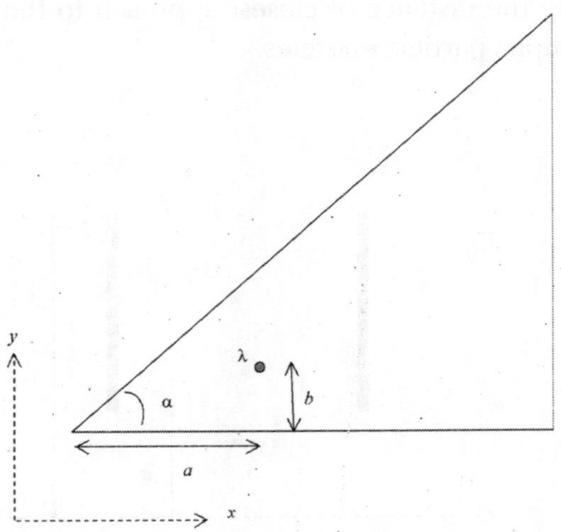

FIGURE E-4.19 **Line charge inside a hollow grounded right triangular conduit. Geometry: *x-y* symmetry.**

4.20 Calculate the potential and electric field distribution of a point charge inside the dielectric sphere shown in Figure E-4.20. Make contour plots of the surface charge distribution on the inner and outer surfaces of the sphere and calculate the total charge on each surface. Compare your results for the point charge located off-center along the z-axis. Also, calculate the force between the point charge and the sphere when the point charge is off-center.

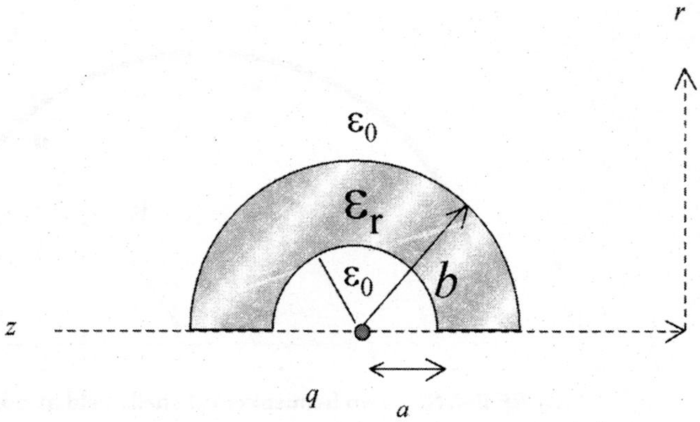

FIGURE E-4.20 **Point charge inside a dielectric sphere. Geometry: axial symmetry.**

4.21 Model a particle accelerator consisting of two washers held at opposite potentials V_- and V_+ as shown in Figure E-4.21. Inject alpha particles with zero initial speed at the center of the rings. Position a point charge equivalent to that of a uranium nucleus on the z-axis outside of the rings. Calculate the distance of closest approach to the uranium nucleus for several alpha particle energies.

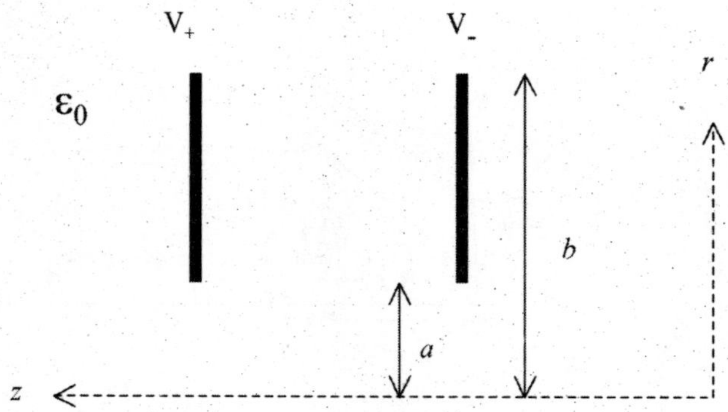

FIGURE E-4.21 **Particle accelerator consisting of two washers held at opposite potentials. Geometry: axial symmetry.**

4.22 Guard rings are used to reduce fringing fields between capacitor plates thereby resulting in greater field uniformity between the plates. The guard ring in Figure E-4.22 is formed by a circular groove in one of the capacitor plates separating an outer ring from the central disk. The guard ring is maintained at the same potential as the rest of the plate. Compare the field uniformity between the plates with and without the guard ring by making a contour plot of the electric field along a radial contour between the capacitor plates.

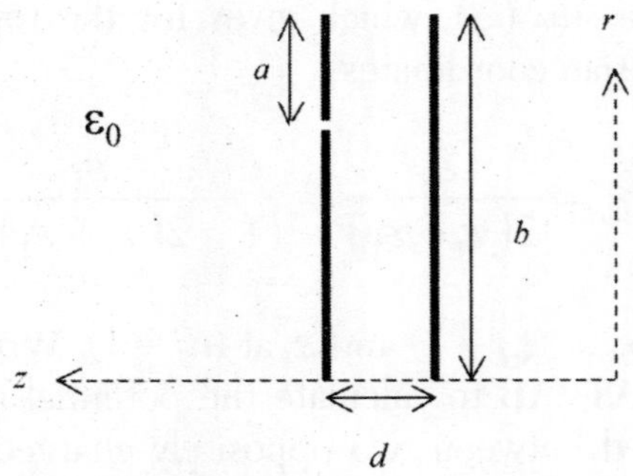

FIGURE E-4.22 **Circular capacitor with a guard ring. Geometry: axial symmetry.**

4.23 Plot the electric field distribution resulting from a dielectric wedge partially inserted into a parallel plate capacitor with the plate held at opposite potential as shown in Figure E-4.23. Calculate the force acting on the dielectric slab varying the position $0 < d < L$ using Label Mover.

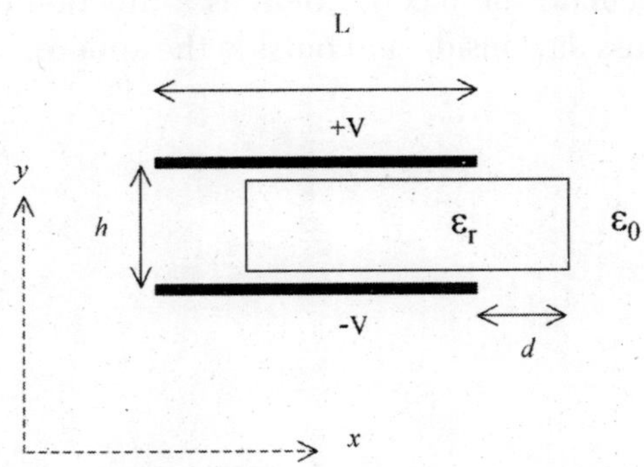

FIGURE E-4.23 **Dielectric wedge partially inserted in a parallel plate capacitor. Geometry: *x-y* symmetry.**

4.24 Two positive charges Q separated by a distance d are held fixed while a third negative charge orbits both charges chaotically. Write a MATLAB program that calculates the orbit of the negative charge. Make a 3-D plot of the orbit. Refer to Appendix 1 for tips on solving the system of three second-order differential equations.

4.25 The finite difference formula for the potential at the boundary between two dielectrics is obtained from the requirement

$$\oint \mathbf{D}\cdot d\ell = \oint \varepsilon \nabla U \cdot d\ell = 0$$

at the site (i,j), which gives for the top surface of the dielectric in Cartesian coordinates

$$U_{i,j} = \frac{\varepsilon_1}{2(\varepsilon_1+\varepsilon_2)}U_{i,j+1} + \frac{\varepsilon_2}{2(\varepsilon_1+\varepsilon_2)}U_{i,j-1} + \frac{1}{4}U_{i-1,j} + \frac{1}{4}U_{i+1,j}$$

with ε_1 at $(i,j+1)$ and ε_2 at $(i,j-1)$. Write a finite difference program in MATLAB to calculate the potential near a square dielectric beam located between two oppositely charged plates. Note that the coefficients in the above expression would also be equal to the Monte Carlo stepping probabilities at the site (i,j). These coefficients depend on the coordinate system one is working in and should sum to one. With a little resilience, they can be worked out for the other coordinate systems.

4.26 An oblate spheroid has a uniform charge density ρ_0 for $\xi \le \xi_0$. Calculate the electric field as a function of ξ using equation (4.14) or Gauss' law.

4.27 A prolate spheroidal shell has a uniform charge density ρ_0 for $\xi_1 \le \xi \le \xi_2$. Calculate the electric field as a function of ξ using equation (4.14) or Gauss' law.

4.28 A sphere of radius R carries a radial charge density $\rho_0 e^{-r}$ for $r \le R$. Calculate the electric field as a function of r using equation (4.14) or Gauss' law inside and outside the sphere.

Chapter 5 MAGNETOSTATICS

In This Chapter

- Biot-Savart Law
- Ampere's Law
- Gauss' Law for Magnetic Field
- Magnetic Scalar Potential
- Magnetic Vector Potential
- QuickField Magnetostatic Analysis
- Inductance Calculations
- Uniform Magnetic Fields
- Dipole Sources
- Shielding Applications
- Magnetic Monopoles

5.1 BIOT-SAVART LAW

While preparing a lecture demonstration in 1820, Ørsted noticed that current flowing through a wire deflected a nearby compass needle. After further investigations, he published his discovery that an electric current flowing through a wire gives rise to a magnetic field.

The strength of the electric field is determined by the magnitude the force exerted on a hypothetical test charge according to $\mathbf{F} = q\mathbf{E}$. In a similar way, the magnetic field $\mathbf{B}$ may be determined by the torque $\boldsymbol{\tau}$ exerted on a magnetic dipole with dipole moment $\boldsymbol{\mu}$

$$\tau = \mu \times \mathrm{B}. \tag{5.1}$$

The continuity equation

$$\frac{\partial \rho}{\partial t} + \nabla \cdot \mathbf{J} = 0 \tag{5.2}$$

is a statement of charge conservation. Magnetostatic fields are produced by steady state currents that are not changing in time so that we have

$$\nabla \cdot \mathbf{J} = 0. \tag{5.3}$$

The field $d\mathbf{B}$ produced by a current dipole $I\, d\ell$ located at the position $\mathbf{r}'$ decreases as the inverse square of the distance from the current source according to

$$d\mathbf{B}(\mathbf{r}) = \frac{\mu_0}{4\pi} I\, d\ell \times \frac{(\mathbf{r} - \mathbf{r}')}{|\mathbf{r} - \mathbf{r}'|^3}. \tag{5.4}$$

This expression is analogous to the Coulomb's law describing the field produced by a single point charge. The Biot-Savart law gives the magnetic field at a position $\mathbf{r}$ resulting from a continuous current distribution $\mathbf{J}(\mathbf{r}')$

$$\mathbf{B}(\mathbf{r}) = \frac{\mu_0}{4\pi} \iiint_{\text{vol}} \mathbf{J}(\mathbf{r}') \times \frac{(\mathbf{r} - \mathbf{r}')}{|\mathbf{r} - \mathbf{r}'|^3}\, dv'. \tag{5.5}$$

5.2 AMPERE'S LAW

Ampere's law relates the line integral of magnetic field around any closed path Γ surrounding a wire carrying a total current I as shown in Figure 5.1.

$$\oint_{\Gamma} \mathbf{B} \cdot d\ell = \mu_0 I. \tag{5.6}$$

The total current threading the loop is given by the surface integral of current density over any capping surface bound by the integration loop

$$I = \iint_{\text{surf}} \mathbf{J} \cdot d\mathbf{a}. \tag{5.7}$$

ON THE CD

Four color images for many of the figures may be found on the CD-ROM.

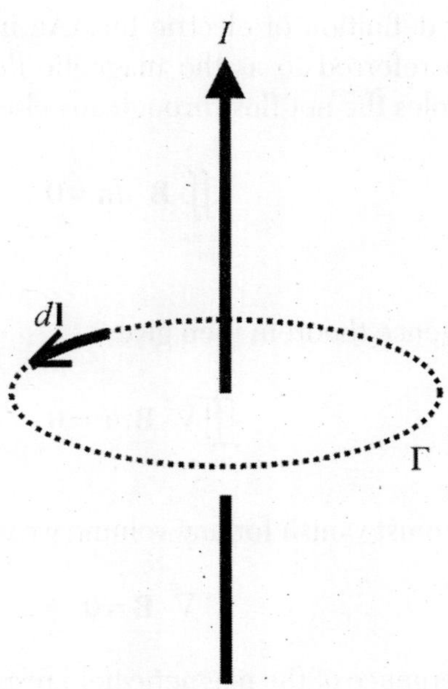

FIGURE 5.1 **Amperian loop Γ surrounding a wire carrying a current *I*.**

Combining (5.6) and (5.7) and applying Stokes' theorem gives

$$\iint_{\text{surf}} \nabla \times \mathbf{B} \cdot d\mathbf{a} = \mu_0 \iint_{\text{surf}} \mathbf{J} \cdot d\mathbf{a}. \tag{5.8}$$

Equating integrands since (5.8) holds for arbitrary capping surfaces gives the Maxwell relation

$$\nabla \times \mathbf{B} = \mu_0 \mathbf{J}. \tag{5.9}$$

5.3 Gauss' Law for Magnetic Fields

Magnetic flux is defined as the surface integral

$$\Phi_m = \iint_{\text{surf}} \mathbf{B} \cdot d\mathbf{a} \tag{5.10}$$

analogous to the definition of electric flux. An incremental flux $d\Phi_m = \mathbf{B} \cdot d\mathbf{a}$ so that $\mathbf{B}$ is also referred to as the magnetic *flux density*. In the absence of magnetic monopoles the net flux through any closed surface must be zero

$$\oiint_{\substack{\text{closed}\\ \text{surf}}} \mathbf{B} \cdot d\mathbf{a} = 0. \tag{5.11}$$

Gauss' divergence theorem then gives

$$\iiint_{\text{vol}} \nabla \cdot \mathbf{B}\, dv = 0 \tag{5.12}$$

This integral must vanish for any volume we choose so that

$$\nabla \cdot \mathbf{B} = 0. \tag{5.13}$$

The zero divergence of the magnetic field results from an absence of isolated magnetic charge.

5.4 MAGNETIC SCALAR POTENTIAL

In source free regions $\nabla \times \mathbf{B} = 0$ so that we may write the magnetic field as the gradient of a scalar function

$$\mathbf{B} = -\nabla \Omega, \tag{5.14}$$

where the scalar potential Ω is not considered to be a physical field as is the electric potential U. Since $\nabla \cdot \mathbf{B} = 0$ the scalar potential satisfies Laplace's equation

$$\nabla^2 \Omega = 0. \tag{5.15}$$

Spherical Permanent Magnet

As an example of the application of the scalar potential in treating magnetostatics problems, we consider the field established by a uniformly magnetized sphere of radius a with magnetization pointing in the z-direction. In spherical coordinates, the solution has symmetry about the z-axis so that $\Omega = \Omega(r,\theta)$. Solving Laplace's equation using separation of variables we have the scalar potential inside the sphere $r \leq a$

$$\Omega(r,\theta)=\sum_{n=0}^{\infty}a_n r^n P_n(\cos\theta). \tag{5.16}$$

The potential outside the sphere $r > a$ is given by

$$\Omega(r,\theta)=\sum_{n=0}^{\infty}c_n r^{-(n+1)} P_n(\cos\theta). \tag{5.17}$$

The boundary conditions for continuity of tangential **H** and normal **B** are

$$\left\{\begin{matrix} H_\theta^< = H_\theta^> \\ B_r^< = B_r^> \end{matrix}\right\}_{r=a}. \tag{5.18}$$

This gives the potential outside the sphere

$$\Omega^>(r,\theta)=\frac{M}{3}\frac{a^3}{r^2}\cos\theta \tag{5.19}$$

for $r \le a$ and inside the sphere

$$\Omega^<(r,\theta)=\frac{M}{3}r\cos\theta \tag{5.20}$$

for $r > a$. The magnetic field components are then given by the gradient in spherical coordinates.

5.5 MAGNETIC VECTOR POTENTIAL

Because the diverge of the curl of a vector field is zero

$$\nabla\cdot(\nabla\times\mathbf{A})=0 \tag{5.21}$$

and the divergence of the magnetic field is zero, we may write the magnetic field as the curl of a vector field

$$\mathbf{B}=\nabla\times\mathbf{A}. \tag{5.22}$$

Whereas the electric field **E** is obtained from the gradient of a scalar potential U, the magnetic field **B** is obtained from the curl of a vector field **A**. In general, the vector potential **A** will point in the same direction as the current source points in the same direction as the current source **J**. The vector potential at a location **r** may be expressed as an integral over the current distribution

$$\mathbf{A}(\mathbf{r}) = \frac{\mu_0}{4\pi} \iiint_{\text{vol}} \frac{\mathbf{J}(\mathbf{r}')}{|\mathbf{r}-\mathbf{r}'|} dv'. \tag{5.23}$$

Taking the curl of this expression and using the vector identity

$$\nabla \times \frac{\mathbf{J}}{|\mathbf{r}-\mathbf{r}'|} = \frac{1}{|\mathbf{r}-\mathbf{r}'|} \nabla \times \mathbf{J} - \mathbf{J} \times \nabla \frac{1}{|\mathbf{r}-\mathbf{r}'|}, \tag{5.24}$$

where

$$\nabla \frac{1}{|\mathbf{r}-\mathbf{r}'|} = -\frac{\mathbf{r}-\mathbf{r}'}{|\mathbf{r}-\mathbf{r}'|^3} \tag{5.25}$$

and we obtain the Biot-Savart law. As an example, consider the vector potential of an elemental current dipole pointing along the z-axis

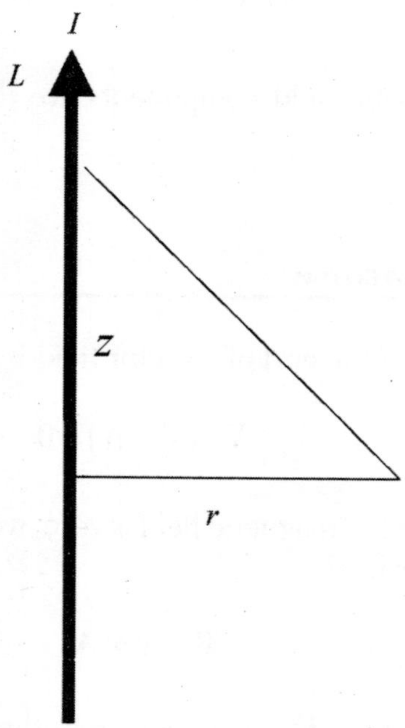

FIGURE 5.2 **Wire of length 2L carrying a current I.**

$$\mathbf{A} = \frac{\mu_0 I}{4\pi} \frac{d\ell}{r} \mathbf{k}. \tag{5.26}$$

The z-component vector potential for a current carrying wire of length $2L$ as shown in Figure 5.2 is given by

$$A_z = \frac{\mu_0 I}{4\pi} \int_{-L}^{L} \frac{dz}{\sqrt{r^2 + z^2}} \tag{5.27}$$

or

$$A_z = \frac{\mu_0 I}{4\pi} \ln\left(\frac{\sqrt{L^2 + r^2} + L}{\sqrt{L^2 + r^2} - L} \right) \tag{5.28}$$

The curl of the vector potential then provides the magnetic field.

5.6 QuickField Magnetostatic Analysis

Magnetostatic problems in QuickField are solved for the vector potential. The constitutive relation

$$\mathbf{H} = \frac{1}{\mu} \mathbf{B} \tag{5.29}$$

relates the magnetic field strength **H** to the flux density **B**. A steady current **J** will give rise to a field **H** with curl according to Maxwell's equation

$$\nabla \times \mathbf{H} = \mathbf{J}. \tag{5.30}$$

In addition, a material with magnetization vector **M** will contribute the field strength

$$\nabla \times \mathbf{H} = \mathbf{J} + \nabla \times \mathbf{M}, \tag{5.31}$$

where **H** is parallel to **B** and antiparallel to **M** and

$$\nabla \times \mathbf{H} = \nabla \times \left(\nabla \times \frac{1}{\mu} A \right). \tag{5.32}$$

For x-y symmetry, where the third dimension is considered infinite in extent, the vector potential $A_z(x,y)$ only has a z-component and is described by

$$\frac{\partial}{\partial x}\left(\frac{1}{\mu_y}\frac{\partial A_z}{\partial x}\right)+\frac{\partial}{\partial y}\left(\frac{1}{\mu_x}\frac{\partial A_z}{\partial y}\right)=-J+\left(\frac{\partial H_{cy}}{\partial x}-\frac{\partial H_{cx}}{\partial y}\right). \tag{5.33}$$

H_{cx} and H_{cy} are the x- and y-components of the coercive force vector $\boldsymbol{H}_c$, treated as the equivalent magnetization surface current on the boundary defining a region of constant magnetization. Once the vector potential is known, the flux density components can be determined from

$$B_x=\frac{\partial A_z}{\partial y}\quad B_y=-\frac{\partial A_z}{\partial x}. \tag{5.34}$$

In axial symmetry, the vector potential $A_\phi(r,z)$ only has a ϕ component and satisfies the equation

$$\frac{\partial}{\partial r}\left(\frac{1}{r\mu_z}\frac{\partial(rA_\phi)}{\partial r}\right)+\frac{\partial}{\partial z}\left(\frac{1}{\mu_r}\frac{\partial A_\phi}{\partial z}\right)=-J+\left(\frac{\partial H_{cr}}{\partial z}-\frac{\partial H_{cz}}{\partial r}\right) \tag{5.35}$$

with H_{cr} and H_{cz} the r- and z-components of the coercive force vector. The magnetic field components are then

$$B_z=\frac{1}{r}\frac{\partial}{\partial r}(rA_\phi)\quad B_r=-\frac{\partial A_\phi}{\partial z}. \tag{5.36}$$

The permeability tensor components (μ_x,μ_y) in equation (5.33) and (μ_r,μ_z) in equation (5.35) are constant in each block of the model, equal for isotropic materials and field dependent for nonlinear magnetics.

In carrying out magnetostatic simulations, the magnetostatics problem type is initially specified along with the designated x-y or r-z symmetry.

Magnetic Material Properties

Figure 5.3 shows the window for editing block label data for magnetostatics problems. Material properties are specified for air, orthotropic materials with constant permeability, ferromagnets, current carrying conductors, and permanent magnets.

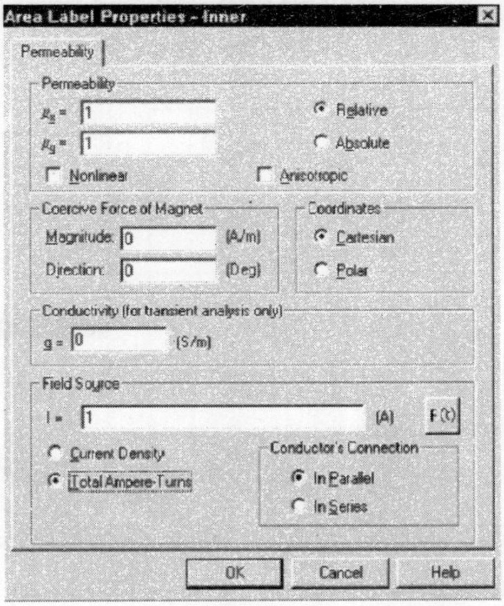

FIGURE 5.3 Window for editing block label data for magnetostatics problems: To open the block label window click on the corresponding block label in the tree diagram on the left of the screen.

Block labels

Block labels in magnetostatic problems can be associated with data containing two components of magnetic permeability tensor, the total current or the current density, and the magnitude and the direction of permanent magnet's coercive force. Source current distributions are specified either by the current density value or by the total current for a given coil. For total current, you may specify the labeled blocks as a single conductor or as several conductors connected in series. Serially connected conductors always carry the same current with calculated current densities inversely proportional to their areas. For axisymmetric coils, the current density may be specified to vary inversely with the radius rather than being distributed uniformly.

Nonlinear magnetics

B-*H* curves for ferromagnets can be specified through the interactive curve editor. Nonlinear magnetic properties may be assigned, where $B(H)$ gives a relation between the permeability and the flux density

$$\mu(B) = \frac{B}{H + H_c}. \tag{5.37}$$

Figure 5.4 illustrates the curve editing window for specifying nonlinear $B(H)$ magnetic properties.

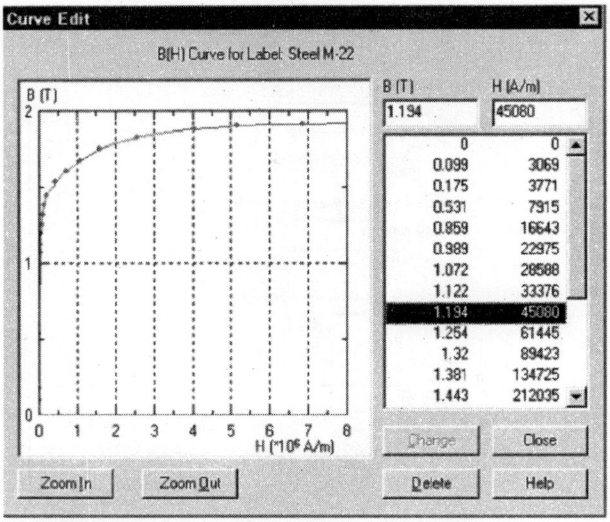

FIGURE 5.4 **Curve editing window for specifying nonlinear $B(H)$ magnetic properties.**

Boundary Conditions

Boundary conditions may be specified on boundaries of the model as well as edges of the solution region. Uniform and non-uniform fields as well as surface current densities can be simulated with assigned values of vector potential and

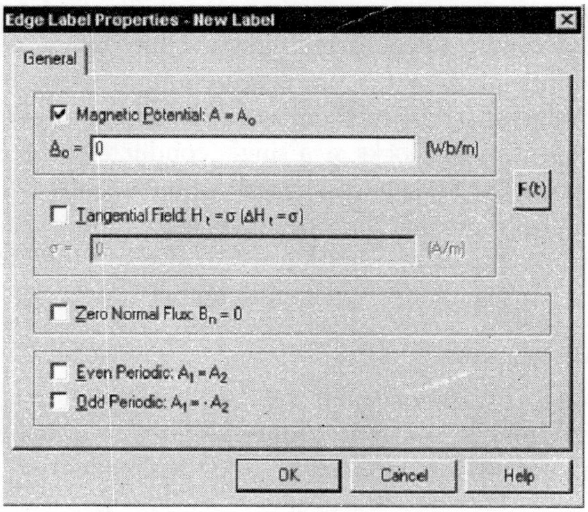

FIGURE 5.5 **Window for editing edge label data for magnetostatics problems: To open the edge label window click on the corresponding edge label in the tree diagram on the left of the screen.**

tangential field strength. Figure 5.5 illustrates the window for editing edge label data for magnetostatics problems.

Dirichlet conditions

Prescribe values of the vector potential $A_0 = a + bx + cy$ in x-y symmetry or the flux function $r\ A_0 = a + bzr + cr^2/2$ in axial symmetry on the boundaries of the model or external edges. a, b, and c are constants for each edge and are determined subject to the above constraints.

Neumann conditions

Specify the tangential flux density. At outward boundaries $H_t = \sigma$, where H_t is the tangential component of the field strength and σ is the linear surface current density. The homogeneous Neumann condition with $\sigma = 0$ is the default condition on external boundaries. At boundaries over interior regions within the model, the Neumann condition specifies a discontinuity in the field strength across the surface $H_t^+ - H_t^- = \sigma$, where the surface normal points in the direction from (−) to (+).

Periodic boundary conditions

Even $A_1 = A_2$ or odd $A_1 = -A_2$ can be applied to opposite edges of a model to simulate periodic structures. Both normal and tangential field components may be present at a periodic boundary.

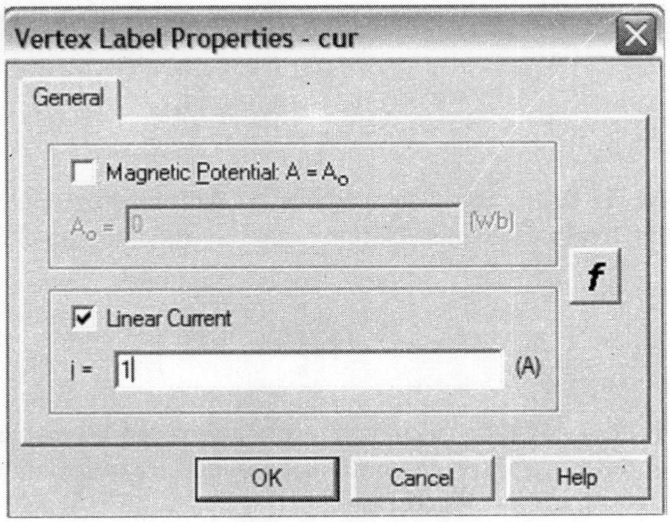

FIGURE 5.6 **Window for editing vertex label data for magnetostatics problems: To open the vertex label window click on the corresponding vertex label in the tree diagram on the left of the screen.**

Vertex Conditions

Vertex labels can be associated with specific values of the vector potential or electric current that may depend on coordinates. For coordinate dependent values, QuickField will calculate the current or vector potential for all vertices linked to a given label. Figure 5.6 shows the window for editing vertex label data for magnetostatics problems.

Integral values

Once a contour has been specified around a body, the magnetic force **F** acting on the body is calculated as

$$\mathbf{F} = \frac{1}{2}\oiint_{\text{surf}} (\mathbf{H}(\mathbf{n}\cdot\mathbf{B}) + \mathbf{B}(\mathbf{n}\cdot\mathbf{H}) - \mathbf{n}(\mathbf{H}\cdot\mathbf{B}))\, d\mathbf{a} \tag{5.38}$$

by choosing the corresponding integral calculator. The torque **T** acting on a given body is

$$\mathbf{T} = \frac{1}{2}\oiint_{\text{surf}} ((\mathbf{r}\times\mathbf{H})(\mathbf{n}\cdot\mathbf{B}) + (\mathbf{r}\times\mathbf{B})(\mathbf{n}\cdot\mathbf{H}) - (\mathbf{r}\times\mathbf{n})(\mathbf{H}\cdot\mathbf{B}))\, d\mathbf{a}. \tag{5.39}$$

Both integrals for the force and torque are evaluated over the surface of the body. The QuickField Post Processor also supports volume integrals. For example, the volume integral for magnetic energy W is given by

$$W = \frac{1}{2}\iiint_{\text{vol}} (\mathbf{H}\cdot\mathbf{B})\, dv, \tag{5.40}$$

where $\mathbf{H}\cdot\mathbf{B}/2$ is the magnetic energy density. For nonlinear magnetics this volume integral becomes (suppressing the triple integrals)

$$W = \int \left(\int_0^B H(B')\, dB' \right) dv. \tag{5.41}$$

Total flux may be calculated by integrating the vector potential over any part of the solution region. In particular, we may also calculate the total flux linking a single coil for the planar case

$$\Psi = \frac{\iint A ds}{S}, \tag{5.42}$$

where this integral is over the cross-sectional area of the coil (not the area of the loop). In the axisymmetric case

$$\Psi = \frac{2\pi \iint rA ds}{S}, \tag{5.43}$$

where S is the cross-sectional area of the coil in these expressions. Note that the calculated flux may then be used in evaluating self-inductance or mutual inductance between one or more current carrying coils.

5.7 INDUCTANCE CALCULATIONS

The self-inductance of a single coil and the mutual inductance of multiple coils may be calculated in the QuickField magnetostatics solver. Note, however, that the electromotive force, or induced voltage, across the terminals of the coil, results from *time-changing* magnetic fields threading the coil according to Faraday's law. The inductance of a single coil with n-turns is calculated as

$$L = \frac{n\Psi}{I}, \tag{5.44}$$

where Ψ is the flux threading the coil and I is the current flowing through each turn. The mutual inductance between two coils (coil-1 and coil-2) is calculated according to

$$L_{12} = \frac{n_2 \Psi_2}{I_1}. \tag{5.45}$$

The above expression is evaluated with current I_1 flowing in coil-1 and evaluating the flux threading coil-2 with I_2 set to zero. The Inductance.Wizard is also available to simplify the calculation of self- and mutual inductance of the coils.

5.8 UNIFORM MAGNETIC FIELDS

Uniform Fields in *x-y* Symmetry

Uniform fields may consist of a superposition of both an axial and a transverse component. Assuming a rectangular solution region with solution boundaries at $x = (0, a)$ and $y = (0, b)$, a uniform field in the y-direction is generated by assigning separate Direchlet boundary conditions on the vector potential to the edges

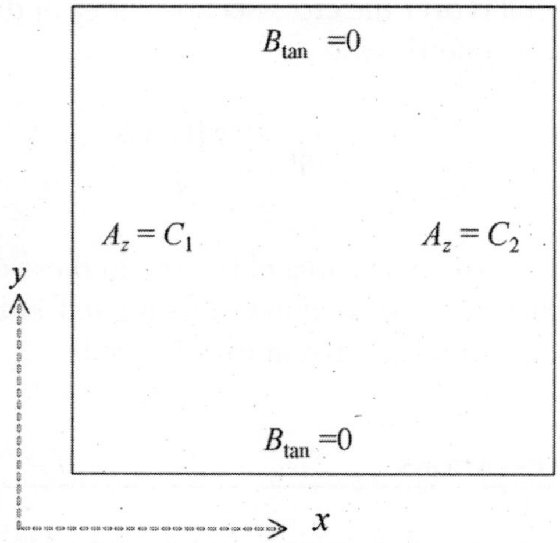

FIGURE 5.7 **Specifying boundary conditions to establish a uniform magnetic induction in the *y*-direction.**

$x = (0, a)$ so that $A_z(x = 0) \neq A_z(x = a)$ and assigning a Neumann boundary condition, where the tangential magnetic field $B_{tan} = 0$ on the edges $y = (0, b)$ as shown in Figure 5.7. Alternatively, a uniform field in the x-direction is generated by assigning Dirichlet conditions to the edges $y = (0,b)$ and Neumann conditions to $x = (0, a)$.

Uniform Fields in Axisymmetry

In r-z axisymmetry, a uniform field along the z-axis is generated by assigning a Dirichlet boundary condition $rA_\phi = const.$ to the edge $r = a$, and Neumann

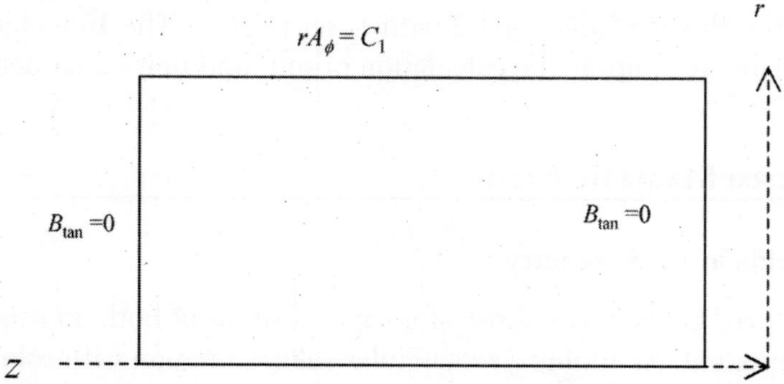

FIGURE 5.8 **Specifying boundary conditions to establish a uniform magnetic induction in the *z*-direction.**

boundary conditions $B_{tan} = 0$ to $z = (0,b)$ as shown in Figure 5.8. The boundary condition at $r = 0$ is $rA_\phi = 0$, by default. For an accurate determination of the flux distribution near a particular point, the edges of the solution region should be removed a sufficient distance so as not to influence the field distribution in the region of interest. A good rule of thumb is to position boundary edges at least ten times further than the dimension of a given region, where an accurate solution is desired.

5.9 DIPOLE SOURCES

In Figure 5.9 we see the results from a line current modeled in planar x-y symmetry: A single node assigned a current of 1.0 A models the current carrying wire.

Modeling a Line Current Source

In x-y geometry, a current source with planar symmetry is modeled as an infinite line source. This is accomplished by either assigning a current value to an individual node in the finite element mesh, or by specifying a current density over a local subregion. A line source cannot be modeled in r-z coordinate axisymmetry, where the vector potential and all current sources have only a ϕ component.

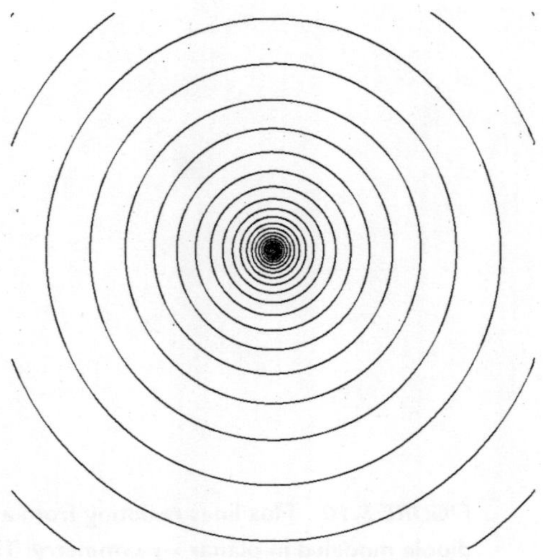

FIGURE 5.9 **Flux lines.**

Current Dipole

The field distribution of a current source of finite length, or current dipole, may be qualitatively reproduced with a single node line current source (see Figure 5.9). A current dipole of length L, carrying a current I, has a dipole moment of magnitude $Q = IL$. Note that the field arising from a line source and a current dipole falls off as $1/r$ and $1/r^2$, respectively.

Current Density

The current density J in a coil with cross-sectional area S is

$$J = \frac{nI}{S}, \tag{5.46}$$

where n is the total number of turns carrying a current I.

Magnetic Dipole

The more familiar magnetic dipole can be modeled in both Cartesian x-y and axisymmetric r-z coordinates. Using x-y coordinates, a magnetic dipole field can be generated by two nearby infinite line sources carrying currents in the opposite direction or as a region of permanent magnetization with a coercive force vector pointing in a given direction (see Figure 5.10). The easiest way to

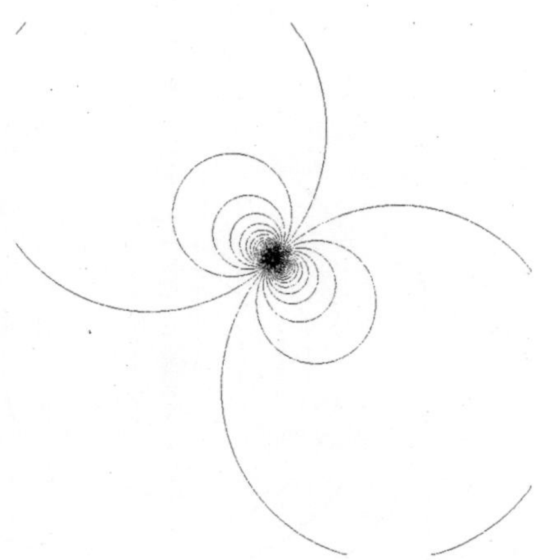

FIGURE 5.10 **Flux lines resulting from a magnetic dipole modeled in planar *x-y* symmetry: The magnetic dipole is simulated as a cylindrical region having a coercive force of 1.0 A/m oriented 45 degrees from the horizontal.**

generate a magnetic dipole field in r-z axisymmetry is with a small current loop with a dipole moment pointing along the z-axis. This is accomplished by specifying current at a single node near the z-axis. The product of the current times the loop area IS gives the magnetic moment of a loop dipole with the direction determined by the right-hand rule.

5.10 SHIELDING APPLICATIONS

Magnetic Shielding by a Highly Permeable Tube

The most common shield design for screening low-frequency magnetic fields is a cylindrical tube made of highly permeable alloy, or mumetal. A mumetal tube

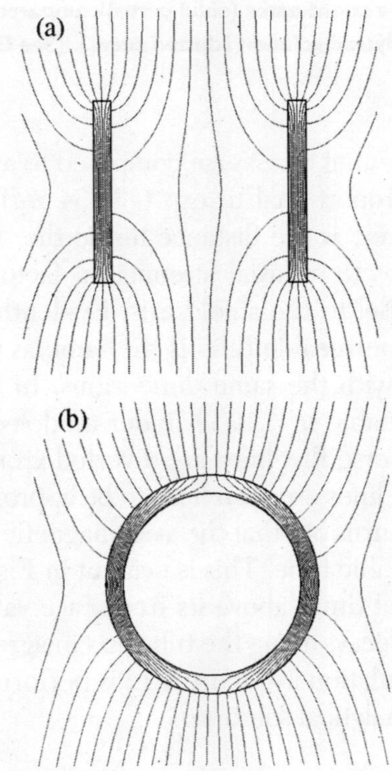

FIGURE 5.11 **Flux lines of a cylindrical shield of high permeability in a previously uniform (a) axial and (b) transverse noise field modeled in *x-y* coordinates: Mumetal attenuates magnetic fields by diverting flux lines around regions enclosed by the material.**

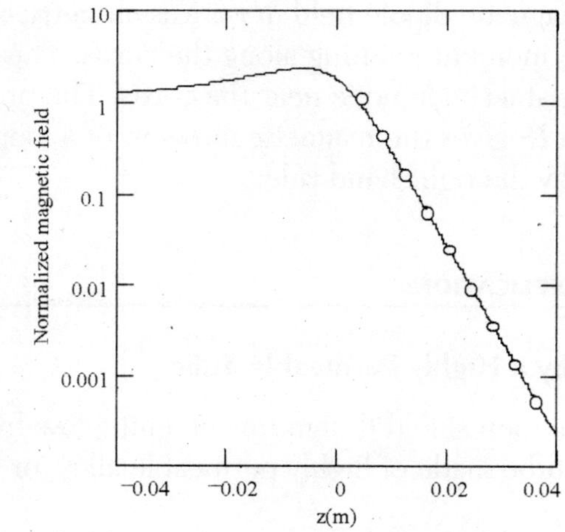

FIGURE 5.12 **Attenuation of the magnetic field along the axis of a highly permeable shield modeled in *r-z* coordinates (solid curve) compared to the analytical solution (dotted curve). (See Exercise 5.11.)**

is a more effective screening transverse compared to axial fields. The transverse attenuation factor is proportional to exp $(-3.834\ z/a)$ for fields perpendicular to the shield axis, where z is the distance inside the shield measured from the opening. This compares to an axial attenuation factor of exp $(-2.405\ z/a)$ for field components parallel to the shield axis. Evidently the transverse attenuation factor of a highly permeable tube is the same as the axial attenuation of a superconducting tube with the same dimensions. In Figure 5.11 (a) axial flux lines penetrate a permeable $\mu_r = 10^4$ cylinder modeled as two parallel strips in x-y coordinates. Transverse flux lines are diverted around the cylindrical shield in Figure 5.11 (b). Flux lines are seen here to be approximately normal to highly permeable surfaces. It turns out that the axial magnetic field actually increases in the opening of a permeable tube. This is evident in Figure 5.12, where the axial field is increased several times above its free space value in the opening of the tube at $z = 0$ and then decays along the tube axis in agreement with the theoretical prediction. The simulation in Figure 5.12 is performed in r-z axial symmetry and compared to the analytical solution.

5.11 MAGNETIC MONOPOLES

In 1931 Dirac demonstrated that the existence of even a single magnetic monopole in the universe is sufficient to require the quantization of electric charge. Coulomb's law gives the electric field established by a single point charge q_e as

$$\mathbf{E}(\mathbf{r}) = \frac{1}{4\pi\varepsilon_0} q_e \frac{\mathbf{r}-\mathbf{r}'}{|\mathbf{r}-\mathbf{r}'|^3}. \tag{5.47}$$

For a hypothetical magnetic charge q_m the analogous expression for the resulting magnetic field is

$$\mathbf{B}(\mathbf{r}) = \frac{\mu_0}{4\pi} q_m \frac{\mathbf{r}-\mathbf{r}'}{|\mathbf{r}-\mathbf{r}'|^3}. \tag{5.48}$$

The field angular momentum density is given by

$$\mathbf{L} = \varepsilon_0 \mathbf{r} \times (\mathbf{E} \times \mathbf{B}), \tag{5.49}$$

where the total angular momentum is given by the integral over all space of which has a constant magnitude independent of **r**

$$L = \frac{\mu_0}{4\pi} q_e q_m \tag{5.50}$$

Since angular momentum is quantized in integral units of Plank's constant $L = n\text{h}/2$ we have quantization of the electric charge, where $n = 1, 2, 3...$

$$q_e = n\frac{h}{\mu_0 q_m} \tag{5.51}$$

For $n = 1$, the product of fundamental electric and magnetic charges is given by Plank's constant divided by the permeability of free space $q_e q_m = h/\mu_0$.

REFERENCES

QuickField User's Guide: Tera Analysis Ltd. (2004). *Chapter 2: Introductory Guide; Chapter 9 Theoretical Description.*

David J. Griffiths. (1999). *Introduction to Electrodynamics 3rd edition*, Prentice-Hall, Inc., New Jersey. *Chapter 5: Magnetostatics.*

Wolfgang K. H. Panofsky and Melba Phillips. (1962). *Classical Electricity and Magnetism 2nd edition,* Addison-Wesley Publishing Company. *Chapter 7: Steady Currents and Their Interaction; Chapter 8: Magnetic Materials and Boundary Value Problems.*

John D. Jackson. (1998). *Classical Electrodynamics 3rd edition*, John Wiley & Sons, Inc. *Chapter 5: Magnetostatics.*

William R. Smythe. (1989). *Static and Dynamic Electricity*, Taylor & Francis. *Chapter VII: Magnetic Interaction of Currents; Chapter IIX: Electromagnetic Induction; Chapter IX: Magnetism.*

EXERCISES

5.1 Model a line current as a single node in *x-y* magnetostatics. Compare Ampere's law by calculating $\oint H_{\tan} d\ell$ over rectangular Γ_1 and circular Γ_2 contours surrounding the line current.

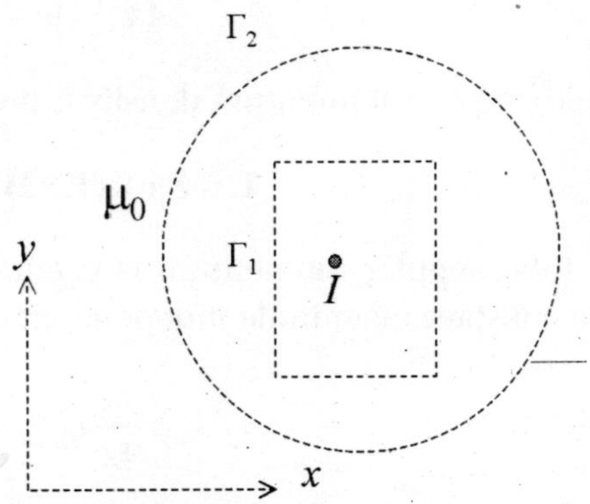

FIGURE E-5.1 **Two contours surrounding a single line current. Geometry: *x-y* symmetry.**

5.2 Calculate the force per unit length between two parallel line currents. Compare to the analytical solution

$$\frac{F}{L} = \frac{\mu_0}{2\pi r} I_1 I_2.$$

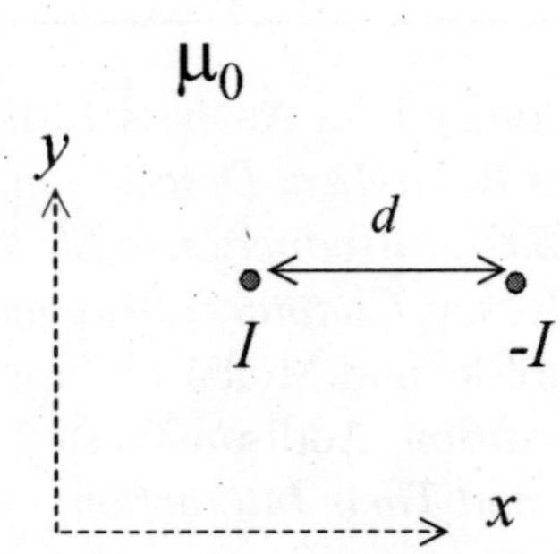

FIGURE E-5.2 **Two oppositely directed line currents. Geometry: *x-y* symmetry.**

5.3 Model a current loop of radius *a* as a single node in axisymmetric geometry. Plot the *z*-component magnetic field along the *z*-axis and the z- and r-field components along the radius of the current loop. Show by comparison that the z-component field along the loop axis falls off as

$$B_z = \frac{\mu_0 I}{2} \frac{a^2}{\left(a^2 + z^2\right)^{3/2}},$$

while the vector potential is given by

$$A_\phi(r,z) = \frac{\mu_0 I a}{2} \int_0^\infty J_1(ka) J_1(kr) \exp(-k|z|)\, dk,$$

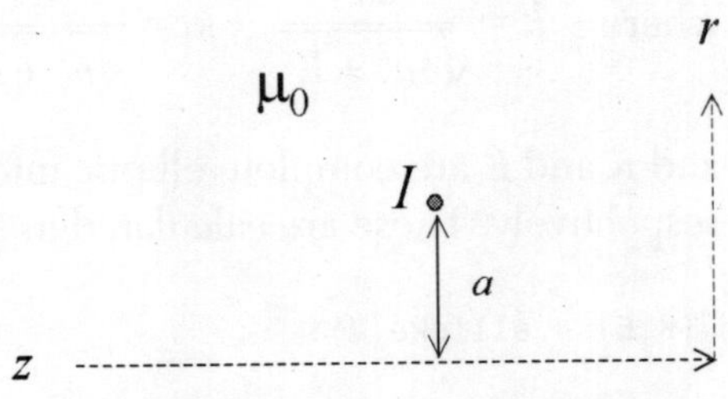

FIGURE E-5.3 **Current loop of radius *a*.**
Geometry: Axial symmetry.

where J_1 is the first-order Bessel function of the first kind.

5.4 Calculate the self-inductance L of the coil above. Compare to the analytical formula

$$L = \mu_0 \pi a^2 \int_0^\infty J_1^2(ka)\, dk.$$

5.5 Model a permeable cylinder of radius a carrying a total current I in the presence of an x-directed magnetic field B_0, perpendicular to the axis of the cylinder. Compare the solution for the vector potential to the analytical expressions

$$A_z = -\frac{\mu_0 I\left(x^2 + y^2\right)}{4\pi a^2} + \frac{2\mu}{\mu + \mu_0} B_0 y$$

inside the cylinder and

$$A_z = \frac{\mu_0 I}{2\pi}\ln\left(\frac{a}{\sqrt{x^2+y^2}}\right) - \frac{\mu I}{4\pi} + \left(1+\frac{\mu-\mu_0}{\mu+\mu_0}\frac{a^2}{x^2+y^2}\right)B_0 y$$

outside the cylinder. Calculate the total force acting on the cylinder. Make a vector plot of the magnetic field distribution near the cylinder.

5.6 Calculate the force between a permanent magnet of length L and a plane permeable boundary. Plot F for several values of L using Label Mover. Compare the analytical solution

$$F = 8\pi a L H_c^2\left[\frac{K(k)-E(k)}{k} - \frac{K(k_1)-E(k_1)}{k_1}\right],$$

where $k = \frac{2a}{\sqrt{4a^2+L^2}}$, $k_1 = \frac{a}{\sqrt{a^2+L^2}}$

and K and E are complete elliptic integrals of the first and second kind, respectively. These are calculated in MATLAB as

```
[K,E] = ellipke(M)
```

where $0 \le M \le 1$.

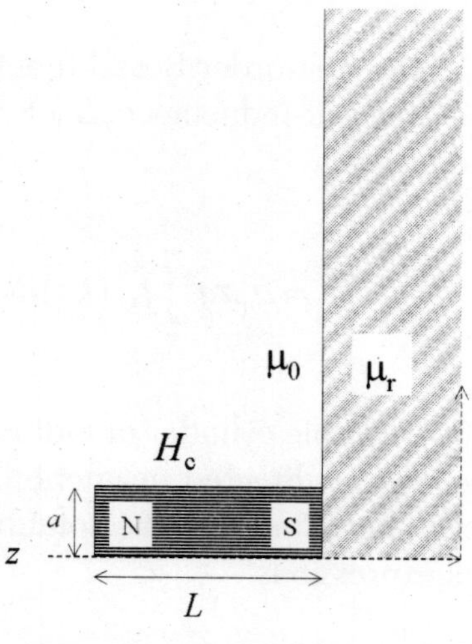

FIGURE E-5.6 **Permanent magnet above a permeable plane. Geometry: Axial symmetry.**

5.7 Calculate the force between a line current I, a height h, above a plane permeable boundary μ. Choose values for the line current and the permeability of the half-space. Compare the value obtained with the analytical solution

$$\frac{F}{L}=\frac{\mu_0}{2\pi}\left(\frac{\mu-1}{\mu+1}\right)\frac{I^2}{2h}.$$

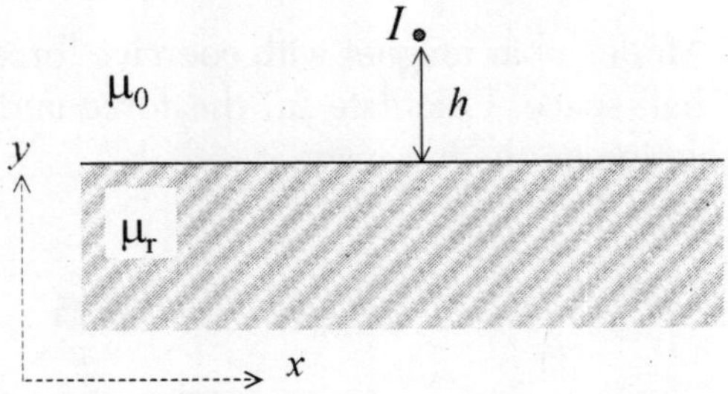

FIGURE E-5.7 **Line current above a permeable plane. Geometry: *x-y* symmetry.**

5.8 Model a uniformly magnetized sphere with coercive force vector H_c. Verify that the magnetic field inside the sphere is uniform with magnitude

$$B=\frac{2}{3}\mu_0 H_c.$$

5.9 Model a hollow permeable sphere in a uniform magnetic field oriented in the z-direction. Use axial symmetry. Choose values for the inner radius a, and outer radius b of the sphere and the relative permeability μ of the sphere. Calculate the attenuation of the field inside the sphere. Compare to analytical result

$$\frac{B_{\text{in}}}{B_{\text{ext}}}=\frac{9}{2\mu_r\left(1-\dfrac{a^3}{b^3}\right)}.$$

Verify that the magnetic field inside the spherical cavity is uniform.

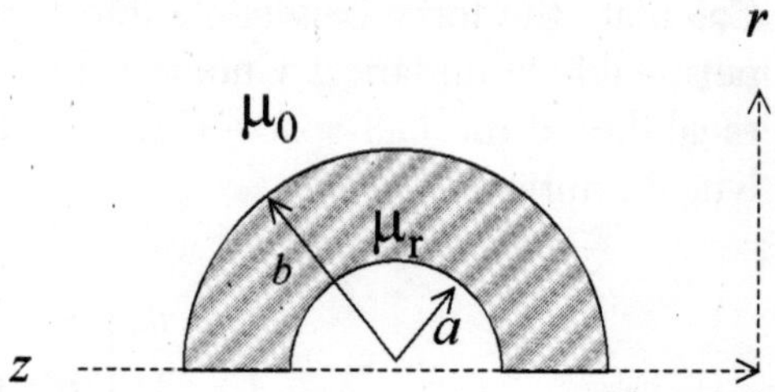

FIGURE E-5.9 **Hollow permeable sphere. Geometry: Axial symmetry.**

5.10 Model a bar magnet with coercive force vector tangential to permeable half-space. Calculate (a) the force and (b) the torque per unit length acting on the bar magnet.

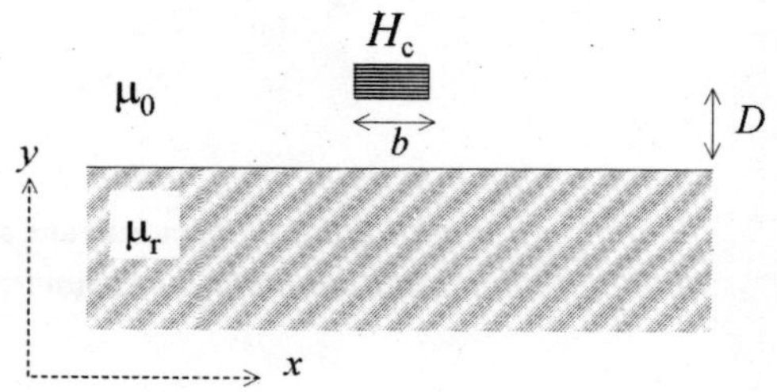

FIGURE E-5.10 **Tangentially oriented permanent magnet above a permeable plane. Geometry: x-y symmetry.**

5.11 Model a permeable tube of radius a in uniform axial field in axial symmetry. Compare the z-component field along the axis of the tube to the analytical solution

$$B_z(z) = A\exp\left(-2.405\frac{z}{a}\right),$$

where A is a constant.

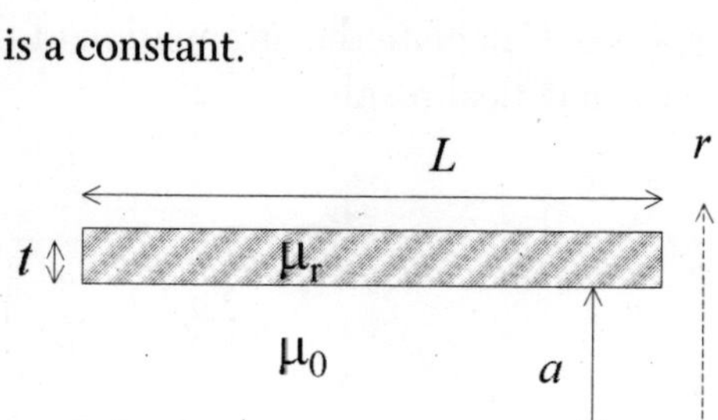

FIGURE E-5.11 **Permeable tube. Geometry: Axial symmetry.**

5.12 Model a line current along the axis of a hollow permeable cylinder in x-y coordinates. Make a contour plot of the magnetic field along a radial contour.

5.13 Model a bar magnet next to the permeable c-section shown in Figure E-5.13. Calculate the total magnetic field energy per unit length in the c-section.

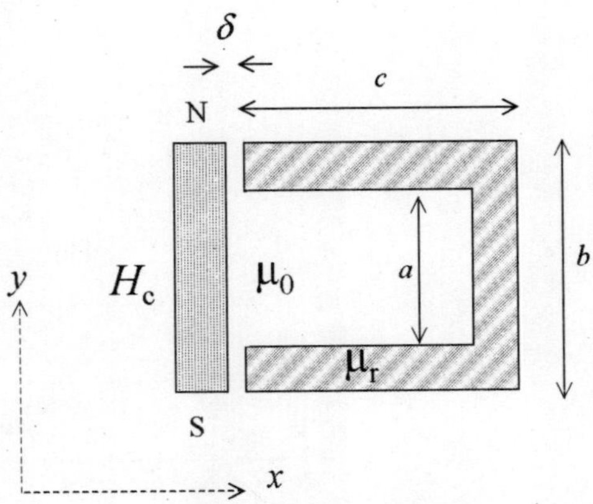

FIGURE E-5.13 **Permanent magnet next to c-shaped permeable conduit. Geometry: *x-y* symmetry.**

5.14 Model a bar magnet next to the permeable reluctance circuit shown in Figure E-5.14. Compare the flux through each section of the circuit. Plot the total magnetic energy stored in the reluctance circuit for several values of δ.

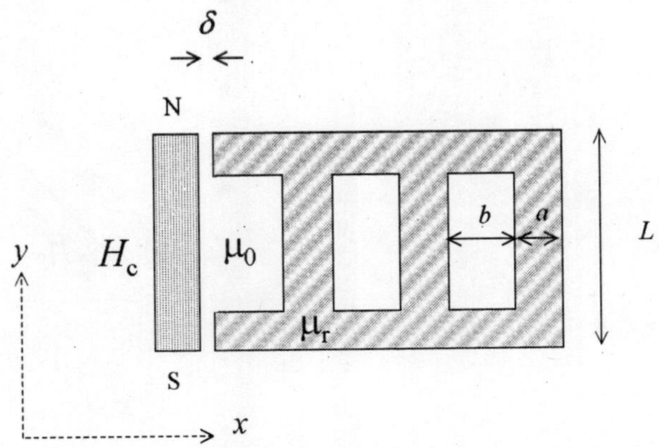

FIGURE E-5.14 **Permanent magnet next to a permeable structure. Geometry: *x-y* symmetry.**

5.15 Model an electromagnet consisting of a current carrying coil wound around a nail in axisymmetry shown in Figure E-5.15. Calculate the force between the electromagnet and an adjacent permeable plane.

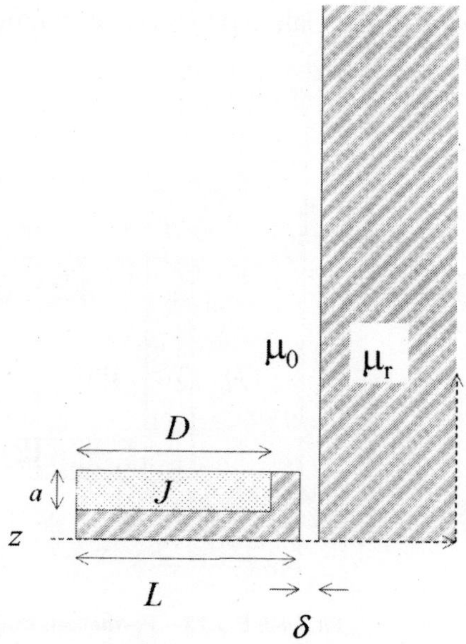

FIGURE E-5.15 **Electromagnet with permeable core above a permeable plane. Geometry: Axial symmetry.**

5.16 Calculate the torque per unit length acting on a bar magnet tilted at an angle θ with respect to the external field shown in Figure E-5.16.

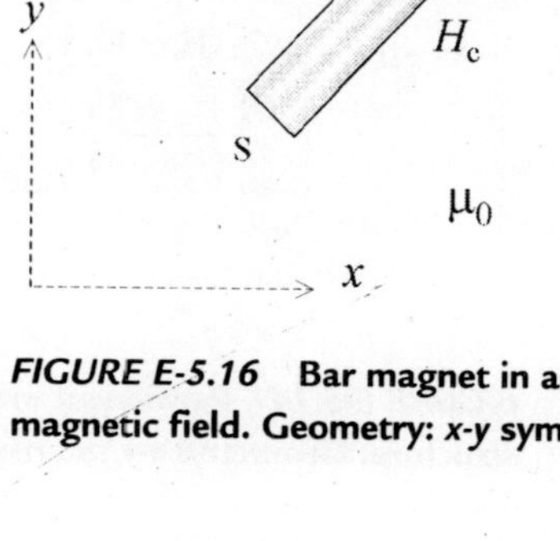

FIGURE E-5.16 **Bar magnet in an external magnetic field. Geometry: *x-y* symmetry.**

5.17 Model a Helmholtz coil consisting of two coaxial current loops of radius R separated by a distance $2d$. Make a contour plot of the z-component field along the axis of the coils. Compare the axial z-component field to the analytical solution

$$B_z = \frac{\mu_0 I R^2}{2}\left(\frac{1}{\left(R^2+(z+d)^2\right)^{3/2}} + \frac{1}{\left(R^2+(z-d)^2\right)^{3/2}}\right).$$

Choose the value of d such that $\frac{\partial^2 B_z}{\partial z^2} = 0$ and $B_z = \frac{8\mu_0 I}{5\sqrt{5}R}$ at the center. Estimate the volume of space in between the coils, where the field is uniform to within 1% by making contour plots of the field distribution. Calculate the force on each coil.

5.18 Calculate the mutual inductance between the two coils in exercise 5.17 and compare to the analytical solution.

$$M_{12} = \mu_0 \pi R^2 \int_0^\infty J_1^2(kR)\exp(-2kd)\,dk.$$

5.19 The inductance between a coil and superconducting plane may be modeled by an oppositely directed image current located an equal distance from the plane of the superconductor in the half-space that would be occupied by the superconductor. Applying this procedure, use the results of exercise 5.17 to calculate mutual inductance between the coil and a superconducting plane a distance d below it.

5.20 Calculate the mutual inductance between the two coplanar, coaxial coils of radii a and b. Compare to the analytical solution

$$M_{12} = 2\mu_0 a\left[\mathrm{K}\left(\frac{b}{a}\right) - \mathrm{E}\left(\frac{b}{a}\right)\right]$$

5.21 Verify that equations 5.19 and 5.20 follow from the boundary conditions given in equation 5.18. Model a uniformly magnetized sphere in QuickField and compare B_z along the z-axis to that obtained from the analytical solution.

5.22 Calculate the scalar magnetic potential and the magnetic field components inside and outside the uniformly magnetized prolate spheroid with uniform magnetization pointing in the z-axis.

5.23 Repeat exercise 5.22 for a uniformly magnetized oblate spheroid with uniform magnetization pointing in the z-axis.

5.24 The finite difference formula in Cartesian coordinates for the scalar magnetic potential Ω at the site (i,j) between two permeable regions is

$$\Omega_{i,j} = \frac{\mu_1}{2(\mu_1+\mu_2)}\Omega_{i,j+1} + \frac{\mu_2}{2(\mu_1+\mu_2)}\Omega_{i,j-1} + \frac{1}{4}\Omega_{i-1,j} + \frac{1}{4}\Omega_{i+1,j}$$

with μ_1 at $(i,j+1)$ and μ_2 at $(i,j-1)$. Write a finite difference program in MATLAB to calculate the magnetic potential near a square permeable beam located in a previously uniform magnetic field. Note that the coefficients in the finite difference expression are also equal to the Monte Carlo stepping probabilities at the site (i,j). These coefficients depend on the coordinate system we are working in and should sum to one. If one of the regions is a superconductor (say region 2), then $\mu_2 = 0$ and the finite difference formula becomes

$$\Omega_{i,j} = \frac{1}{2}\Omega_{i,j+1} + \frac{1}{4}\Omega_{i-1,j} + \frac{1}{4}\Omega_{i+1,j}.$$

In the Monte Carlo analysis, the random walker will not step into a superconducting region.

5.25 Repeat exercise 5.24 using the Monte Carlo method for the case of a superconducting beam in a previously uniform field. Note that a uniform field may be established by two surfaces held at different values of Ω. Compare the Monte Carlo simulation using random walkers to that using the Exodus method.

Chapter

6 TIME-HARMONIC MAGNETICS

In This Chapter

- Faraday's Law and Maxwell's Displacement Current
- Time-harmonic Maxwell's Equations
- Vector Potential Formulation
- Time-harmonic Analysis in QuickField
- Eddy Current Nondestructive Testing Applications
- Metal Detectors
- Geophysical Applications
- Transformers

6.1 FARADAY'S LAW AND MAXWELL'S DISPLACEMENT CURRENT

In 1831 Michael Faraday discovered that a time-changing magnetic field would induce a current flow in an electric circuit. The resulting voltage, or electromotive force (*emf* or *E*), is proportional to the time rate of change of magnetic flux threading the circuit. The magnetic flux is defined as

$$\Phi = \iint_{\text{surf}} \mathbf{B} \cdot d\mathbf{a}, \tag{6.1}$$

where the integral is over any capping surface bound by the current loop. The *emf* is then

$$E = -\frac{\partial \Phi}{\partial t} \tag{6.2}$$

The minus sign in this equation indicates that the induced current flow opposes the change in magnet flux through the circuit according to Lenz's law. The *emf* is also equal to the line integral of electric field around the circuit

$$E = \oint_{\Gamma} \mathbf{E} \cdot d\boldsymbol{\ell}. \tag{6.3}$$

Combining the above relations we have that

$$\oint_{\Gamma} \mathbf{E} \cdot d\boldsymbol{\ell} = -\frac{\partial}{\partial t} \iint_{\text{surf}} \mathbf{B} \cdot d\mathbf{a}. \tag{6.4}$$

From Stokes' theorem, the line integral of the electric field becomes the surface integral of its curl

$$\iint_{\text{surf}} \nabla \times \mathbf{E} \cdot d\mathbf{a} = -\frac{\partial}{\partial t} \iint_{\text{surf}} \mathbf{B} \quad d\mathbf{a}. \tag{6.5}$$

Equating the integrands gives the differential form of Faraday's law

$$\nabla \times \mathbf{E} = -\frac{\partial \mathbf{B}}{\partial t}. \tag{6.6}$$

Since a time-changing magnetic field gives rise to an electric field with curl, we might suspect that a time-changing electric field would produce a magnetic field with curl. Maxwell, in fact, discovered that time-changing electric fields give rise to magnetic fields. If we take the time derivative of Gauss' law

$$\nabla \cdot \mathbf{E} = \frac{\rho}{\varepsilon_0} \tag{6.7}$$

and substitute the continuity equation

$$\nabla \cdot \mathbf{J} = \frac{\partial \rho}{\partial t}, \tag{6.8}$$

we obtain $\mathbf{J} = \varepsilon \partial \mathbf{E}/\partial t$ eliminating $\partial \rho/\partial t$. Note, however, that displacement currents result from time-changing electric fields and not charge transport.

Ampere's law is thus modified by the addition of a second term proportional to the time rate of change of electric flux

$$\oint_{\Gamma} \mathbf{B} \cdot d\ell = \mu \iint_{\text{surf}} \mathbf{J}_e \cdot d\mathbf{a} + \mu\varepsilon \iint_{\text{surf}} \frac{\partial \mathbf{E}}{\partial t} \cdot d\mathbf{a}, \tag{6.9}$$

where $\mathbf{J}_e = \sigma\mathbf{E} + \mathbf{J}_{\text{source}}$ and $(\mu,\varepsilon) = (\mu_r\mu_0,\ \varepsilon_r\varepsilon_0)$. Stokes' theorem transforms the line integral of the magnetic field to a surface integral of its curl (over an arbitrary capping surface bounded by the contour Γ), so that we may equate the integrands

$$\nabla \times \mathbf{B} = \mu\mathbf{J}_e + \mu\varepsilon\frac{\partial \mathbf{E}}{\partial t} \tag{6.10}$$

In terms of the constitutive relations $\mathbf{B} = \mu\mathbf{H}$ and $\mathbf{D} = \varepsilon\mathbf{E}$, equation (6.10) may be written in regions with homogeneous and isotropic permeability

$$\nabla \times \mathbf{H} = \mathbf{J}_e + \frac{\partial \mathbf{D}}{\partial t}, \tag{6.11}$$

where $\partial\mathbf{D}/\partial t$ is Maxwell's displacement current. In the following sections we consider fields and currents with sinusoidal variation and with arbitrary time dependence in Chapter 7.

6.2 TIME-HARMONIC MAXWELL'S EQUATIONS

For fields with time variation of $\exp(i\omega t)$ we may replace the $\partial/\partial t$ operator by $i\omega$ so Maxwell's curl equations may be written as

$$\nabla \times \mathbf{E} = -i\omega\mathbf{B} \tag{6.12}$$

$$\nabla \times \mathbf{B} = \mu\sigma\mathbf{E} + i\omega\mu\varepsilon\mathbf{E}, \tag{6.13}$$

where $\mathbf{E}$ and $\mathbf{B}$ are now complex phasors. The conduction current will be much greater than the displacement current for frequencies $\omega = \sigma/\varepsilon$. In biological tissue the conductivity is $\sigma \approx$ S/m with effective relative dielectric constants $\varepsilon \approx 10^6\varepsilon_0$. In metals $\sigma = 10^7$–10^8 S/m with $\varepsilon \approx \varepsilon_0$ displacement currents are negligible up to 10 MHz.

Current Sheet Above a Conducting Half-Space

The simplest example of AC current induction is that produced by a time-harmonic magnetic field tangential to conducting half-space with $\mu_r = 1$. This field can be thought of as being produced by an infinite sheet with a uniform surface current density that is parallel to the conducting half-space shown in Figure 6.1. Neglecting the displacement current we have

$$\nabla \times \mathbf{B} = \mu_0 \sigma \mathbf{E}, \tag{6.14}$$

where $\mathbf{J}_e = \sigma\mathbf{E}$ inside the metal. Taking the curl of this equation

$$\nabla \times \nabla \times \mathbf{B} = \mu_0 \sigma \nabla \times \mathbf{E} \tag{6.15}$$

and using the identity

$$\nabla \times \nabla \times \mathbf{B} = \nabla(\nabla \cdot \mathbf{B}) - \nabla^2\mathbf{B} \tag{6.16}$$

with $\nabla \cdot \mathbf{B} = 0$ and $\nabla \times \mathbf{E} = -i\omega\mathbf{B}$ we have

$$\nabla^2\mathbf{B} = i\mu_0 \sigma\omega\mathbf{B}. \tag{6.17}$$

Four color images for many of the figures may be found on the CD-ROM.

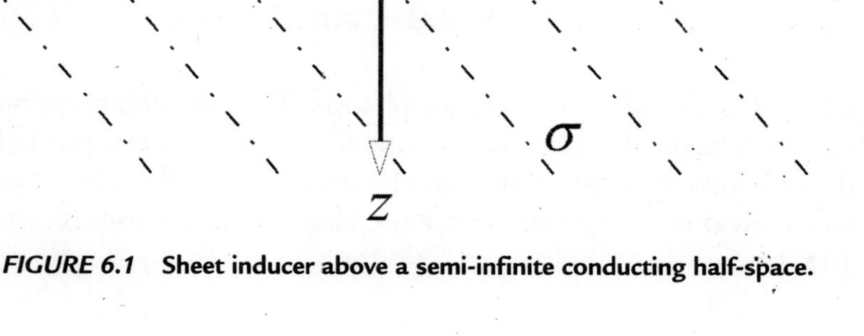

FIGURE 6.1 **Sheet inducer above a semi-infinite conducting half-space.**

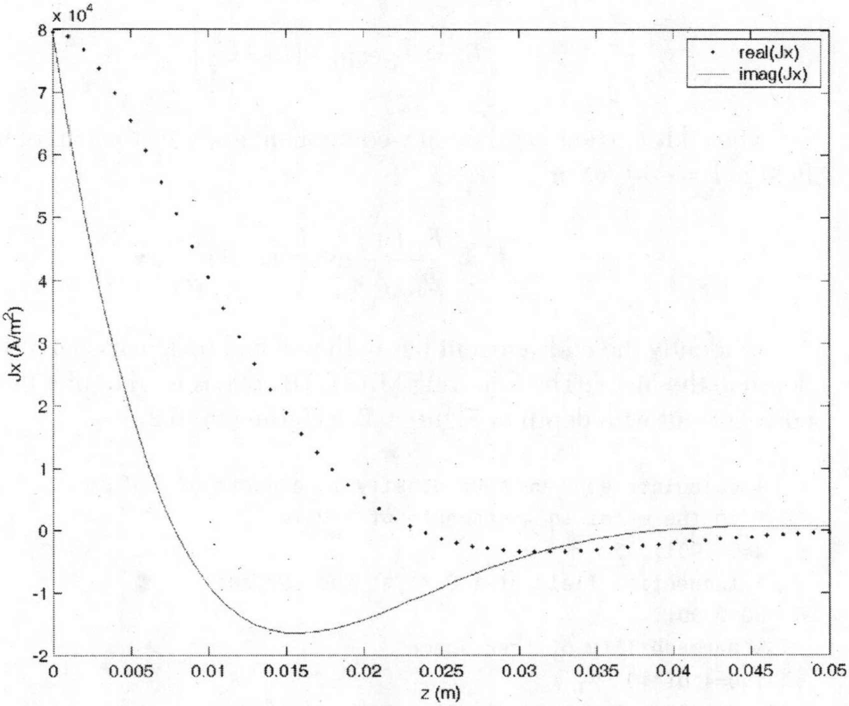

FIGURE 6.2 **Real and imaginary components of eddy current density resulting from a time-harmonic magnetic field tangent to a semi-infinite conducting half-space.**

This may be written as

$$\nabla^2 \mathbf{B} = i\frac{2}{\delta^2}\mathbf{B}, \tag{6.18}$$

where the effective field penetration depth is given by

$$\delta = \sqrt{\frac{2}{\mu\sigma\omega}}. \tag{6.19}$$

For a y-directed external field $\mathbf{B} = B_0\mathbf{j}$ so that the field satisfies

$$\frac{\partial^2 B_y}{\partial z^2} = i\frac{2}{\delta^2}B_y \tag{6.20}$$

inside the metal with solutions of the form $\exp\pm\sqrt{2i}z/\delta$ subject to the boundary conditions $B_y = B_0$ at $z = 0$ and $B_y = 0$ at $z = \infty$, respectively. Taking $\sqrt{i} = (1+i)/2$

$$B_y = B_0 \exp\left(-(i+1)\frac{z}{\delta}\right). \tag{6.21}$$

The eddy current has only an x-component given by the curl of the magnetic field $\mu_0 J_x = -\partial B_y/\partial z$ or

$$J_x = \frac{B_0}{\mu_0}\frac{1+i}{\delta}\exp\left(-(i+1)\frac{z}{\delta}\right). \tag{6.22}$$

Evidently the eddy current has both real and imaginary parts that vary with depth in the metal. The following MATLAB code is used to plot the variation of eddy current with depth in Figure 6.2 according to (6.22)

```
% calculate eddy current density to a depth of 5.0 cm
% in the metal in increments of 1.0 mm
z=0:.001:.05;
% tangential field of 1.0 mT at the surface
B0=0.001;
% permeability of free space
mu0=4*pi*10^-7;
% electrical skin depth of 1.0 cm
d=0.01;
%eddy current density
Jx=(B0/mu0/d)*(1+i)*exp(-(i+1)*z/d);

%plot real and imaginary parts of
plot(z,real(Jx),'.',z,imag(Jx));

% include axis labels and legend
xlabel('z (m)')
ylabel('Jx (A/m^2)')
legend('real(Jx)','imag(Jx)');
```

In the following sections we solve time-harmonic magnetic problems by calculating the vector potential instead of the fields initially.

6.3 VECTOR POTENTIAL FORMULATION

In Chapter 5 we expressed the magnetic field as the curl of a vector potential $\mathbf{B} = \nabla \times \mathbf{A}$. Taking the time derivative of this expression

$$\frac{\partial \mathbf{B}}{\partial t} = \nabla \times \frac{\partial \mathbf{A}}{\partial t} \tag{6.23}$$

and from Faraday's law $\nabla \times \mathbf{E} = -\partial \mathbf{B}/\partial t$ we have

$$\nabla \times \left(\mathbf{E} + \frac{\partial \mathbf{A}}{\partial t} \right) = 0. \tag{6.24}$$

The quantity in brackets may be written as the gradient of a scalar potential

$$\mathbf{E} + \frac{\partial \mathbf{A}}{\partial t} = -\nabla U \tag{6.25}$$

since the curl of the gradient is zero. If the electric potential is taken to be constant, as in a conductor, then

$$\mathbf{E} = -\frac{\partial \mathbf{A}}{\partial t} \tag{6.26}$$

For time-harmonic fields $\mathbf{E} = -i\omega\mathbf{A}$, so that Maxwell's equation (6.10) becomes

$$\nabla \times \nabla \times \mathbf{A} = -i\omega\mu\sigma\mathbf{A} + \omega^2 \mu\varepsilon\mathbf{A}. \tag{6.27}$$

taking the Coulomb gauge, where $\nabla \cdot \mathbf{A} = 0$ we obtain the vector Helmholtz equation

$$\nabla^2\mathbf{A} = i\omega\mu\sigma\mathbf{A} - \omega^2 \mu\varepsilon\mathbf{A} \tag{6.28}$$

with an additional term $-\mu_0 \mathbf{J}_{\text{source}}$ in the presence of sources. Once the vector potential is solved for in a conducting region, the eddy current density can be immediately obtained

$$\mathbf{J}_{\text{eddy}} = \sigma\mathbf{E} = -i\omega\sigma\mathbf{A}. \tag{6.29}$$

Analytical solutions to eddy current induction problems are often quite complex involving infinite integrals or summations over Bessel, or other special functions. However, the procedure for calculating eddy current distributions is quite similar for different conducting geometries. Analytical results are useful for the verification of computational methods and can often be extended to model more complex eddy current distributions using superposition integrals and summations as we demonstrate in the following sections.

Circular Coil Above a Conducting Half-Space

Dodd, Deeds, and Liquire treated eddy current induction problems in multi-layered planar, cylindrical, and spherical conductors in a series of papers

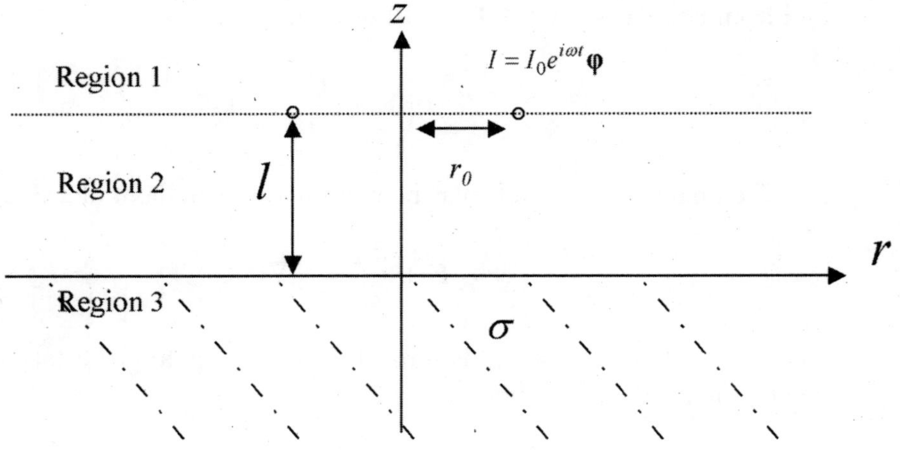

FIGURE 6.3 **Circular excitation coil above a semi-infinite conducting half-space.**

beginning in 1968. For simplicity, we consider a circular AC current carrying loop of radius r_0 located a height ℓ above and coplanar to a semi-infinite, permeable conducting half-space in Figure 6.3.

In the air region above the conducting half-space, only the source current term appears neglecting displacement currents

$$\nabla^2 \mathbf{A} = -\mu_0 \mathbf{J}_{\text{source}}. \tag{6.30}$$

Inside the conductor only conduction currents appear

$$\nabla^2 \mathbf{A} = i\omega\mu\sigma\mathbf{A}, \tag{6.31}$$

where $\mu = \mu_r \mu_0$. The problem has axial symmetry so the vector potential only has a ϕ-component and the vector Laplacian has a term in addition to the scalar Laplacian

$$\left(\nabla^2 \mathbf{A}\right)_\phi = \frac{\partial^2 A_\phi}{\partial r^2} + \frac{1}{r}\frac{\partial A_\phi}{\partial r} - \frac{A_\phi}{r^2} + \frac{\partial^2 A_\phi}{\partial z^2} = \nabla^2 A_\phi - \frac{A_\phi}{r^2}. \tag{6.32}$$

Above the conducting half-space, the drive current may be written as a product of delta functions

$$\frac{\partial^2 A_\phi}{\partial r^2} + \frac{1}{r}\frac{\partial A_\phi}{\partial r} - \frac{A_\phi}{r^2} + \frac{\partial^2 A_\phi}{\partial z^2} = \begin{cases} -\mu_0 I \delta(r - r_0)\delta(z - \ell) & z > 0 \\ i\omega\mu\sigma A_\phi & z \leq 0 \end{cases}. \tag{6.33}$$

Solving the above equations by separation of variables, we first substitute $A_\phi(r,z) = R(r) \cdot Z(z)$ into the differential equation and then divide by $R \cdot Z$

$$\frac{1}{R}\frac{\partial^2 R}{\partial r^2} + \frac{1}{rR}\frac{\partial R}{\partial r} - \frac{1}{r^2} + \frac{1}{Z}\frac{\partial^2 Z}{\partial z^2} - i\omega\mu\sigma = 0. \tag{6.34}$$

Introducing the separation constant α^2, we set all r-dependent terms equal to $-\alpha^2$ and the remaining terms equal to α^2. The differential equation for $R(r)$ becomes

$$\frac{1}{R}\frac{\partial^2 R}{\partial r^2} + \frac{1}{rR}\frac{\partial R}{\partial r} - \frac{1}{r^2} = -\alpha^2, \tag{6.35}$$

which has the linearly independent solutions

$$R(r) = A \cdot J_1(\alpha r) + B \cdot Y_1(\alpha r). \tag{6.36}$$

In constructing a solution for the vector potential, we take B to be zero because $Y_1(\alpha r)$ diverges at $r = 0$. Setting the remaining terms equal to α^2 gives the differential equation

$$\frac{1}{Z}\frac{\partial^2 Z}{\partial z^2} - i\omega\mu\sigma = \alpha^2 \tag{6.37}$$

with solutions

$$Z(z) = C \cdot \exp\left(\sqrt{i\omega\mu\sigma + \alpha^2}\, z\right) + D \cdot \exp\left(-\sqrt{i\omega\mu\sigma + \alpha^2}\, z\right). \tag{6.38}$$

Since there are no restrictions on α, it must be summed over continuously in constructing a solution for $A_\phi(r,z)$. Relabeling the constants with subscripts corresponding to region number, the solution for the vector potential above the drive coil $z \geq \ell$ is

$$A_\phi^1(r,z) = \int_0^\infty B_1 e^{-\alpha z} J_1(\alpha r)\, d\alpha. \tag{6.39}$$

In the region between the drive coil and the conducting half-space $0 < z < \ell$

$$A_\phi^2(r,z) = \int_0^\infty \left(B_2 e^{-\alpha z} + C_2 e^{\alpha z}\right) J_1(\alpha r)\, d\alpha. \tag{6.40}$$

Inside the conducing half-space $z \le 0$

$$A_\varphi^3(r,z) = \int_0^\infty C_3 e^{\beta z} J_1(\alpha r)\, d\alpha, \tag{6.41}$$

where $\beta = \sqrt{\alpha^2 + i\omega\mu\sigma}$. The constants B_1, B_2, C_2, and C_3 are obtained requiring continuity in the tangential fields **E** and **H** at $z = 0$ and $z = \ell$. The tangential electric boundary condition is straightforward since $E_\phi = -i\omega A_\phi$ we just require continuity in A_ϕ

$$A_\varphi^1\Big|_{z=\ell} = A_\varphi^2\Big|_{z=\ell} \rightarrow B_1 \cdot e^{-\alpha\ell} = B_2 \cdot e^{-\alpha\ell} + C_2 \cdot e^{\alpha\ell} \tag{6.42}$$

$$A_\varphi^2\Big|_{z=0} = A_\varphi^3\Big|_{z=0} \rightarrow C_3 = B_2 + C_2. \tag{6.43}$$

Requiring tangential H_r at $z = 0$

$$\frac{1}{\mu_0}\frac{\partial A_\varphi^2}{\partial z}\bigg|_{z=0} = \frac{1}{\mu}\frac{\partial A_\varphi^3}{\partial z}\bigg|_{z=0} \rightarrow C_2 - B_2 = \frac{1}{\mu_r}\frac{\beta}{\alpha} \cdot C_3. \tag{6.44}$$

The final boundary condition takes into account the discontinuity in H_r at the position of the drive coil

$$-\frac{\partial A_\varphi^1}{\partial z}\bigg|_{z=\ell} + \frac{\partial A_\varphi^2}{\partial z}\bigg|_{z=\ell} = \mu_0 I \delta(r - r_0). \tag{6.45}$$

This condition is written as

$$\int_0^\infty \alpha\left(C_2 \cdot e^{\alpha\ell} + (B_1 - B_2)\cdot e^{-\alpha\ell}\right) J_1(\alpha r)\, d\alpha = \mu_0 I \delta(r - r_0). \tag{6.46}$$

Multiplying both sides of this equation by $r \cdot J_1(\alpha' r)$ and integrating over r

$$\begin{aligned}&\int_0^\infty \alpha\left(C_2 \cdot e^{\alpha\ell} + (B_1 - B_2)\cdot e^{-\alpha\ell}\right)\int_0^\infty r J_1(\alpha' r) J_1(\alpha r)\, dr\, d\alpha \\ &\quad = \mu_0 I \int_0^\infty r J_1(\alpha' r)\delta(r - r_0)\, dr.\end{aligned} \tag{6.47}$$

Using the relation

$$\int_0^\infty r J_1(\alpha' r) J_1(\alpha r)\, dr = \frac{\delta(\alpha - \alpha')}{\alpha} \tag{6.48}$$

collapses the integral over α and we have

$$B_1 e^{-\alpha'\ell} - B_2 e^{-\alpha'\ell} + C_2 e^{\alpha'\ell} = \mu_0 I r_0 J_1(\alpha' r_0), \tag{6.49}$$

where the primes are dummy and can be erased so that the four equations and four unknowns are

$$\begin{bmatrix} e^{-\alpha\ell} & -e^{-\alpha\ell} & -e^{\alpha\ell} & 0 \\ 0 & 1 & 1 & -1 \\ 0 & -1 & 1 & -\frac{1}{\mu_r}\frac{\beta}{\alpha} \\ e^{-\alpha\ell} & -e^{-\alpha\ell} & e^{\alpha\ell} & 0 \end{bmatrix} \begin{bmatrix} B_1 \\ B_2 \\ C_2 \\ C_3 \end{bmatrix} = \begin{bmatrix} 0 \\ 0 \\ 0 \\ \mu_0 I r_0 J_1(\alpha r_0) \end{bmatrix}. \tag{6.50}$$

Solving the above system for the constants in MATLAB

```
% Solve for the constants B1, B2, C2 and C3
syms B1, B2, C2, C3, a, l, b, ur, u0,I, r0, J1
Eq1 = 'exp(a*l)*B1-exp(-a*l)*B2-exp(a*l)*C2 =0'
Eq2 = 'B2 + C2 - C3 = 0'
Eq3 = '-B2+C2 -b/a/ur *C2 =0'
Eq4 = 'exp(-a*l)*B1-exp(-a*l)*B2+exp(a*l)*C2=u0*I*r0*J1'
[B1,B2,C2,C3]=Solve(Eq1,Eq2,Eq3,Eq4)
```

and substituting into the integral expressions for the vector potential yields

$$A_\phi^1(r,z) = \frac{\mu_0 I r_0}{2}\int_0^\infty J_1(\alpha r) J_1(\alpha r_0) e^{-\alpha z}\left(e^{\alpha\ell} + e^{-\alpha\ell}\frac{\alpha\mu_r - \beta}{\alpha\mu_r + \beta}\right) d\alpha \tag{6.51}$$

$$A_\phi^2(r,z) = \frac{\mu_0 I r_0}{2}\int_0^\infty J_1(\alpha r) J_1(\alpha r_0) e^{-\alpha\ell}\left(e^{\alpha z} + e^{-\alpha z}\frac{\alpha\mu_r - \beta}{\alpha\mu_r + \beta}\right) d\alpha \tag{6.52}$$

$$A_\phi^3(r,z) = \frac{\mu_0 I r_0}{2}\int_0^\infty J_1(\alpha r) J_1(\alpha r_0) e^{-\alpha\ell}\left(e^{\beta z}\frac{2\alpha\mu_r}{\alpha\mu_r + \beta}\right) d\alpha. \tag{6.53}$$

The eddy current density in the conducting plane is then $-i\omega\sigma$ times the vector potential in the third region

$$\mathbf{J}_{\text{eddy}} = -i\omega\sigma A_\phi^3(r,z)\hat{\varphi}. \tag{6.54}$$

If $\sigma = 0$, we have $\beta = \alpha$ and the vector potential for a single current loop in free space is obtained

$$A_\phi(r,z) = \frac{\mu_0 I r_0}{2} \int_0^\infty J_1(\alpha r) J_1(\alpha r_0) e^{-\alpha|z-\ell|} d\alpha. \tag{6.55}$$

By comparing equations 6.51, 6.52, and 6.55, the contribution to the vector potential above the conducting half-space $z > 0$ due solely to eddy currents can be identified as

$$\Delta A_\phi(r,z) = \frac{\mu_0 I r_0}{2} \int_0^\infty J_1(\alpha r) J_1(\alpha r_0) e^{-\alpha|z-\ell|} \frac{\mu_r \alpha - \beta}{\mu_r \alpha + \beta} d\alpha \tag{6.56}$$

and $\beta \to \infty$ for a perfectly conducting half-space and the factor $(\mu_r \alpha - \beta)/(\mu_r \alpha + \beta) = -1$. The above expressions may be modified to include the displacement current by replacing $\beta \to \sqrt{\alpha^2 + i\omega\mu\sigma - \omega^2 \mu\varepsilon}$.

The vector potential resulting from an excitation coil with a finite cross-section can be obtained by integrating the expression for a single filamentary current loop over the cross-section of the coil

$$A_\phi(r,z) = \frac{1}{\Delta\ell \Delta r} \int_{\ell_1}^{\ell_2} \int_{r_1}^{r_2} A_\phi(r,z,r_0,\ell) dr_0 d\ell, \tag{6.57}$$

where $\Delta \ell = \ell_2 - \ell_1$ and $\Delta r = r_2 - r_1$. For a thin pancake, or spiral coil, the integral is performed only over r_0 and divided by Δr. For a thin solenoid with a stand-off ℓ_1 above the conducting plane, the integral is performed only over ℓ and divided by $\Delta \ell$.

Circular Coil Above a Conducting Sphere

Consider a circular AC excitation coil located above a conducting sphere in Figure 6.4 with $\mu_r = 1$, for simplicity.

In spherical coordinates with axial symmetry, the vector Laplacian has an additional term

$$\left(\nabla^2 \mathbf{A}\right)_\phi = \nabla^2 A_\phi - \frac{1}{r^2 \sin^2\theta} A_\phi, \tag{6.58}$$

where $\nabla^2 A_\phi$ is the scalar Laplacian. Outside the sphere we have

$$\nabla^2 A_\phi - \frac{1}{r^2 \sin^2\theta} A_\phi = -\mu_0 I \frac{\delta(r - r_0)}{r_0} \delta(\theta - \theta_0). \tag{6.59}$$

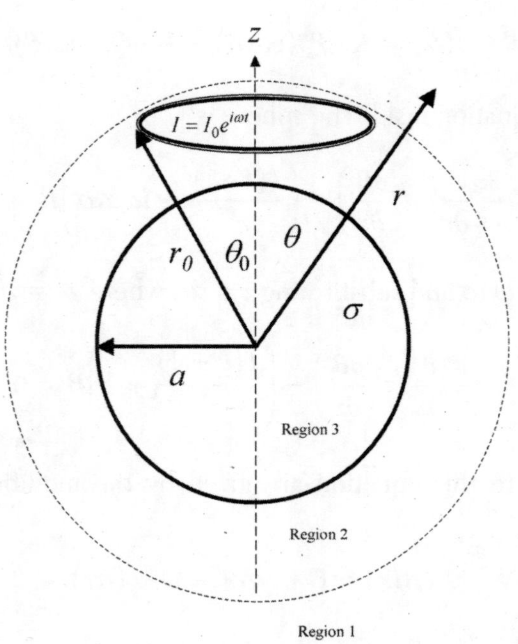

FIGURE 6.4 **Circular excitation coil above a conducting sphere.**

Inside the conducting sphere

$$\nabla^2 A_\phi - \frac{1}{r^2 \sin^2\theta} A_\phi = i\omega\mu_0\sigma A_\phi, \tag{6.60}$$

which has the form

$$\frac{1}{r^2}\frac{\partial}{\partial r}\left(r^2\frac{\partial A_\phi}{\partial r}\right) + \frac{1}{r^2\sin\theta}\frac{\partial}{\partial\theta}\left(\sin\theta\frac{\partial A_\phi}{\partial\theta}\right) - \frac{1}{r^2\sin^2\theta}A_\phi - i\omega\mu_0\sigma A_\phi = 0. \tag{6.61}$$

Solving this equation by separation of variables we replace $A \to R(r)\Theta(\theta)$ in the differential equation and divide by $R \cdot \Theta$

$$\frac{1}{R}\frac{\partial}{\partial r}\left(r^2\frac{\partial R}{\partial r}\right) + \frac{1}{\Theta\sin\theta}\frac{\partial}{\partial\theta}\left(\sin\theta\frac{\partial\Theta}{\partial\theta}\right) - \frac{1}{\sin^2\theta} - i\omega\mu_0\sigma r^2 = 0. \tag{6.62}$$

The separation constant in this case is $\ell(\ell + 1)$ so that the Θ equation is

$$\frac{1}{\Theta\sin\theta}\frac{\partial}{\partial\theta}\left(\sin\theta\frac{\partial\Theta}{\partial\theta}\right) - \frac{1}{\sin^2\theta} = -\ell(\ell + 1), \tag{6.63}$$

which has the solution

$$\Theta(\theta) = A \cdot P_\ell^1(\cos\theta) + B \cdot Q_\ell^1(\cos\theta). \tag{6.64}$$

The radial equation inside the sphere is

$$\frac{1}{r^2}\frac{\partial}{\partial r}\left(r^2\frac{\partial R}{\partial r}\right) - \left(\frac{\ell(\ell+1)}{r^2} + i\omega\mu_0\sigma\right)R = 0. \tag{6.65}$$

Dividing by $i\omega\mu\sigma$ and substituting $z = \alpha r$, where $\alpha = \sqrt{i\omega\mu_0\sigma}$

$$\frac{1}{z^2}\frac{\partial}{\partial z}\left(z^2\frac{\partial R}{\partial z}\right) - \left(\frac{\ell(\ell+1)}{z^2} + 1\right)R = 0, \tag{6.66}$$

where solutions to this equation are given by the modified spherical Bessel functions

$$R(r) = C\, i_\ell(\alpha r) + D\, k_\ell(\alpha r). \tag{6.67}$$

In terms of the modified cylindrical Bessel functions

$$i_\ell(z) = \sqrt{\frac{\pi}{2z}} I_{\ell+\frac{1}{2}}(z) \quad \text{and}\; k_\ell(z) = \sqrt{\frac{\pi}{2z}} K_{\ell+\frac{1}{2}}(z). \tag{6.68}$$

Finally, the radial equation outside the sphere is

$$\frac{1}{r^2}\frac{\partial}{\partial r}\left(r^2\frac{\partial R}{\partial r}\right) - \frac{\ell(\ell+1)}{r^2}R = 0, \tag{6.69}$$

which has the solutions

$$R(r) = E \cdot r^\ell + F \cdot r^{-(\ell+1)}. \tag{6.70}$$

In forming the general solution we discard the $Q_\ell^1(\cos\theta)$ since they are ill-behaved for $\cos\theta = \pm 1$. Also, $k_\ell(\alpha r)$ are undefined at $r = 0$ and are discarded. The vector potential in each region is given by

$$A_\phi^3(r,\theta) = \sum_{\ell=1}^{\infty} b_3 \cdot i_\ell(\alpha r) \cdot P_\ell^1(\cos\theta) \tag{6.71}$$

$$A_\phi^2(r,\theta) = \sum_{\ell=1}^{\infty} \left(b_2 r^\ell + c_2 r^{-(\ell+1)}\right) \cdot P_\ell^1(\cos\theta) \tag{6.72}$$

$$A_\phi^1(r,\theta) = \sum_{\ell=1}^{\infty} b_1 \cdot r^{-(\ell+1)} \cdot P_\ell^1(\cos\theta), \tag{6.73}$$

where the constants are now labeled according to region number. Applying the tangential electric boundary condition at $r = a$ and $r = r_0$ gives the two conditions

$$b_3 \cdot i_\ell(\alpha \cdot a) = b_2 \cdot a^\ell + c_2 \cdot a^{-(\ell+1)} \tag{6.74}$$

$$b_1 \cdot r_0^{-(\ell+1)} = b_2 \cdot r_0^\ell + c_2 \cdot r_0^{-(\ell+1)}. \tag{6.75}$$

Requiring tangential H_θ at $r = a$

$$-\frac{1}{r}\frac{\partial}{\partial r} rA_\phi^3\Big|_{r=a} = -\frac{1}{r}\frac{\partial}{\partial r} rA_\phi^2\Big|_{r=a} \tag{6.76}$$

gives

$$b_3 \cdot \left[i_\ell(\alpha \cdot a) + a \cdot i_\ell'(\alpha \cdot a)\right] = b_2 \cdot a^\ell \cdot (\ell + 1) - c_2 \cdot \ell \cdot a^{-(\ell+1)}. \tag{6.77}$$

The discontinuity in H_θ at $r = r_0$ gives

$$-\frac{1}{r}\frac{\partial}{\partial r} rA_\phi^2\Big|_{r=r_0} + \frac{1}{r}\frac{\partial}{\partial r} rA_\phi^1\Big|_{r=r_0} = -\frac{\mu_0 I}{r_0}\delta(r - r_0)\delta(\theta - \theta_0). \tag{6.78}$$

Multiplying both sides by r and integrating we have

$$\sum_{\ell=0}^{\infty}\left[-b_1 \cdot \ell \cdot r_0^{-(\ell+1)} - b_2 \cdot r_0^\ell \cdot (\ell + 1) + c_2 \cdot \ell \cdot r_0^{-(\ell+1)}\right] P_\ell^1(\cos\theta) = -\mu_0 I \delta(\theta - \theta_0). \tag{6.79}$$

Multiplying both sides of this result by $P_{\ell'}^1(\cos\theta)\sin\theta$

$$-b_1 \cdot \ell \cdot r_0^{-(\ell+1)} - b_2 \cdot r_0^\ell \cdot (\ell + 1) + c_2 \cdot \ell \cdot r_0^{-(\ell+1)} = -\mu_0 I \frac{2\ell + 1}{2\ell(\ell + 1)} P_\ell^1(\cos\theta_0)\sin\theta_0, \tag{6.80}$$

where use has been made of the orthogonality condition

$$\int_{-1}^{1} P_\ell^m(x) P_{\ell'}^m(x)\, dx = \frac{2}{2\ell + 1}\frac{(\ell - m)!}{(\ell + m)!}\delta_{\ell\ell'}. \tag{6.81}$$

We now have the following system

$$\begin{bmatrix} i_\ell(\alpha\cdot a) & -a^\ell & -a^{-(\ell+1)} & 0 \\ 0 & -r_0^\ell & -r_0^{-(\ell+1)} & r_0^{-(\ell+1)} \\ \left[i_\ell(\alpha\cdot a)+a\cdot i_\ell'(\alpha\cdot a)\right] & -a^\ell(\ell+1) & \ell\cdot a^{-(\ell+1)} & 0 \\ 0 & -r_0^\ell(\ell+1) & \ell\cdot r_0^{-(\ell+1)} & -\ell\cdot r_0^{-(\ell+1)} \end{bmatrix}\begin{bmatrix} b_3 \\ b_2 \\ c_2 \\ b_1 \end{bmatrix}$$

$$= \begin{bmatrix} 0 \\ 0 \\ 0 \\ -\mu_0 I\dfrac{2\ell+1}{2\ell(\ell+1)}P_\ell^1(\cos\theta_0)\sin\theta_0 \end{bmatrix}. \tag{6.82}$$

After solving for the constants in MATLAB as before, the vector potential becomes

$$A_\phi^3(r,\theta) = \frac{\mu_0 I\sin\theta_0}{2}\sum_{\ell=1}^{\infty}\frac{P_\ell^1(\cos\theta)P_\ell^1(\cos\theta_0)}{\ell(\ell+1)}\cdot\frac{(2\ell+1)a^\ell i_\ell(\alpha r)}{r_0^\ell\left[(\ell+1)i_\ell(\alpha a)+a\alpha\cdot i_\ell'(\alpha a)\right]} \tag{6.83}$$

$$A_\phi^2(r,\theta) = \frac{\mu_0 I\sin\theta_0}{2}\sum_{\ell=1}^{\infty}\frac{P_\ell^1(\cos\theta)P_\ell^1(\cos\theta_0)}{\ell(\ell+1)}\left(\frac{r^\ell}{r_0^\ell}-\frac{a^{2\ell+1}r^{-(\ell+1)}}{r_0^\ell}\cdot\frac{a\alpha\cdot i_\ell'(\alpha a)-\ell\cdot i_\ell(\alpha a)}{(\ell+1)i_\ell(\alpha a)+a\alpha\cdot i_\ell'(\alpha a)}\right) \tag{6.84}$$

$$A_\phi^1(r,\theta) = \frac{\mu_0 I\sin\theta_0}{2}\sum_{\ell=1}^{\infty}\frac{P_\ell^1(\cos\theta)P_\ell^1(\cos\theta_0)}{\ell(\ell+1)}\left(\frac{r_0^\ell}{r^\ell}-\frac{a^{2\ell+1}r^{-(\ell+1)}}{r_0^\ell}\cdot\frac{a\alpha\cdot i_\ell'(\alpha a)-\ell\cdot i_\ell(\alpha a)}{(\ell+1)i_\ell(\alpha a)+a\alpha\cdot i_\ell'(\alpha a)}\right). \tag{6.85}$$

The eddy current inside the sphere is given by $\mathbf{J}_{\text{eddy}} = i\sigma\omega A_\phi^3(r,\theta)\hat{\phi}$ The first term in brackets gives the vector potential in absence of the sphere in regions 1 and 2. For a dielectric and conducting sphere, the factor $\alpha = \sqrt{i\omega\mu_0\sigma + \omega^2\mu_0\varepsilon}$ may be substituted in the above expressions. Despite the complex nature of the eddy current examples above, the resulting closed form solutions gives us the dependence of physical properties such as eddy current density, power loss, and forces on parameters such as lift-off, drive frequency, and material conductivity.

6.4 TIME-HARMONIC ANALYSIS IN QUICKFIELD

QuickField enables the solution of time-harmonic magnetic problems with planar or axial symmetry. The permeability is anisotropic if $\mu_x \neq \mu_y$ in x-y symmetry or $\mu_r \neq \mu_z$ in axial symmetry. In x-y symmetry the vector potential is calculated according to

$$\frac{\partial}{\partial x}\left(\frac{1}{\mu_y}\frac{\partial A_z}{\partial x}\right)+\frac{\partial}{\partial y}\left(\frac{1}{\mu_x}\frac{\partial A_z}{\partial y}\right)-i\omega\sigma A_z=-J_{\text{source}} \qquad (6.86)$$

and in axial symmetry

$$\frac{\partial}{\partial r}\left(\frac{1}{r\mu_z}\frac{\partial (rA_\phi)}{\partial r}\right)+\frac{\partial}{\partial z}\left(\frac{1}{\mu_r}\frac{\partial A_\phi}{\partial z}\right)-i\omega\sigma A_\phi=-J_{\text{source}}. \qquad (6.87)$$

The magnetic field components are calculated from the curl of the vector potential, as in the previous chapter. The electric field only has a ϕ component given by $E_\phi = -i\omega A_\phi$ in axial symmetry or a z-component in x-y symmetry $E_z = -i\omega A_z$. The eddy current density is proportional to the vector potential $J_\phi = -i\omega\sigma A_\phi$ in axial symmetry and $J_z = -i\omega\sigma A_z$ in x-y symmetry.

Boundary Conditions

Boundary conditions are assigned in AC Magnetics in the same way as they are assigned in DC Magnetics except that they are understood to vary harmonically in time. The window for editing edge label data is shown in Figure 6.5. Vector potential and tangential field boundary conditions may be assigned spatial dependencies. The zero normal field condition, even and odd periodic boundary conditions may also be assigned in this window.

Dirichlet Conditions

Prescribe values of the vector potential A_z in x-y symmetry and rA_ϕ in axial symmetry.

Neumann Conditions

Specify the tangential field strength numerically equal to the surface current density. At outward boundaries $H_t = \sigma$ at inner boundaries $H_t^+ - H_t^- = \sigma$, where the surface normal points in the direction from (–) to (+).

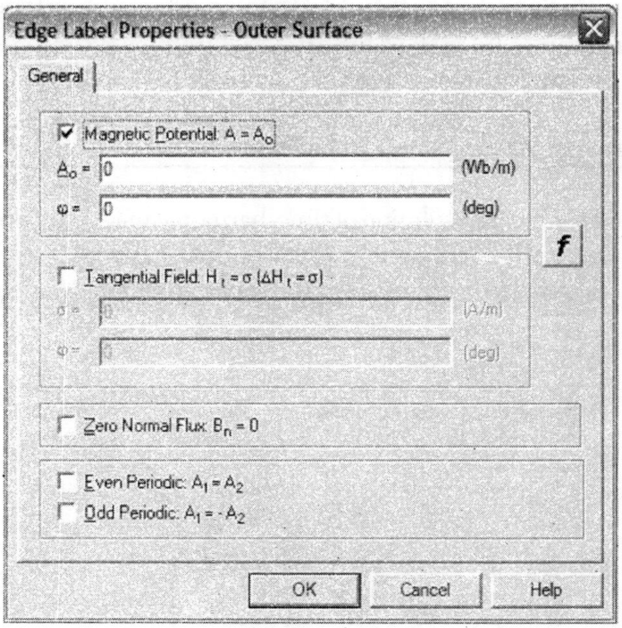

FIGURE 6.5 **Window for editing edge label data in AC Magnetics: To open the edge label window, click on the corresponding edge label in the tree diagram on the left of the screen.**

Periodic Boundary Conditions

Even $A_1 = A_2$ or odd $A_1 = -A_2$ periodicity can be specified on opposite edges of a model to simulate periodic structures. Both normal and tangential field components may be present at a periodic boundary.

Material Properties

Block Labels

Figure 6.6 shows the window for editing block label data.

Block labels in AC Magnetic problems can be associated with data containing two components of magnetic permeability tensor. The total current or the current density may be specified as a function of coordinates. Permanent magnets cannot be simulated because only fields that vary harmonically can be treated in AC Magnetics.

Vertex Labels

Figure 6.7 shows the window for editing vertex label data in AC Magnetics. Specific vector potential and current values may be assigned to individual nodes in the model. These values may also be coordinate dependent.

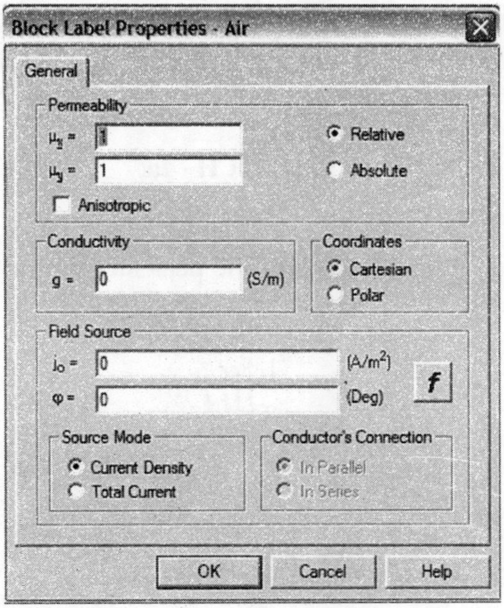

FIGURE 6.6 Window for editing block label data in AC Magnetics: To open the block label window click on the corresponding block label in the tree diagram on the left of the screen.

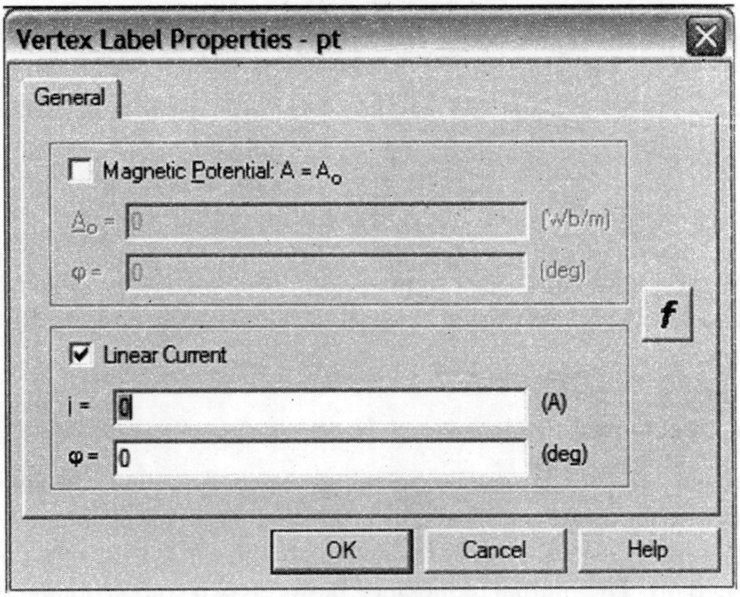

FIGURE 6.7 Window for editing vertex label data in AC Magnetics: To open the vertex label window click on the corresponding vertex label in the tree diagram on the left of the screen.

Integral Quantities

Integral quantities are calculated in the Post Processing module. These include the peak and time average power

$$S = \iint_{surf} \mathbf{E} \times \mathbf{H} \cdot d\mathbf{a}, \tag{6.88}$$

where the Poynting vector $\mathbf{S} = \mathbf{E} \times \mathbf{H}$. The peak and time average Joule heat Q is calculated according to

$$Q = \frac{1}{\sigma} \iiint_{vol} J^2 dv. \tag{6.89}$$

The peak and time average magnetic field energy is

$$W = \frac{1}{2} \iiint_{vol} \mathbf{B} \cdot \mathbf{H}\, dv. \tag{6.90}$$

The peak and time average Lorentz force acting on current carrying conductors is

$$\mathbf{F} = \oint\!\!\oint_{surf} \mathbf{J} \times \mathbf{B} da. \tag{6.91}$$

The peak and time average Lorentz torque is then calculated as

$$\mathbf{T} = \oint\!\!\oint_{surf} \mathbf{r} \times (\mathbf{J} \times \mathbf{B}) da, \tag{6.92}$$

where $\mathbf{r}$ is taken from the origin of the coordinate system. Peak and time average Maxwell forces and torques are also calculated according to the formulas in Chapter 5.

Inductance Calculations

Inductance calculations may be performed in AC Magnetics as in the Magnetostatics solver. The self-inductance of a coil is obtained by dividing the total flux through the coil by the current in the coil with all other magnetic sources turned off. The mutual inductance between two coils is obtained by dividing the flux threading one coil by the current flowing in the second coil with all other sources turned off. The Inductance Wizard is also available in AC Magnetics to systematically calculate self- and mutual inductances.

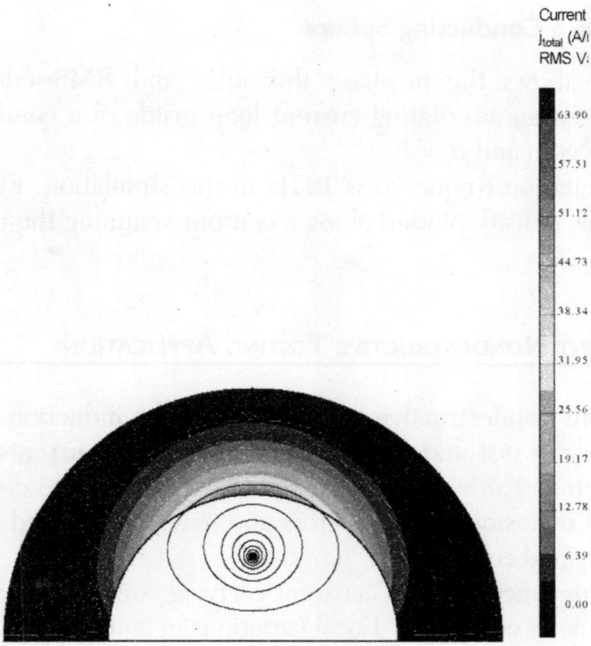

FIGURE 6.8 **Current loop inside of a conducting sphere: In-phase flux lines and RMS eddy current density plot.**

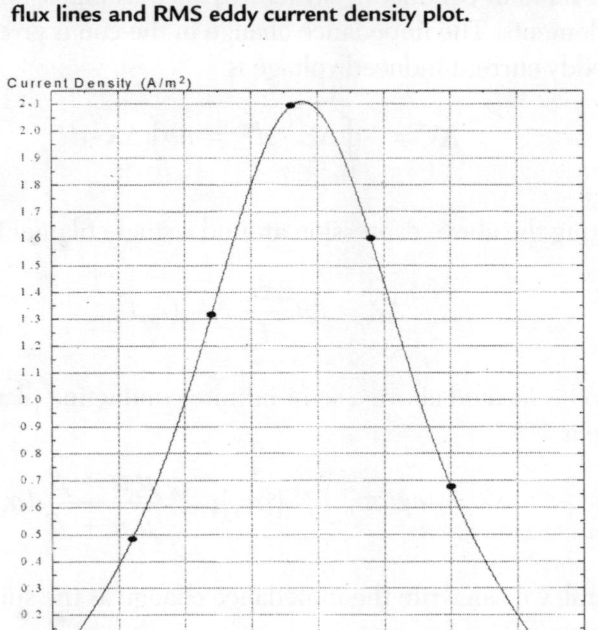

FIGURE 6.9 **Eddy current density plotted along the outer spherical surface in Figure 6.8.**

Current Loop Inside a Conducting Sphere

Figure 6.8 shows the in-phase flux lines and RMS eddy current density resulting from an alternating current loop inside of a conducting sphere with $\sigma = 3.0 \cdot 10^8$ S/m and $\mu_r = 1$.

The excitation frequency is 10 Hz in this simulation. Figure 6.9 shows the eddy current density plotted along a contour spanning the outer surface of the sphere.

6.5 EDDY CURRENT NONDESTRUCTIVE TESTING APPLICATIONS

Eddy current nondestructive testing involves the induction of eddy currents in an object under test and the detection of induced currents by pickup coils or magnetometers. Eddy current inspection may be used to detect defects including pits and inclusions in conductors and may also be used to sort materials by permeability and conductivity.

The impedance of an AC current carrying coil will be affected by the presence of a nearby conductor. Local variations in conductivity will also contribute to the impedance change in the coil. Monitoring the coil impedance enables the detection of flaws in conducting structures such as pipes, aircraft, and metallic structural elements. The impedance change in the coil is given by Ohm's law, where the eddy current induced voltage is

$$\Delta V = -\oint_{\Gamma} \Delta \mathbf{E} \cdot d\ell = i\omega \oint_{\Gamma} \Delta \mathbf{A} \cdot d\ell. \tag{6.93}$$

Integrating the above expression around a single filamentary loop gives

$$\Delta Z = i\omega \frac{2\pi r_0}{I} \Delta A_{\phi}(r_0, \ell). \tag{6.94}$$

If the coil is located above a semi-infinite conducting plane, the coil impedance change is

$$\Delta Z = i\omega\mu_0 \pi r_0^2 \int_0^{\infty} J_1^2(\alpha r_0) e^{-2\alpha \ell} \frac{\mu_r \alpha - \beta}{\mu_r \alpha + \beta} d\alpha. \tag{6.95}$$

In general, we can write the impedance change as the sum of a resistive and a reactive term

$$\Delta Z = \Delta R + i\Delta X. \tag{6.96}$$

ΔZ is typically plotted in the impedance plane plotting resistance versus reactance.

Taking the real and imaginary parts of expression (6.95) gives the coil resistance

$$\Delta R = -\omega\mu_0\pi r_0^2\int_0^\infty J_1^2\,(\alpha r_0)e^{-2\alpha l}\,\mathrm{Im}\left(\frac{\mu_r\alpha-\beta}{\mu_r\alpha+\beta}\right)d\alpha \qquad (6.97)$$

and reactance

$$\Delta X = \omega\mu_0\pi r_0^2\int_0^\infty J_1^2\,(\alpha r_0)e^{-2\alpha l}\,\mathrm{Re}\left(\frac{\mu_r\alpha-\beta}{\mu_r\alpha+\beta}\right)d\alpha. \qquad (6.98)$$

In the case of infinite conductivity the impedance change is purely reactive since Im(−1) = 0 giving

$$\Delta X = -\omega\mu_0\pi r_0^2\int_0^\infty J_1^2\,(\alpha r_0)e^{-2\alpha l}\,d\alpha \qquad (6.99)$$

so that there is no power loss in a perfect conductor with no change in drive coil resistance. Expressions for ΔX and ΔR of an AC current loop above a conducting sphere may be obtained in the same manner.

Change in Magnetic Field and Coil Impedance Due to a Defect in Pipe Inspection

Once the unperturbed vector potential is known, the change in local magnetic field and coil impedance due to a small flaw may be calculated. If the current density inside a flaw is much smaller than the surrounding conductor, it can be treated as the superposition of an unperturbed eddy current and an oppositely directed current $\mathbf{J}_{\text{flaw}}$ canceling within the volume of the flaw. This approximation neglects the complex current flow around the sides of the defect. The Biot-Savart law gives the change in magnetic field due to the perturbation of current flow by a small flaw

$$\Delta\mathbf{B}_{\text{flaw}}(\mathbf{r}) = \frac{\mu_0}{4\pi}\iiint_{\text{vol}}\frac{\mathbf{J}_{\text{flaw}}(\mathbf{r}')\times(\mathbf{r}-\mathbf{r}')}{|\mathbf{r}-\mathbf{r}'|^3}dv', \qquad (6.100)$$

where the flaw $\mathbf{J}_{\text{flaw}} = -\mathbf{J}_{\text{eddy}}$. If the flaw is sufficiently small so that the unperturbed eddy current does not vary appreciably over V_{flaw}, the integral may be approximated

$$\Delta\mathbf{B}_{\text{flaw}}(\mathbf{r}) = \frac{\mu_0}{4\pi}\frac{\mathbf{J}_{\text{flaw}}(\mathbf{r}')\times(\mathbf{r}-\mathbf{r}')}{|\mathbf{r}-\mathbf{r}'|^3}V_{\text{flaw}}, \qquad (6.101)$$

which can be used to calculate the magnetometer response to the flaw. This formula may be implemented with the unperturbed eddy current density obtained from analytical or numerical results. The change in vector potential at the location of an induction or pickup coil may also be obtained

$$\Delta \mathbf{A}_{\text{flaw}}(\mathbf{r}) = \frac{\mu_0}{4\pi} \frac{\mathbf{J}_{\text{flaw}}(\mathbf{r}')}{|\mathbf{r} - \mathbf{r}'|} V_{\text{flaw}}. \tag{6.102}$$

The result may be combined with the formula

$$\Delta Z = i\omega \frac{2\pi r_0}{I} \Delta A_\phi(r_0, \ell). \tag{6.103}$$

to calculate the coil impedance change due to the flaw. The flaw response may then be plotted in the impedance plane. Burrows first modeled the impedance change due to a small flaw obtaining the formula

$$\Delta Z = \frac{3}{2} \omega^2 \sigma V_{\text{flaw}} \left(\frac{A_{\text{flaw}}}{I_{\text{coil}}} \right)^2, \tag{6.104}$$

where A_{flaw} is unperturbed vector potential at the location of the flaw of volume V_{flaw}.

Pipe Inspection

As an example of eddy current nondestructive testing we simulate the eddy currents induced in a section of pipe containing defects with both rectangular and circular cross-sections. Figure 6.10 shows the finite element mesh of a pair of induction coils surrounding a section of pipe containing a notched defect.

The drive coils and pipe cross-sections are highlighted in this figure. The pipe has an inner radius of 2.0 cm, a thickness of 1.0 cm, and a conductivity of 10^8 S/m. The coils have an inner radii of 3.1 cm and are 0.5 cm square with a stand-off between the coil and pipe of 1.0 mm. The flaw has a 1.0 mm thickness, a width of 4.0 mm, and is nonconducting.

The flaw response can be recorded either by monitoring the impedance change in the primary or secondary coils for one- or two-coil configurations, respectively. Alternatively, the change in magnetic field may be monitored directly between two counter-wound coils. From symmetry, the magnetic field generated by the excitation coils, as well as the defect-free eddy current field, will be zero at this location, so that only the flaw response will be recorded. The flux lines and RMS eddy current density resulting from a counter-wound induction coil configuration surrounding the pipe with a notched defect is shown in Figure 6.11. The excitation frequency is 50 Hz in this figure.

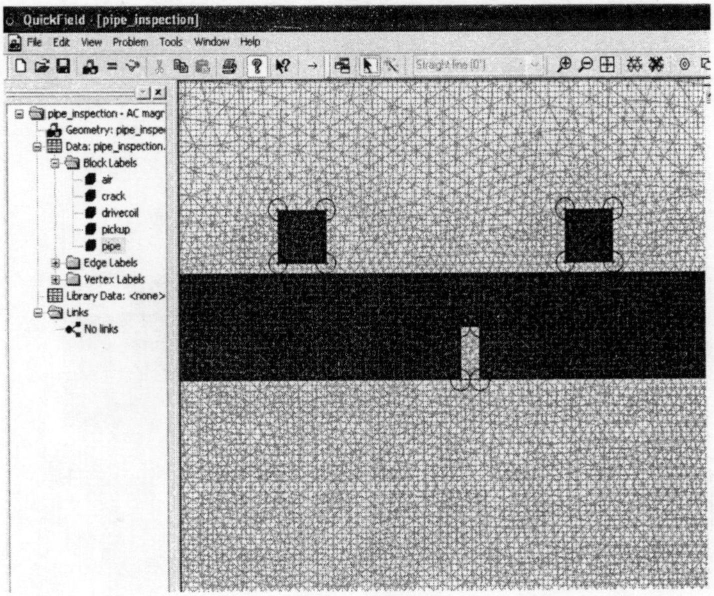

FIGURE 6.10 **Finite element mesh of a pair of induction coils surrounding a section of pipe containing a notched defect: The drive coils and pipe sections are highlighted in this figure.**

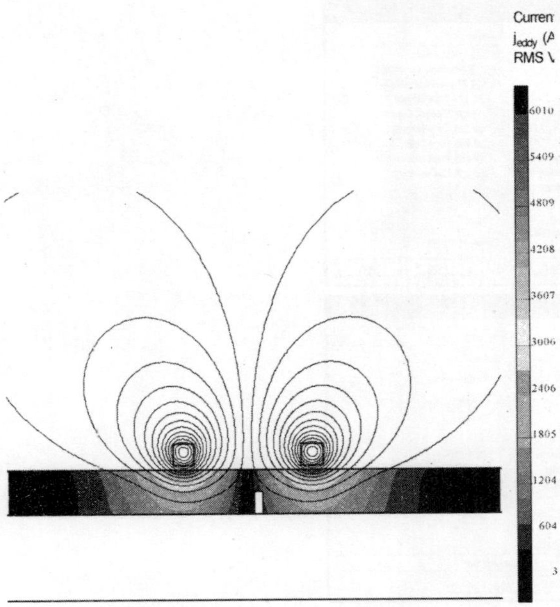

FIGURE 6.11 **Flux lines and RMS eddy current density resulting from a pair of counter-wound induction coils surrounding a section of pipe containing a notched defect. The excitation frequency is 50 Hz in this figure.**

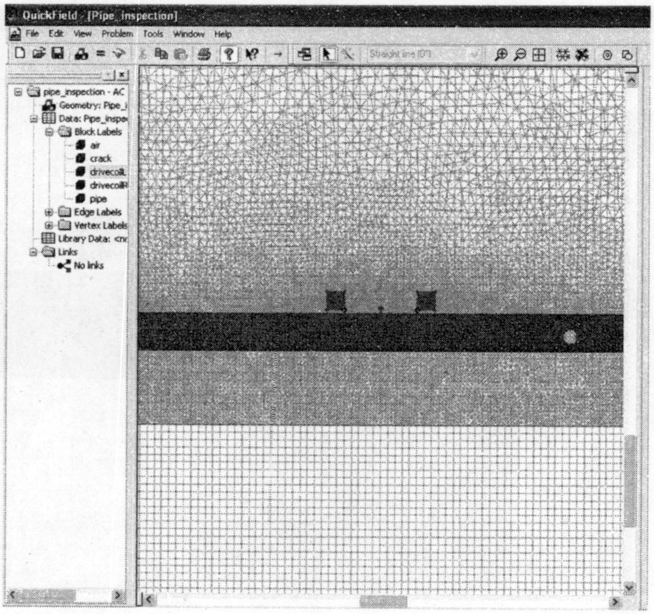

FIGURE 6.12 **Finite element mesh of a pair of induction coils surrounding a section of pipe containing a ring-shaped defect. The drive coils and pipe sections are highlighted in this figure.**

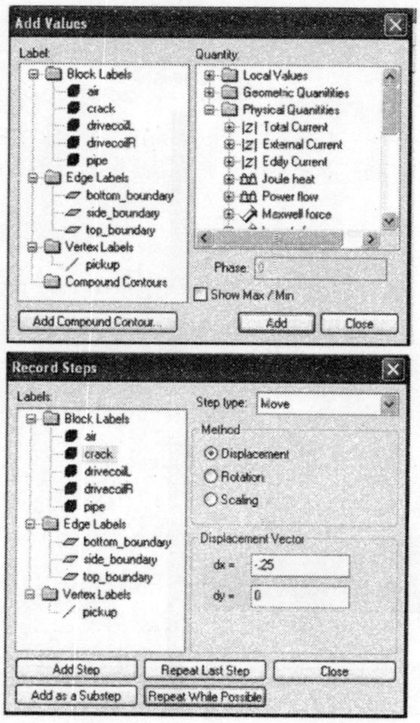

FIGURE 6.13 **Specifying data in the Add Values window and variation of parameters using Record Steps within Label Mover. The crack in the pipe is selected to be translated 0.25 cm to the left each step.**

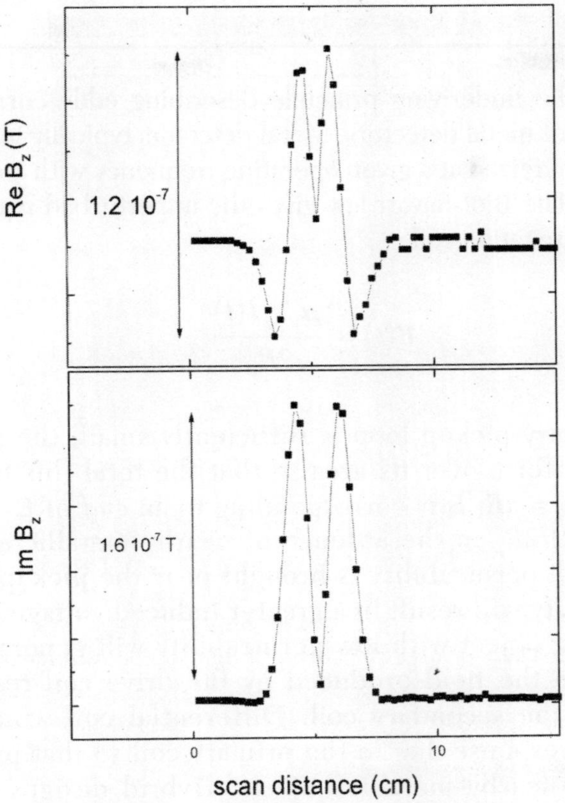

FIGURE 6.14 **Variation in the real part of the *z*-component magnetic field between the induction coils as the defect is moved along the conducting pipe in steps of 0.25 cm. The excitation frequency is 100 Hz in this figure.**

A finite element mesh of a pair of induction coils surrounding a section of pipe containing a ring-shaped defect is shown in Figure 6.12. The drive coils and pipe sections are highlighted in this figure with the same dimensions as in the previous example. Label Mover is used to translate the flaw position along the pipe simulating a measurement scan. Alternatively, the drive coils or the entire pipe containing the defect could be translated. Figure 6.13 shows the Label Mover windows for recording data and changing model parameters after each solution. In this figure, the defect is selected to be translated 0.25 cm to the left each step. Figure 6.14 shows the variation in the real and imaginary parts of the *z*-component magnetic field directly between the induction coils (carrying 1.0 A, 100 Hz AC current) as the defect is moved along the conducting pipe. The signal peaks when the flaw is located in a region of maximum eddy current density and is zero when the flaw is directly between the coils, where the eddy currents produced by each coil cancels.

6.6 METAL DETECTORS

Faraday's law is the underlying principle describing eddy current induction and the operation of metal detectors. Metal detectors typically feature an outer drive coil that is energized at a given operating frequency with an inner pickup, or detection coil. The Biot-Savart law gives the unperturbed magnetic field at the center of the excitation coil

$$B(t) = \frac{\mu_0 N_p I(t)}{2R_p}. \tag{6.105}$$

If the secondary pickup loop is sufficiently small, the primary field will be nearly uniform over its area so that the total flux threading the pickup loop is $\Phi_m = \pi R_s^2 B(t)$ corresponding to an *emf* of $E = -N_s \partial\Phi/\partial t = -i\omega\mu_0 \pi R_S^2 N_S N_P I(t)/2R_p$ in the absence of nearby metallic objects. If an object with a high permeability is brought near the pickup coil, the increased flux density will result in a greater induced voltage in the coil. A highly conducting object with low permeability will generate eddy currents that oppose the field produced by the drive coil resulting in an *emf* decrease in the secondary coil. Differential coil winding may be used to null the response due to the primary coil so that pickup signals result only from nearby metallic objects. Hybrid designs may also be employed, where a magnetometer is used instead of a secondary coil. Various magnetic sensors available include Hall effect sensors, fluxgate, and Superconducting Quantum Interference Device (SQUID) magnetometers. The primary field will be zeroed if the magnetometer is positioned on the axis of the primary coil and oriented so as to detect the in-plane field component. Alternatively, the out-of-phase z-component magnetometer response resulting solely from eddy currents may be monitored as in the next example.

Figure 6.15 shows the flux lines and eddy current density generated by a metal detector coil above a buried Civil War musket ball with a 1.0 cm diameter and a conductivity of 10^7 S/m. The excitation coil consists of a single filamentary current loop carrying a 1.0-Ampere, AC current. Figure 6.16 shows the imaginary, or out-of-phase, z-component magnetic field that would be measured by a magnetometer located at the center of the drive coil for musket ball depths ranging between 0 and 70 cm below the surface. Two curves in this figure correspond to driving frequencies of 50 and 500 Hz. Note that the out-of-phase response will be solely due to the musket ball and not the drive coil. The ground conductivity has been neglected in this calculation. In the next section we investigate current induction in the earth by time-changing fields.

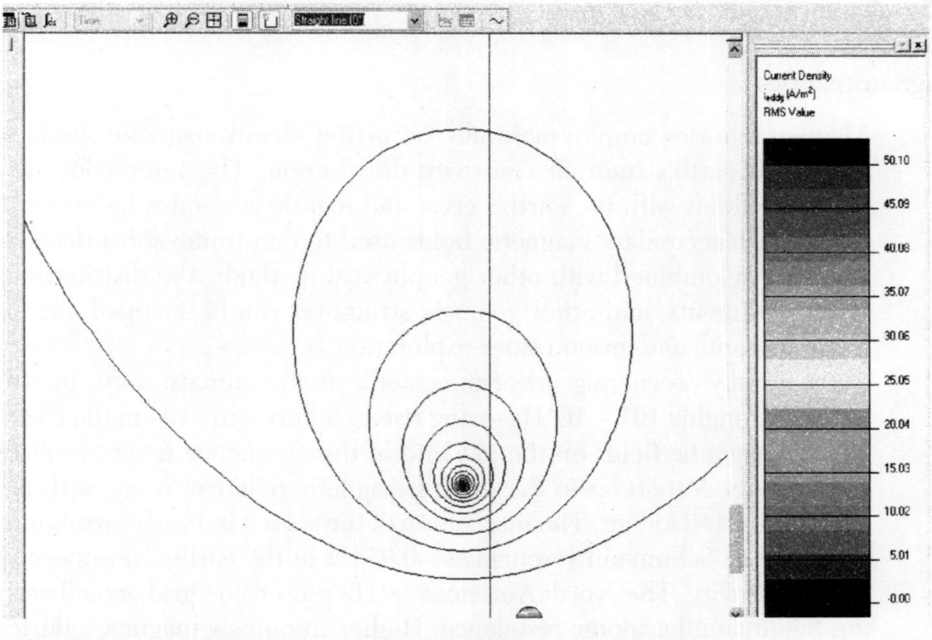

FIGURE 6.15 **Flux lines and eddy current density generated by a metal detector coil above a buried Civil War musket ball.**

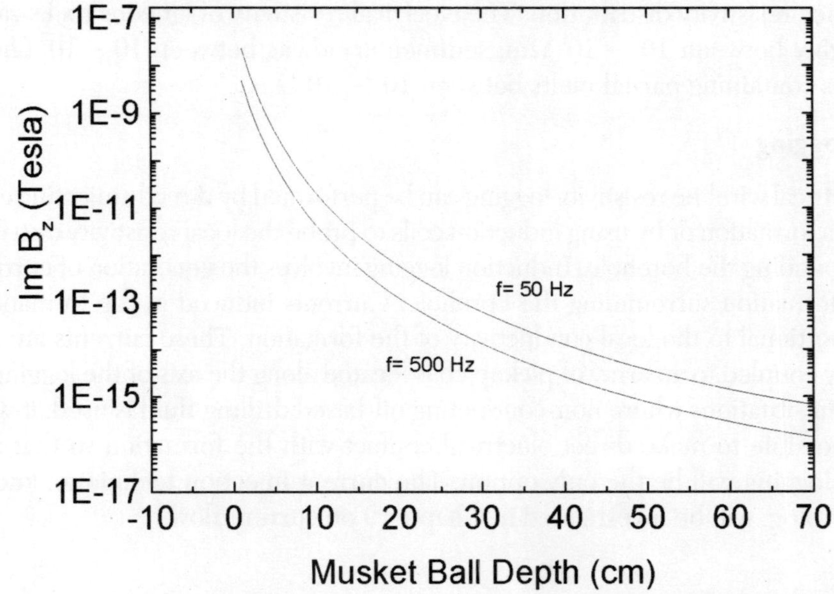

FIGURE 6.16 **Out-of-phase magnetometer response at 50 Hz and 500 Hz for musket ball depths ranging between 0 and 70 cm below the surface: Note that the out-of-phase response will be solely due to the musket ball and not the drive coil.**

6.7 GEOPHYSICAL APPLICATIONS

Magnetotellurics

Magnetotellurics employ naturally occurring electromagnetic fields to investigate the earth's internal resistivity distribution. The interaction of electromagnetic fields with the earth's crust and mantle generates induction currents resulting in secondary magnetic fields used to determine subsurface resistivity $\rho \sim (E/H)^2$. Combined with other geophysical methods, the distribution of rock types, sediments, and other geologic structures can be mapped out for petroleum, mineral, and groundwater exploration.

Naturally occurring electromagnetic field sources vary in frequency between roughly $10^{-4} - 10^4$ Hz giving rise to telluric currents in the earth's crust. Electromagnetic fields on the low end of this frequency range result from sun spot activity. Variations in the earth's magnetic field can occur with periods of about 100 s and longer. The interaction of the solar wind with earth's ionosphere gives rise to Schumann resonances ~0.05 Hz in the earth-ionosphere spherical resonant cavity. The North American 60 Hz power line grid actually excites the 9th Schumann harmonic resonance. Higher frequency magnetotelluric sources above the kHz range result from lightning in the atmosphere.

Because the electromagnetic penetration depth varies with frequency, penetration depths may vary between $10^2 - 10^5$ m, also depending on the subsurface resistivity distribution. The electrical resistivity of igneous rocks can vary roughly between $10^3 - 10^6$ Ωm, sedimentary rocks between $10 - 10^3$ Ωm, and rocks containing partial melts between $10^{-2} - 10$ Ωm.

Induction Logging

Electrical wireline resistivity logging can be performed by directly injecting current into a formation or by using induction coils to probe the local resistivity distribution surrounding the borehole. Induction logging involves the generation of currents in the formation surrounding the borehole. Currents induced in the formation are proportional to the local conductivity of the formation. These currents are inductively coupled to an array of pickup coils located along the axis of the logging tool.

In situations where non-conducting oil-based drilling fluid is used, it will not be possible to make direct electrical contact with the formation so that induction logging will be the only option. The current injection technique, known as Laterolog, will be investigated in Chapter 9 on current flow.

6.8 TRANSFORMER APPLICATIONS

The first practical transformers appeared roughly 50 years after Faraday discovered his law of induction. A typical transformer consists of two coils that are inductively coupled. For greater flux coupling, the coils may be wound on a highly permeable core with $\mu_r \approx 10^4$. Transformers can either convert high voltage low current to low voltage high current signals (step down) or vice versa (step up). Electrical power lines carry high AC voltage, low current electricity so as to minimize I^2R Joule heating losses.

For an ideal transformer with primary and secondary turn numbers N_p and N_s, the AC voltages across the primary and secondary turns are related to the time rate of change of magnetic flux threading the coils $V_p = -N_p \partial\Phi/\partial t$ and $V_s = -N_s \partial\Phi/\partial t$. Since $\partial\Phi/\partial t$ is the same for both coils we have

$$\frac{V_p}{V_s} = \frac{N_p}{N_s}. \tag{6.106}$$

The primary voltage is stepped down for $V_p > V_s$ and stepped up for $V_p < V_s$. If there are no power losses, the input power is equal to the output $I_p V_p = I_s V_s$. Figure 6.17 shows a transformer simulation carried out in x-y symmetry. The primary coil is located on the right of this figure and the secondary coil on the left. Flux lines are channeled through the central ferrite μ-metal core with $\mu_r = 10^4$. The current in the primary transformer coil on the right is 1.0 A with an operating frequency of 1.0 kHz.

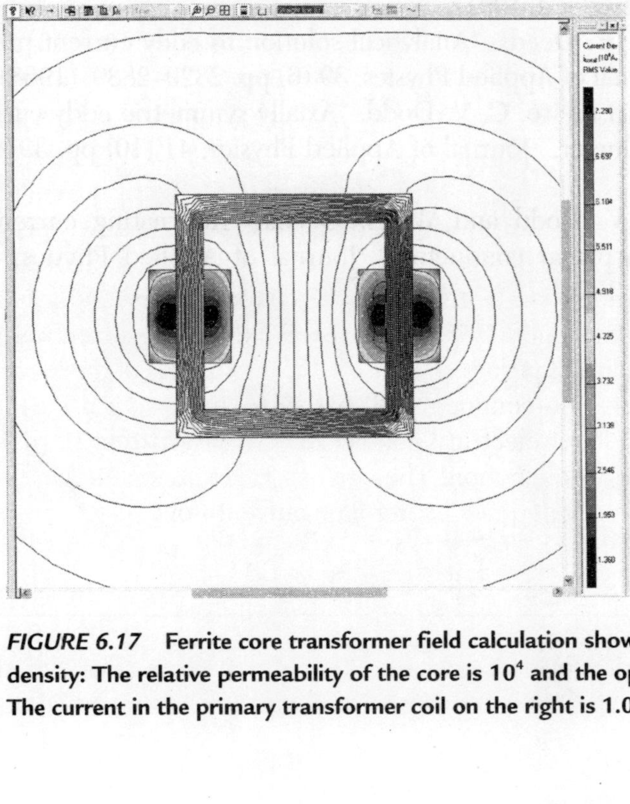

FIGURE 6.17 **Ferrite core transformer field calculation showing flux lines and current density: The relative permeability of the core is 10^4 and the operating frequency is 1.0 kHz. The current in the primary transformer coil on the right is 1.0 A.**

In this chapter, we developed and applied the time-harmonic Maxwell equations to treat AC magnetics problems. The analysis may be carried out initially by solving for the primary fields **E** and **B** or the magnetic vector potential **A**. Applications of Faraday's law explored here include eddy current nondestructive testing, metal detectors, magnetotellurics, induction logging, and transformers. In the next chapter, we extend the time-harmonic analysis developed here to treat transient magnetic problems given sources with arbitrary time dependence.

REFERENCES

QuickField User's Guide: Tera Analysis Ltd. (2007) *Chapter 2: Introductory Guide; Chapter 10: Theoretical Description.*

David J. Griffiths, *Introduction to Electrodynamics* 3rd *ed.*, Prentice-Hall, Inc., New Jersey (1999). *Chapter 7: Electrodynamics.*

William R. Smythe, *Static and Dynamic Electricity*, Taylor & Francis. (1989) *Chapter XIII: Electromagnetic Induction; Chapter X: Eddy Currents.*

J. A. Tegopoulos and E. E. Kriezis, *Eddy Currents in Linear Conducting Media*, (Elsevier, New York, 1985).

Jack Blitz, *Electrical and Magnetic Methods of Nondestructive Testing*, (Adam Hilger, Bristol England, 1991).

C. V. Dodd and W. E. Deeds, "Analytical solution to eddy current probe coil problems," Journal of Applied Physics, 39 (6) pp. 2829–2839, (1968).

W. E. Deeds, J. W. Liquire, C. V. Dodd, "Axially symmetric eddy currents in a spherical conductor," Journal of Applied Physics, 41 (10) pp. 3976–3981, (1970).

J. W. Liquire, C. V. Dodd and W. E. Deeds, "Alternating current distri bution between planar conductors," Journal of Applied Physics, 41 (10) pp. 3983–3985, (1970).

EXERCISES

6.1 For an AC magnetic field tangential to a conducting half-space, show that the real and imaginary parts of the eddy current density are proportional to $\mathrm{Re}(J_x) \propto \cos\frac{z}{\delta} + \sin\frac{z}{\delta}$ and $\mathrm{Im}(J_x) \propto \cos\frac{z}{\delta} - \sin\frac{z}{\delta}$. Show that $\mathrm{Re}(J_x)$ changes sign (is zero) at characteristic depths $z = (4n+3)\pi\delta/4$ inside the metal while $\mathrm{Im}(J_x)$ changes sign at $z = (4n+1)\pi\delta/4$, where $n = 0,1,2\ldots$

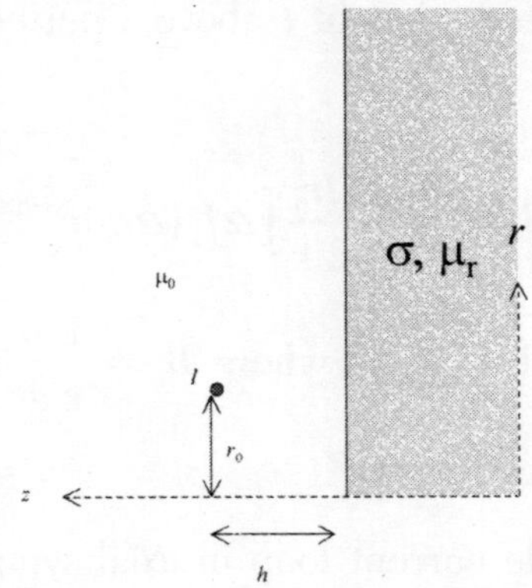

FIGURE E-6.2 **AC current loop above a semi-infinite conducting half-space. Geometry: Axial symmetry.**

6.2 Model a current loop of radius r_0 as a single node in axial symmetric geometry a height λ above a semi-infinite, permeable conducting half-space in Figure E-6.2. Plot the imaginary z-component magnetic field at the center of the current loop as a function of frequency using Label Mover.

6.3 Model a current loop of radius r_0 as a single node in axisymmetric geometry a height λ above a plate of thickness d, conductivity σ, and permeability μ_0. Compare the eddy current density along a radial contour on each surface and at the center of the plate to the analytical expression $\boldsymbol{J}_{\text{eddy}} = -i\omega\sigma A_\phi$, where

$$A_\phi(r,z) = \mu_0 I r_0 \int_0^\infty \alpha J_1(\alpha r) J_1(\alpha r_0) e^{-\alpha\ell} \times \left(\frac{(\alpha+\beta)e^{\beta(z+d)} - (\alpha-\beta)e^{-\beta(z+d)}}{(\alpha+\beta)^2 e^{\beta d} - (\alpha-\beta)^2 e^{-\beta d}} \right) d\alpha,$$

where $\beta = \sqrt{\alpha^2 + i\omega\mu_0\sigma}$ and the upper and lower surfaces of the plate are located at $z = 0$ and $z = -d$, respectively.

6.4 Show that for a current loop of radius r_0 above a perfectly conducting semi-infinite half-space the vector potential has the form

$$A_\phi(r,z) = \mu_0 I r_0 \int_0^\infty J_1(\alpha r) J_1(\alpha r_0) \exp(-\alpha z_>) \sinh(\alpha z_<) d\alpha,$$

where $z_>$ or $z_<$ is the greater or lesser of z and ℓ.

6.5 Show that the magnetic field resulting from an AC current loop of radius r_0 located a height ℓ above a perfectly conducting semi-infinite half-space is

$$B_z(z) = \frac{\mu_0 I r_0}{2} \int_0^\infty \alpha J_1(\alpha r_0) e^{-\alpha(z+\ell)} \frac{\mu_r \alpha - \beta}{\mu_r \alpha + \beta} d\alpha$$

along the z-axis for $z \geq 1$, where $B_z = \frac{1}{r}\frac{\partial}{\partial r}(rA_\phi)$ and using the fact that $\lim_{r \to 0} \frac{1}{r}\frac{\partial}{\partial r}(r \cdot J_1(\alpha r)) = \alpha.$

6.6 Model a single current loop in axial symmetry above the conducting half-space shown in Figure E-6.2. Plot the real and imaginary components of eddy current density along radial contours at several depths inside the conductor and compare to the analytical solution evaluated in MATLAB. Calculate the total power dissipated by eddy currents in the conducting half-space.

6.7 Model a current loop inside the hollow conducting, permeable sphere shown in Figure E-6.7. Create contour plots of the in-phase and out-of-phase eddy current densities and field values along the inner and outer spherical surfaces. Calculate the Joule heating in sphere as a function of frequency using Label Mover.

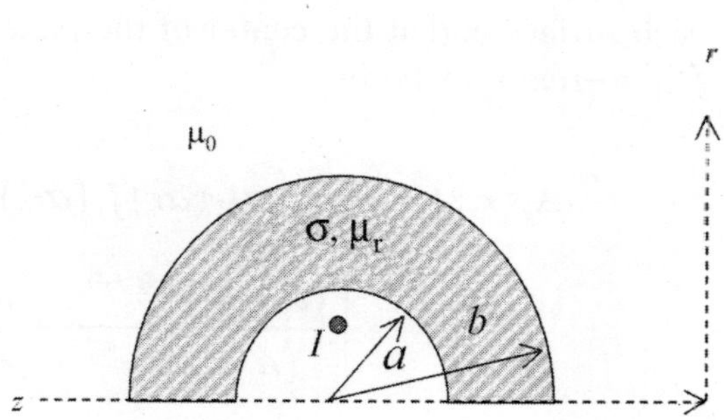

FIGURE E-6.7 **AC current loop inside a permeable conducting sphere. Geometry: Axial symmetry.**

6.8 Analytically model current induction in the human brain by a single filamentary current loop. Assume the brain is roughly spherical with conductivity $\sigma = 0.06$ S/m and effective permittivity $\varepsilon = 10^6 \varepsilon_0$. Numerically evaluate the resulting summations in MATLAB.

6.9 Write analytical expressions for the resistance and reactance of a drive coil above a conducting sphere.

6.10 Model an AC current loop surrounding the permeable conducting sphere shown in Figure E-6.10. Use Label Mover to plot the eddy current distribution for various current loop radii.

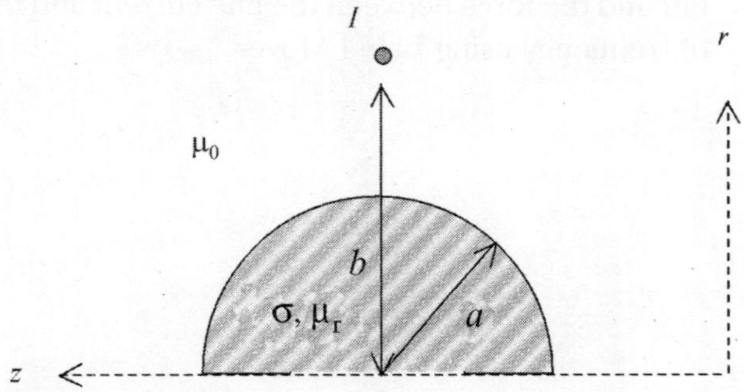

FIGURE E-6.10 **AC current loop of radius b surrounding a solid conducting, permeable sphere of radius *a*. Geometry: Axial symmetry.**

6.11 The conducting, permeable sphere shown in Figure E-6.11 is positioned in a *z*-directed AC magnetic field. Make a contour plot of the radial eddy current density from the center of the sphere to its edge. Determine the total power dissipated by eddy currents in the sphere.

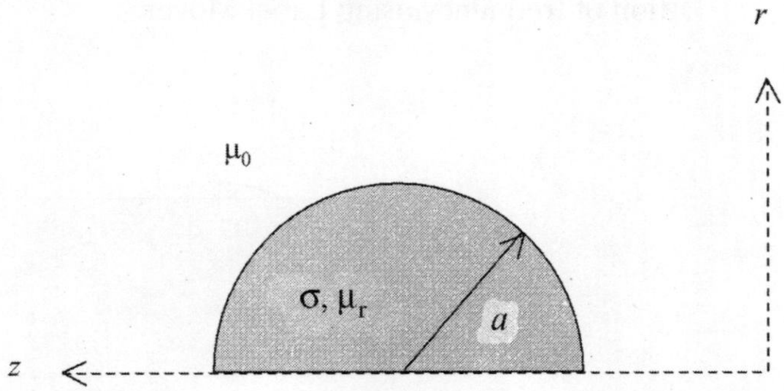

FIGURE E-6.11 **Conducting sphere in an external AC magnetic field. Geometry: Axial symmetry.**

6.12 Use the small flaw model to map the impedance of a circular drive coil above a semi-infinite conducting half-space containing a small spherical void. Repeat for several flaw depths in the conductor.

6.13 Model a sheet inducer above a semi-infinite conducting half-space containing a defect. Use the small flaw model to create magnetic field maps of the z-component flaw field for several flaw depths in the conductor. Choose flaw depths between locations where the imaginary eddy current density vanishes.

6.14 Model a line current parallel to a conducting, permeable cylinder in Figure E-6.14. Plot the power dissipated by eddy currents in the cylinder and the force between the line current and the cylinder as a function of frequency using Label Mover.

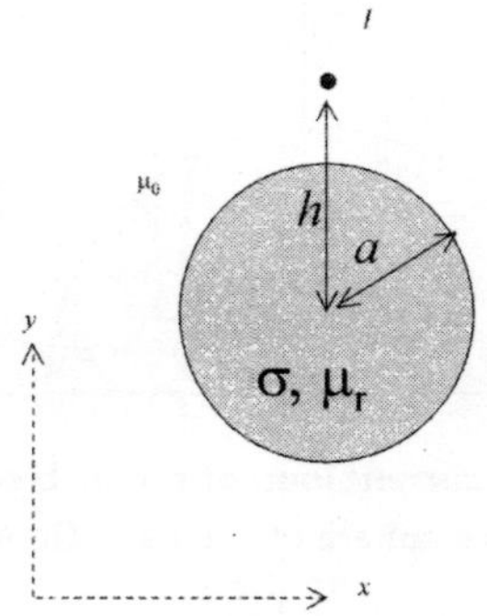

FIGURE E-6.14 **Line current parallel to a conducting, permeable cylinder. Geometry: *x-y* symmetry.**

6.15 Model a line current parallel to a semi-infinite conducting, permeable half-space in Figure E-6.15. Create contour plots of the eddy current density parallel to the surface of the metal at various depths in the metal. Plot the power dissipated by eddy currents in the half-space as a function of frequency using Label Mover.

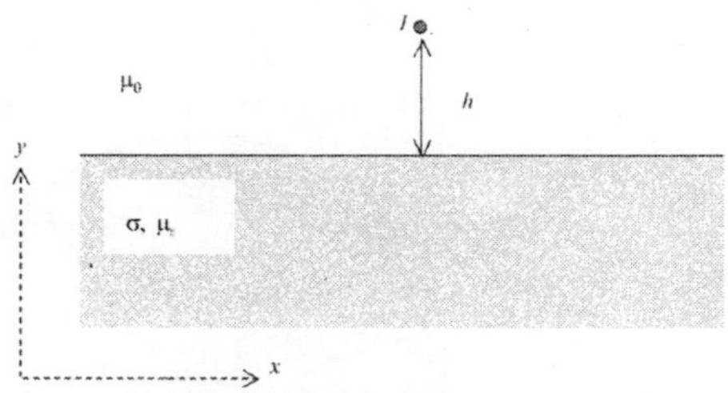

FIGURE E-6.15 **Line current parallel to a conducting, permeable half-space. Geometry: *x-y* symmetry.**

6.16 Model an AC current loop of radius r_0 surrounding a conducting, permeable cylinder of radius b containing a spherical defect of radius δ at the center of the cylinder in Figure E-6.16. Calculate the drive coil impedance while translating the coil over the flaw using Label Mover. Compare impedance plane plots of Im(Z) vs. Re(Z) performing the scan at several different operating frequencies.

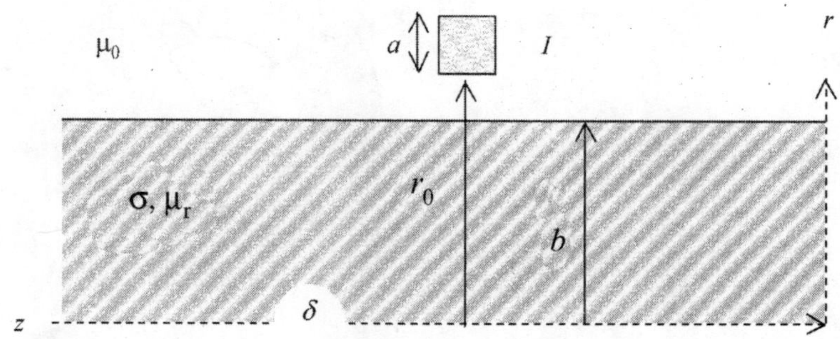

FIGURE E-6.16 **AC current loop of radius r_0 surrounding a conducting permeable cylinder of radius *b* containing a spherical defect of radius δ at the center of the cylinder. Geometry: Axial symmetry.**

6.17 Model a hollow conducting, permeable sphere in the uniform AC magnetic field in Figure E-6.17. Plot the 0-degree and 90-degree phase magnetic field at the center of the sphere as a function of frequency using Label Mover.

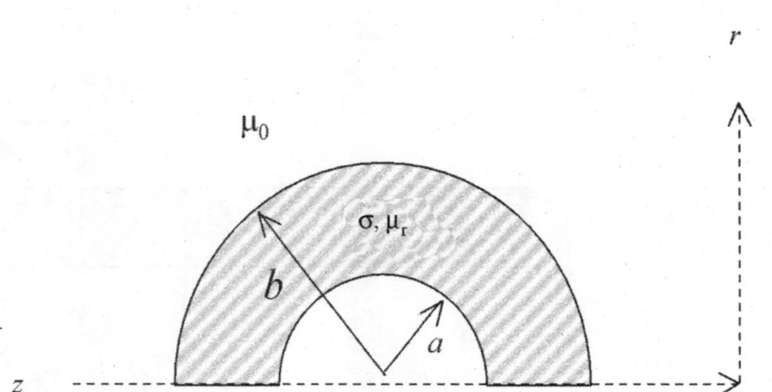

FIGURE E-6.17 **Hollow permeable, conducting sphere in an AC magnetic field. Geometry: Axial symmetry.**

6.18 Model an AC line current at the center of a hollow conducting, permeable cylinder containing the line defect of diameter δ shown in Figure E-6.18. Plot the normal magnetic field component along a circular contour surrounding the cylinder. Note that the normal field component would be zero without the defect from symmetry.

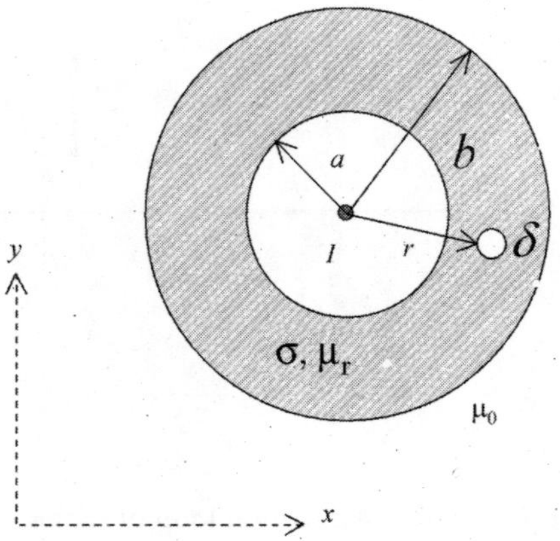

FIGURE E-6.18 **AC line current inside a permeable conducting cylinder with a defect. Geometry: *x-y* symmetry.**

6.19 Simulate the eddy current nondestructive testing of a cylindrical conducting, permeable pipe with the notch defect in Figure E-6.19. Plot the impedance in the secondary coil translating both coils along the pipe using Label Mover for several operating frequencies.

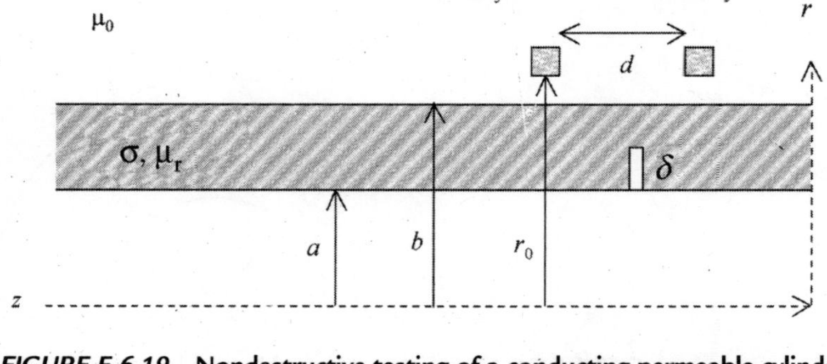

FIGURE E-6.19 **Nondestructive testing of a conducting permeable cylinder of inner radius *a*, outer radius *b*, and notch defect of width *δ*. Geometry: Axial symmetry.**

6.20 Simulate an eddy current drive coil above a semi-infinite conducting, permeable half-space with the cylindrical defect in Figure E-6.20. Calculate the change in impedance of the eddy current drive coil as a function of flaw depth using Label Mover for excitation frequencies of 1 kHz, 10 kHz, 100 kHz, and 1 MHz. Create an impedance plane plot of the inductive and reactive components of the coil impedance using Label Mover to vary the flaw depth at a single frequency.

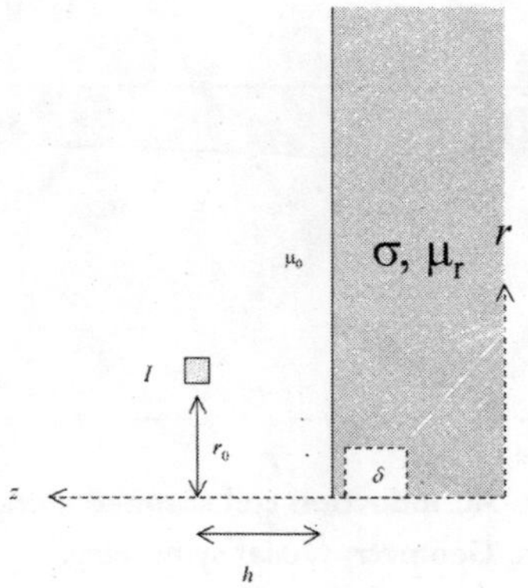

FIGURE E-6.20 **Eddy current drive coil above a semi-infinite conducting, permeable half-space with a cylindrical defect. Geometry: Axial symmetry.**

6.21 Compare the results obtained in 6.20 to the impedance obtained from the small flaw model calculating the required integrals in MATLAB.

6.22 Model a transformer with primary and secondary coils wound on the cylindrical, conducting permeable core shown in Figure E-6.22. Calculate the power loss in the transformer and the induced voltage in the secondary coil as a function of frequency of the primary coil using Label Mover.

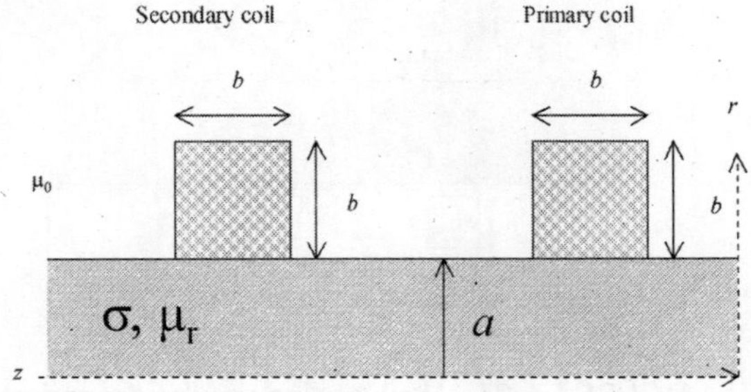

FIGURE E-6.22 **Transformer with primary and secondary coils wound on a permeable conducting cylinder. Geometry: Axial symmetry.**

6.23 Model the eddy current nondestructive testing of a section of pipe with thinning due to corrosion in Figure E-6.23. Calculate the impedance of the drive coil as it is translated along the length of the thinning pipe using Label Mover.

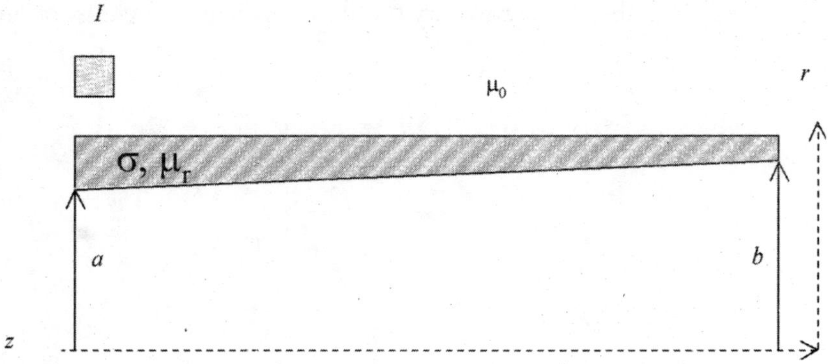

FIGURE E-6.23 **AC induction coil scanned along the length of a pipe with thinning due to corrosion. Geometry: Axial symmetry.**

6.24 Perform a well logging simulation of a logging tool consisting of excitation and receiver coils translated down a borehole passing through geological formations with differing conductivity as shown in Figure E-6.24. Using Label Mover, calculate the impedance change in the secondary coil as the logging tool is translated in the z-direction.

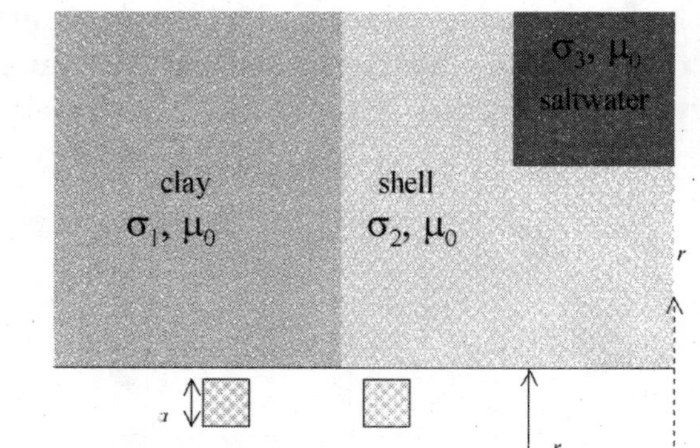

FIGURE E-6.24 **Well logging simulation: excitation and receiver coils are translated through a borehole passing through geological formations with differing conductivity. Geometry: Axial symmetry.**

6.25 Model the magnetotelluric currents resulting from variations in the earth's magnetic field of 0.5 μT with a period of about 100 seconds. Experiment with both uniform field and current loop excitations. Choose realistic values for the conductivity, permeability, and radii of the solid and liquid core regions, mantle, and the earth's crust according to Figure E-6.25. Calculate the magnitude of the induced current density in the earth and calculate the total power generated by magnetotelluric currents in the crust. Make plots of E_ϕ and H_n along a circular contour extending form N Pole to S Pole. Calculate the average ratio of $(H_n/E_\phi)^2$ along the contour. Use the following data: radius of earth = 6371 km, radius of inner core = 1216 km, thickness of outer core = 2270 km, thickness of mantle = 2885 km.

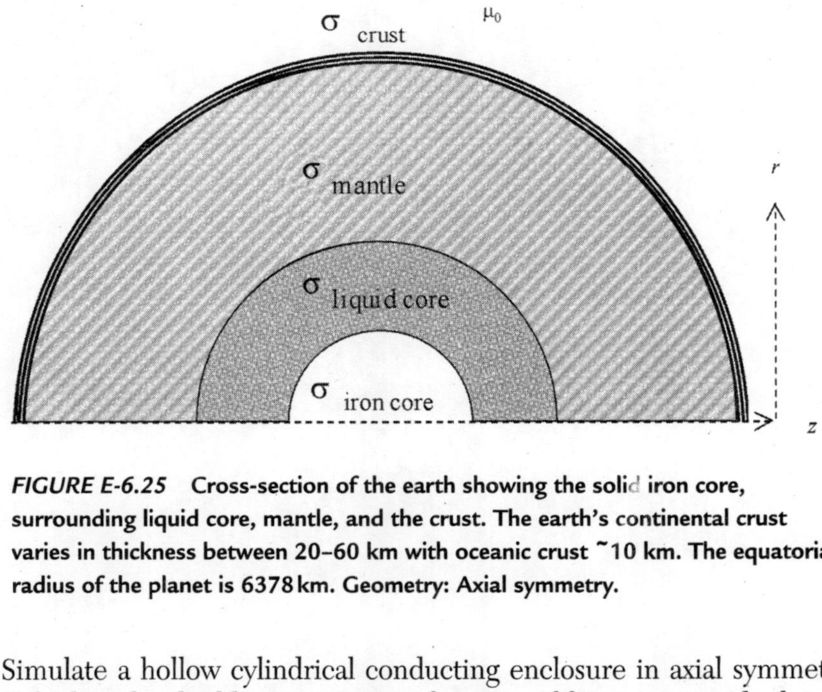

FIGURE E-6.25 **Cross-section of the earth showing the solid iron core, surrounding liquid core, mantle, and the crust. The earth's continental crust varies in thickness between 20–60 km with oceanic crust ~10 km. The equatorial radius of the planet is 6378 km. Geometry: Axial symmetry.**

6.26 Simulate a hollow cylindrical conducting enclosure in axial symmetry. Calculate the shielding B_{in}/B_{out} as a function of frequency inside the can for a z-directed external field with magnitude B_{out}. Use Label Mover to sweep the frequency while recording the real and imaginary field components inside the shield.

Chapter 7 TRANSIENT MAGNETICS

In This Chapter

- Time-dependent Maxwell's Equations
- Finite Difference Time-domain Method
- Vector Potential Formulation
- Time-dependent Magnetics in QuickField
- Pulsed Eddy Current Nondestructive Testing Applications
- Maxwell's Equations with Monopole Source Term

7.1 TIME-DEPENDENT MAXWELL'S EQUATIONS

In the previous chapter we considered the special case of Maxwell's equations with time-harmonic sources. If the field sources have arbitrary time dependence, then we must work with the time-dependent curl and divergence equations

$$\nabla \times \mathbf{E} = -\frac{\partial \mathbf{B}}{\partial t} \quad \nabla \times \mathbf{B} = \mu \mathbf{J}_e + \mu\varepsilon\frac{\partial \mathbf{E}}{\partial t}$$
$$\nabla \cdot \mathbf{E} = \frac{\rho}{\varepsilon} \quad \nabla \cdot \mathbf{B} = 0, \tag{7.1}$$

where $\mathbf{J}_e = \sigma\mathbf{E} + \mathbf{J}_{\text{source}}$ and $(\mu, \varepsilon) = (\mu_r\mu_0, \varepsilon_r\varepsilon_0)$.

Current Sheet Above a Conducting Half-Space

The simplest example of transient current induction is that produced by a magnetic field tangential to a conducting half-space that is switched on at $t = 0$.

This field can be thought of as being produced by an infinite sheet with a uniform surface current density that is parallel to the conducting half-space. Neglecting the displacement current we have

$$\nabla \times \mathbf{B} = \mu_0 \sigma \mathbf{E} \tag{7.2}$$

inside the metal and taking the curl of this equation and using

$$\nabla \times \mathbf{E} = -\frac{\partial \mathbf{B}}{\partial t} \tag{7.3}$$

we obtain

$$\nabla^2 \mathbf{B} = \mu_0 \sigma \frac{\partial \mathbf{B}}{\partial t}. \tag{7.4}$$

For an external field $\mathbf{B}(t) = B_0 \theta(t)\mathbf{j}$, the differential equation with initial conditions becomes

$$\frac{\partial^2 B_y}{\partial z^2} = \mu_0 \sigma \frac{\partial B_y}{\partial t} \quad \begin{Bmatrix} B_y(z,0) \\ B_y(0,t) \\ B_y(\infty,t) \end{Bmatrix} = \begin{Bmatrix} 0 \\ B_0 \theta(t) \\ 0 \end{Bmatrix}, \tag{7.5}$$

where $\theta(t)$ is zero for $t < 0$ and 1 for $t \geq 0$. Taking the Laplace transform of both the differential equation and initial conditions gives

$$\frac{\partial^2 b_y(s)}{\partial z^2} = \mu_0 \sigma \left(s b_y(s) - B_y(0)\right). \quad \begin{Bmatrix} b_y(z,0) \\ b_y(0,s) \\ b_y(\infty,s) \end{Bmatrix} = \begin{Bmatrix} 0 \\ \dfrac{B_0}{s} \\ 0 \end{Bmatrix}, \tag{7.6}$$

where $B_y(0) = 0$. The solution to this equation is

$$b_y(s) = A \cdot \exp\left(\sqrt{\mu_0 \sigma s} z\right) + B \cdot \exp\left(-\sqrt{\mu_0 \sigma s} z\right). \tag{7.7}$$

Applying the boundary conditions we have $A = 0$ and $B = \frac{B_0}{s}$ giving

$$b_y(s) = \frac{B_0}{s} \cdot \exp\left(-\sqrt{\mu_0 \sigma s} z\right). \tag{7.8}$$

Performing the inverse Laplace transform in MATLAB

```
%MATLAB code for performing the inverse Laplace transform
% of exp(-a*sqrt(s))
% where a = z*sqrt(mu_0*sigma)
syms s a positive
% requiring a>0
ilaplace(exp(-a*sqrt(s))/s,s,t)
```

gives the field in the time-domain

$$B_y(z,t) = B_0 \mathrm{erfc}\left(\frac{1}{2}\sqrt{\frac{\mu_0\sigma}{t}}z\right), \tag{7.9}$$

where the complementary error function is given by

$$\mathrm{erfc}(z) = \frac{2}{\sqrt{\pi}}\int_z^\infty e^{-t^2}dt. \tag{7.10}$$

The eddy current is given by $\mu_0 J_x = -\partial B_y/\partial z$ so that using the identity

$$\frac{d}{dz}\mathrm{erfc}(z) = -\frac{2}{\sqrt{\pi}}e^{-z^2} \tag{7.11}$$

finally gives the eddy current

$$J_x(z,t) = \frac{B_0}{\mu_0}\frac{2}{\sqrt{\pi}}\exp\left(-\frac{\mu_0\sigma}{4t}z^2\right). \tag{7.12}$$

Equation 7.5 may be solved numerically using an explicit finite difference numerical method

$$\frac{B_y^n(i+1) - 2B_y^n(i) + B_y^n(i-1)}{\Delta z^2} = \mu_0\sigma\frac{B_y^{n+1}(i) - B_y^n(i)}{\Delta t}, \tag{7.13}$$

where the spatial and time derivatives are evaluated using central difference and forward difference schemes, respectively. The magnetic field is then advanced in time at each grid point according to

$$B_y^{n+1}(i) = B_y^n(i) + \frac{\Delta t}{\mu_0\sigma}\frac{B_y^n(i+1) - 2B_y^n(i) + B_y^n(i-1)}{\Delta z^2}. \tag{7.14}$$

This explicit method is stable for $\Delta t \leq \mu_0 \sigma \Delta z^2/2$. Once the field is known, the current density may be calculated according to

$$J_x^n(i) = -\frac{1}{\mu_0}\left(B_y^n(i+1) - B_y^n(i)\right). \tag{7.15}$$

Implicit methods, such as the Crank-Nicholson method, may also be used for greater stability and accuracy. The following MATLAB program calculates the penetration of magnetic fields and the induced currents inside the metal:

```
%MATLAB program for solving for the field and
%current penetration into a semi-infinite conducting half-space
%B-field of 1.0 T is switched on at t=0

%specify parameters
dz=10^-4; mu_0=4*pi*10^-7; sigma=10^8;
dt=(mu_0*sigma*dz^2)/2.;
r=dt/(mu_0*sigma*dz^2);
imax=50; % number of spatial grid points
nmax=200; % number of time steps

% initialize the By and Jx arrays
By=zeros(nmax,imax);
Jx=zeros(nmax,imax);
% require that the magnetic field is 1.0 at the surface
By(:,1)=1.0;

% perform the explicit finite difference scheme
for n=2:nmax-1                %time loop
    for i=2:imax-1            %space loop

      %calculate the field distribution
      By(n+1,i)=By(n,i)+r*(By(n,i+1)-2*By(n,i)+By(n,i-1));

      %calculate the current distribution
      Jx(n,i)=(By(n,i)-By(n,i+1))/mu_0;

    end
end

% create two mesh plots of Jx and By side-to-side
subplot(1,2,1);mesh(Jx)
subplot(1,2,2);mesh(By)
```

The output of this program is shown in Figure 7.1. Axis labels and the figure perspectives were adjusted in the window show.

COLOR images for many of the figures may be found on the CD-ROM.

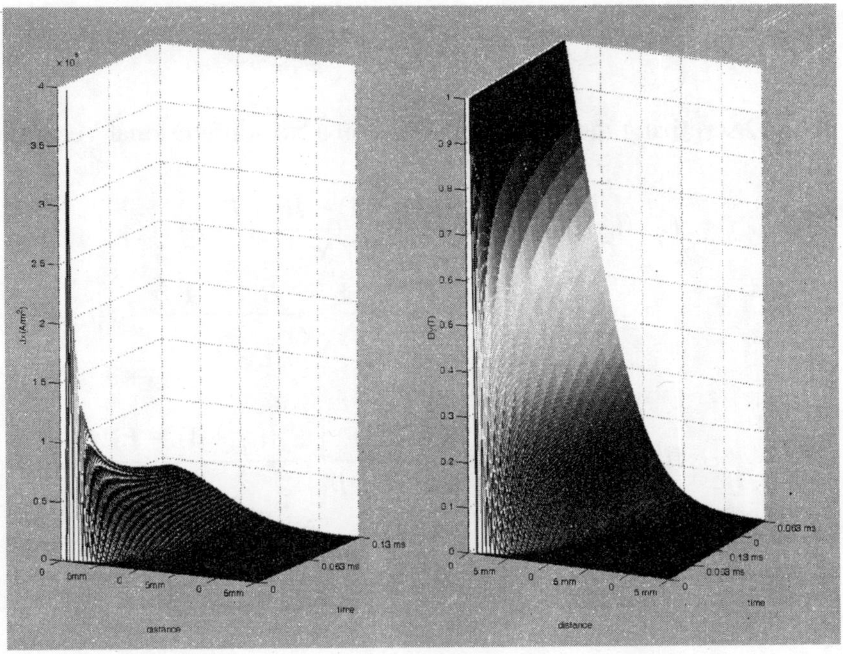

FIGURE 7.1 **MATLAB simulation of the spatial and time-dependent current (left) and magnetic field (right) distributions inside a conducting half-space.**

7.2 FINITE DIFFERENCE TIME-DOMAIN METHOD

Fields Inside a Cylindrical Cavity

The Finite Difference Time-domain (FDTD) method involves the discretization and numerical solution of the first-order Maxwell equations for the primary field **E** and **H** according to Yee's algorithm. In general, this will involve six first-order equations, three for each field component, depending on the coordinate system. Here, we illustrate a special field distribution in cylindrical coordinates, where the electric field has only a ϕ-component so that $\mathbf{E} = E_\phi \hat{\varphi}$ and $\mathbf{H} = H_r \hat{\mathbf{r}} + H_z \hat{\mathbf{z}}$. Maxwell's equations then reduce to the three time-dependent equations for the electric and magnetic fields

$$\varepsilon \frac{\partial E_\phi}{\partial t} = \left(\frac{\partial H_r}{\partial z} - \frac{\partial H_z}{\partial r} \right) - \sigma E_\phi \tag{7.16}$$

$$\mu \frac{\partial H_r}{\partial t} = \frac{\partial E_\phi}{\partial z} \tag{7.17}$$

$$\mu\frac{\partial H_z}{\partial t} = -\frac{1}{r}\frac{\partial}{\partial r}rE_\phi. \tag{7.18}$$

Discretizing the above equations on a two-dimensional Yee lattice yields

$$E_\phi^{n+1}(i,j) = E_\phi^{n-1}(i,j) + \frac{\Delta t}{\varepsilon(i,j)}\left[\frac{H_r^n(i,j+1) - H_r^n(i,j-1)}{\Delta z} - \frac{H_z^n(i+1,j) - H_z^n(i-1,j)}{\Delta r} - \sigma(i,j)E_\phi^n(i,j)\right] \tag{7.19}$$

$$H_r^{n+1}(i,j) = H_r^{n-1}(i,j) + \frac{\Delta t}{\mu(i,j)}\left[\frac{E_\phi^n(i,j+1) - E_\phi^n(i,j-1)}{\Delta z}\right] \tag{7.20}$$

$$H_z^{n+1}(i,j) = H_z^{n-1}(i,j) - \frac{\Delta t}{\mu(i,j)}\left[\frac{(i+1)E_\phi^n(i+1,j) - (i-1)E_\phi^n(i-1,j)}{i\Delta r}\right], \tag{7.21}$$

where $r = i\Delta r$ $z = j\Delta z$ $t = n\Delta t$. A central difference scheme is employed in calculating the space and time derivatives in equations (7.19 through 7.21). Space derivatives and time derivatives are calculated at alternate time steps according to the leapfrog scheme. For greater stability, especially if the solution region contains highly conductive elements, E_ϕ^n in equation (7.19) is replaced by the average $(E_\phi^{n+1} + E_\phi^{n-1})/2$ so that this equation becomes

$$E_\phi^{n+1}(i,j) = \frac{\left(1 - \frac{\sigma(i,j)\Delta t}{2\varepsilon(i,j)}\right)}{\left(1 + \frac{\sigma(i,j)\Delta t}{2\varepsilon(i,j)}\right)}E_\phi^{n+1}(i,j) + \frac{\frac{\Delta t}{\varepsilon(i,j)}}{\left(1 + \frac{\sigma(i,j)\Delta t}{2\varepsilon(i,j)}\right)} \times\left[\frac{H_r^n(i,j+1) - H_r^n(i,j-1)}{\Delta z} - \frac{H_z^n(i+1,j) - H_z^n(1-1,j)}{\Delta r}\right] \tag{7.22}$$

The time step $\Delta t \le \Delta x/c\sqrt{n}$ is chosen, where c is the speed of light and n is the number of spatial dimensions. E_ϕ and H_r are equal to zero along the z-axis so that H_z must be treated separately there. Discretizing the integral form of Maxwell's equation

$$\oint \mathbf{E} \cdot d\ell = -\mu \iint \frac{\partial \mathbf{H}}{\partial t} \cdot d\mathbf{a} \tag{7.23}$$

along a circular contour of radius Δr from the z-axis gives

$$H_z^{n+1}(0,j) = H_z^{n-1}(0,j) - \frac{4\Delta t}{\mu(i,j)\Delta r} E_\phi^n(1,j) \tag{7.24}$$

along the z-axis.

Resonant Frequencies

Time series obtained by Yee's algorithm may be used to determine resonant frequencies inside a cavity with an arbitrary distribution of dielectrics and conductors. The cavity is given an E_ϕ pulse (or other field pulse) at $t = 0$ and a time series of one field component is then recorded at a given lattice site. A Fourier transform algorithm can then be used to determine cavity resonant modes. Once a resonant frequency is identified, the cavity may be driven at that particular frequency and the resulting field distributions recorded and plotted in MATLAB. Alternatively, the Finite Difference Frequency Domain (FDFD) may be used to obtain the cavity modes.

7.3 VECTOR POTENTIAL FORMULATION

Instead of solving for the electric and magnetic fields directly, we may solve for the vector potential and then obtain the fields from the curl and time derivative of **A**. Substituting $\nabla \times \mathbf{A} = \mathbf{B}$ and $\mathbf{E} = -\partial \mathbf{A}/\partial t$ into equation (7.1) results in a second-order equation for **A**

$$\nabla^2 \mathbf{A} = -\mu_0 \sigma \frac{\partial \mathbf{A}}{\partial t} + \mu_0 \varepsilon_0 \frac{\partial^2 \mathbf{A}}{\partial t^2} \tag{7.25}$$

with an additional term $-\mu_0 \mathbf{J}_{\text{source}}$ in the presence of sources.

Circular Coil Above a Conducting Half-Space

As an example of the vector potential formulation, where displacement currents are neglected, we consider the example in the previous chapter of a drive coil above and coplanar to a conducting half-space. The drive current written as a product of delta functions now has an arbitrary time dependence so that $A_\phi = A_\phi(r,z,t)$

$$\frac{\partial^2 A_\phi}{\partial r^2} + \frac{1}{r}\frac{\partial A_\phi}{\partial r} - \frac{A_\phi}{r^2} + \frac{\partial^2 A_\phi}{\partial z^2} = \begin{cases} -\mu_0 I(t)\delta(r - r_0)\delta(z - \ell) & z > 0 \\ \mu\sigma \dfrac{\partial A_\phi}{\partial t} & z \le 0 \end{cases} \tag{7.26}$$

Taking the Laplace transform of both sides of this equation gives

$$\frac{\partial^2 a_\phi}{\partial r^2} + \frac{1}{r}\frac{\partial a_\phi}{\partial r} - \frac{a_\phi}{r^2} + \frac{\partial^2 a_\phi}{\partial z^2} = \begin{cases} -\mu_0 I(s)\delta(r - r_0)\delta(z - \ell) & z > 0 \\ i\omega\mu \cdot s \cdot \sigma a_\phi & z \leq 0 \end{cases}, \quad (7.27)$$

where $L A_\phi(r,z,t) = a_\phi(r,z,s)$ and

$$L\left\{\frac{\partial A_\phi(r,z,t)}{\partial t}\right\} = s \cdot a_\phi(r,z,s) - A_\phi(r,z,0). \quad (7.28)$$

The current source is switched on at $t = 0$ so $a_\phi(r,z,0) = 0$. Solving the above equations by separation of variables as before we take

$$a_\phi(r,z,s) = R(r,s) \cdot Z(z,s). \quad (7.29)$$

Substituting (7.29) into equation (7.27) and then dividing by $R \cdot Z$ gives

$$\frac{1}{R}\frac{\partial^2 R}{\partial r^2} + \frac{1}{rR}\frac{\partial R}{\partial r} - \frac{1}{r^2} + \frac{1}{Z}\frac{\partial^2 Z}{\partial z^2} - s\mu\sigma = 0, \quad (7.30)$$

which has the same form as equation (6.31). Evidently the solution in the s-domain has the same form as obtained in Chapter 6 with the replacements $I(t) \rightarrow I(s)$ and $i\omega \rightarrow s$ so that we may write in each region

$$a_\phi^1(r,z,s) = \frac{\mu_0 I(s) r_0}{2}\int_0^\infty J_1(\alpha r) J_1(\alpha r_0) e^{-\alpha z}\left(e^{\alpha\ell} + e^{-\alpha\ell}\frac{\alpha\mu_r - \sqrt{s\mu\sigma + \alpha^2}}{\alpha\mu_r + \sqrt{s\mu\sigma + \alpha^2}}\right) d\alpha \quad (7.31)$$

$$a_\phi^2(r,z,s) = \frac{\mu_0 I(s) r_0}{2}\int_0^\infty J_1(\alpha r) J_1(\alpha r_0) e^{-\alpha\ell}\left(e^{\alpha z} + e^{-\alpha z}\frac{\alpha\mu_r - \sqrt{s\mu\sigma + \alpha^2}}{\alpha\mu_r + \sqrt{s\mu\sigma + \alpha^2}}\right) d\alpha \quad (7.32)$$

$$a_\phi^3(r,z,s) = \frac{\mu_0 I(s) r_0}{2}\int_0^\infty J_1(\alpha r) J_1(\alpha r_0) e^{-\alpha\ell}\left(\frac{2\alpha\mu_r e^{\sqrt{s\mu\sigma + \alpha^2}\, z}}{\alpha\mu_r + \sqrt{s\mu\sigma + \alpha^2}}\right) d\alpha. \quad (7.33)$$

Taking the inverse Laplace transform with $I(t) = L^{-1} I(s)$,

$$q(t) = L^{-1}\left\{\frac{\alpha\mu_r - \sqrt{s\mu\sigma + \alpha^2}}{\alpha\mu_r + \sqrt{s\mu\sigma + \alpha^2}}\right\} \text{ and } w(t) = L^{-1}\left(\frac{2\alpha\mu_r e^{\sqrt{s\mu\sigma+\alpha^2}z}}{\alpha\mu_r + \sqrt{s\mu\sigma + \alpha^2}}\right) \quad (7.34)$$

may now be performed in MATLAB. Note that $z < 0$ in $w(t)$ so care must be taken in evaluating the inverse Laplace transform:

```
%MATLAB program for calculating inverse Laplace transforms
syms a b c z s t positive
% calculate the inverse Laplace transform of q(t)
ilaplace( (a-sqrt(b*s+c))/(a+sqrt(b*s+c)),s,t)
%for greater readability use the pretty command
pretty(ans)
% calculate the inverse Laplace transform of w(t)
% note that the argument of the
% exponential in w(t) is negative
ilaplace( (2*a*exp(-z*sqrt(b*s+c))/(a+sqrt(b*s+c)),s,t)
pretty(ans)
```

The resulting time-dependent equations are

$$A_\phi^1(r,z,t) = \frac{\mu_0 r_0}{2}\int_0^\infty J_1(\alpha r)J_1(\alpha r_0)\left\{e^{-\alpha(z-\ell)}I(t) + e^{-\alpha(z+\ell)}I(t)\otimes q(t)\right\}d\alpha \quad (7.35)$$

$$A_\phi^2(r,z,t) = \frac{\mu_0 r_0}{2}\int_0^\infty J_1(\alpha r)J_1(\alpha r_0)\left\{e^{-\alpha(\ell-z)}I(t) + e^{-\alpha(z+\ell)}I(t)\otimes q(t)\right\}d\alpha \quad (7.36)$$

$$A_\phi^3(r,z,t) = \frac{\mu_0 r_0}{2}\int_0^\infty J_1(\alpha r)J_1(\alpha r_0)e^{-\alpha\ell}I(t)\otimes w(t)d\alpha, \quad (7.37)$$

where the convolution integrals are given by

$$I(t)\otimes q(t) = \int_0^t I(t-u)q(t)du \text{ and } I(t)\otimes w(t) = \int_0^t I(t-u)w(t)du, \quad (7.38)$$

which may now be performed using MATLAB. (See Appendix 2 for the inverse Laplace transform of pulsed and periodic signals.)

7.4 TIME-DEPENDENT MAGNETICS IN QUICKFIELD

Time-dependent magnetic analysis in QuickField involves the solution of the second-order equation

$$\nabla \times \left(\frac{1}{\mu} \nabla \times \mathbf{A} \right) = \mathbf{J} + \nabla \times \mathbf{H}_c \tag{7.39}$$

for the vector potential. The current density is given by a sum of source current and eddy current $\mathbf{J} = \mathbf{J}_{\text{source}} + \mathbf{J}_{\text{eddy}}$, where the eddy current is

$$\mathbf{J}_{\text{eddy}} = -\sigma \frac{\partial \mathbf{A}}{\partial t}. \tag{7.40}$$

Given an anisotropic permeability tensor in x-y symmetry

$$\frac{\partial}{\partial x}\left(\frac{1}{\mu_y} \frac{\partial A_z}{\partial x} \right) + \frac{\partial}{\partial y}\left(\frac{1}{\mu_x} \frac{\partial A_z}{\partial y} \right) - \sigma \frac{\partial A_z}{\partial t} = -J_{\text{source}} + \left(\frac{\partial H_{cy}}{\partial x} - \frac{\partial H_{cx}}{\partial y} \right). \tag{7.41}$$

For an anisotropic permeability in axial symmetry

$$\frac{\partial}{\partial r}\left(\frac{1}{r\mu_z} \frac{\partial (rA_\varphi)}{\partial r} \right) + \frac{\partial}{\partial z}\left(\frac{1}{\mu_r} \frac{\partial A_\varphi}{\partial z} \right) - \sigma \frac{\partial A_z}{\partial t} = -J_{\text{source}} + \left(\frac{\partial H_{cr}}{\partial z} - \frac{\partial H_{cz}}{\partial r} \right). \tag{7.42}$$

The coercive force vector $\mathbf{H}_c$ is as defined as in Magnetostatics analysis. Note that magnetic material properties can be simulated in Transient Magnetic Analysis, unlike in AC Magnetics, where all fields and sources are required to vary harmonically.

Integral Quantities

Integral quantities may be calculated in Transient Magnetics including the total current through a surface

$$I = \iint_{\text{surf}} \mathbf{J} \cdot d\mathbf{a}. \tag{7.43}$$

The Joule heating in a given volume

$$Q = \iiint_{\text{vol}} \sigma^{-1} J^2 dV. \tag{7.44}$$

Other integral quantities may be calculated as in Magnetostatics including the total magnetic field energy, flux linkages, total magnetic forces, and torques acting on bodies contained in a volume.

Material Properties and Boundary Conditions

The Block Label Properties window is the same as in Magnetostatics except now material conductivity can be specified in addition to permeability, coercive force of magnets, and current sources that may have spatial dependence.

The Edge Label Properties window is also the same as in Magnetostatics. Boundary conditions can be applied including specified vector potential or tangential fields that may have spatial dependence, zero normal flux, and even and odd periodic boundary conditions.

The Vertex Label Properties window is the same as before, where the magnetic potential and linear current can be specified at a point with values that may depend on spatial coordinates.

Time-dependent Quantities

The time dependence of various sources and boundary conditions in Transient Magnetic Analysis can be entered using a library of specific functions. Non-periodic functions include exponential exp(t), logarithm log(t), square root sqrt(t), and absolute value abs(t). The most useful non-periodic functions for transient analysis are the step and impulse functions defined by

$$step(t) = \begin{cases} 1 & \text{if } t \geq 0 \\ 0 & \text{if } t < 0 \end{cases} \quad \text{and } \text{impulse}(t,\ell,r) = \begin{cases} 0 & \text{if } t < \ell \\ 1 & \text{if } \ell \leq t \leq r. \\ 0 & \text{if } t > r \end{cases} \tag{7.45}$$

Periodic functions include: sine sin(t), cosine cos(t), tangent tan(t), arcsine asin(t), arccosine acos(t), and arctangent atan(t). The most useful periodic functions for transient analysis include the saw waves saw(t,p) and saw(t,p,$p0$), where $p0$ is a specified time interval between the individual saw tooth waveforms. A square wave with positive square pulses may be constructed as sign (saw(t,p,$p0$)) and a square wave with alternating sign may be constructed as sign (sin(t)), where the sign function is ±1 depending on the sign of the argument.

If no explicit time dependence is applied to current sources or boundary conditions in the model, these are treated as if switched on at $t = 0$. This is equivalent to specifying the current or boundary condition times the step function $step(t)$.

The Set Timing Parameters Tab located in the Problem Properties window is shown in Figure 7.2. The total integration time and integration time step is

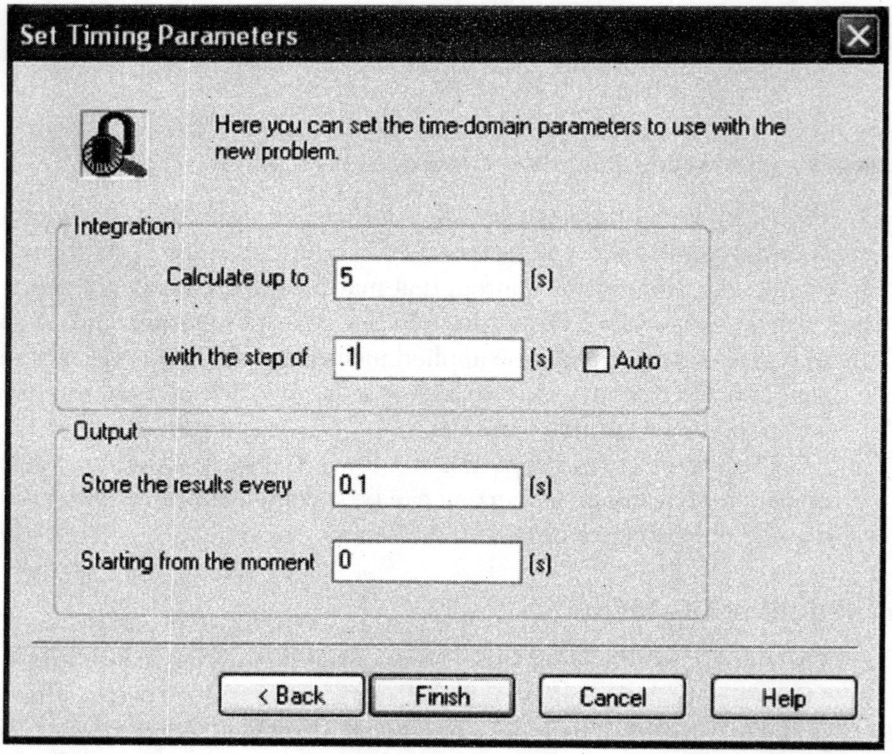

FIGURE 7.2 **Window for setting time-domain properties including total time span, time step, data sampling interval, and initial recording time.**

specified in the upper half of this window. The output time intervals and start time is specified in the lower half of this window.

In QuickField, the time step is adjusted in Transient Magnetics depending on properties of the mesh and potential variation. The initial time step Δt_0 is chosen $\Delta t_0 = \min(\xi^2/4\alpha)$, where ξ is a characteristic size of the mesh elements and $\alpha = 1/\mu\sigma$. Subsequent time steps are updated according to $\Delta t_{n+1} = k\Delta t_n$, where k is a scaling factor that is chosen to be 0.25, 0.5, 1.0, 2.0, or 4.0 depending on the spatial and temporal behavior of the potential, sources, and boundary conditions.

Conducting Cylinder in a Transient Magnetic Field

As an example of transient field analysis, we consider a conducting $\sigma = 10^8$ S/m cylinder with a 10.0 cm radius in a transverse magnetic field of 2.0 T that is switched on at $t = 0$. No explicit time dependence is specified in this example so the boundary conditions are treated as step functions in time. In order to specify a magnetic field in the y-direction, the vector potential A_z is assigned different values on the right- and left-hand sides of the solution region. The meshed cylinder and solution region is shown in Figure 7.3. A finer mesh near the cylinder has been created, where the mesh spacing is smaller, as seen by circles on opposite sides of the cylinder, compared to the larger circles on the corners of the square solution region.

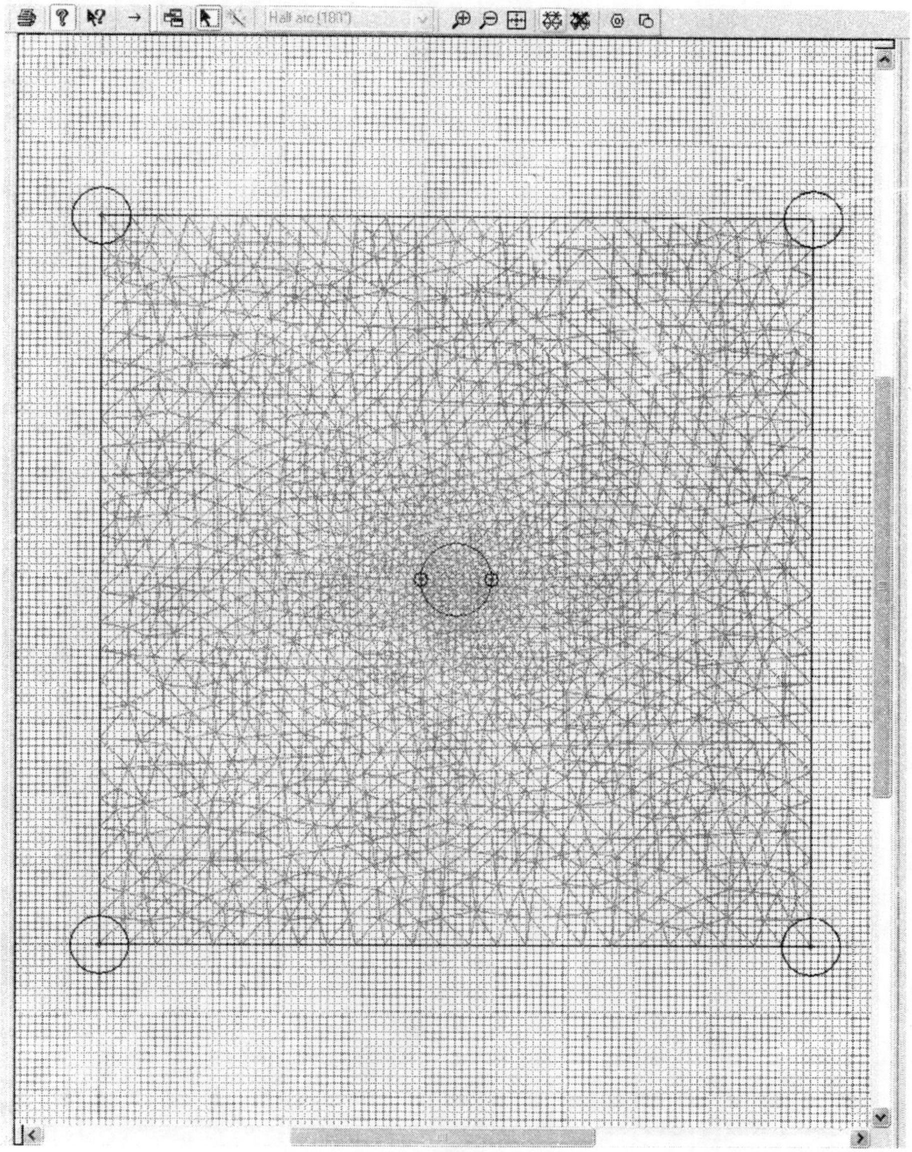

FIGURE 7.3 **Finite element mesh of cylindrical conductor in an air region. The graded mesh is apparent at the location of the conducting cylinder.**

Transient field calculation will require a longer solution time compared to Magnetostatics and AC Magnetics problems. The window displayed during the solution of the problem is shown in Figure 7.4 showing the time moment in the calculation, iteration number with estimated remaining iterations, and the total elapsed time. Figure 7.5 shows the window for specifying quantities at a specific location to plot as a function of time. The magnetic flux lines and transient eddy current densities are plotted at 0.01, 0.03, 0.05, and 0.07 seconds in Figure 7.6. This first figure shows the initial expulsion of field lines for larger induced current values at 0.1 seconds. The field lines are seen to penetrate the cylinder more each increment in time, with a subsequent reduction in current density. In Figure 7.7, the eddy current density is plotted around a circular contour surrounding the conducting cylinder at 0.58 seconds. Eddy current density along the contour may be viewed at each time selected from the pull-down window shown. Other physical quantities along the same contour may be plotted including vector potential, flux density, field strength, permeability, and energy density.

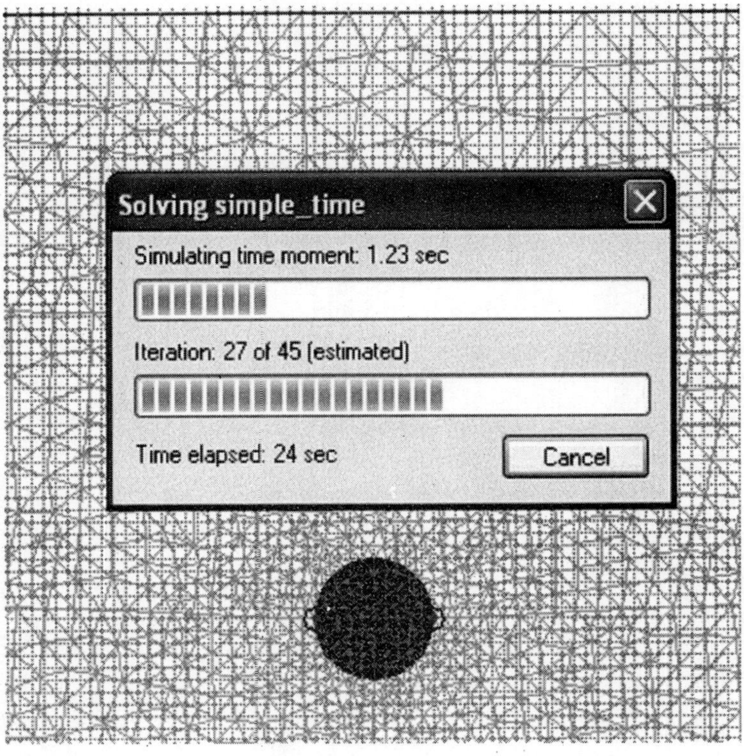

Figure 7.4 Window showing progress during the transient magnetic problem execution.

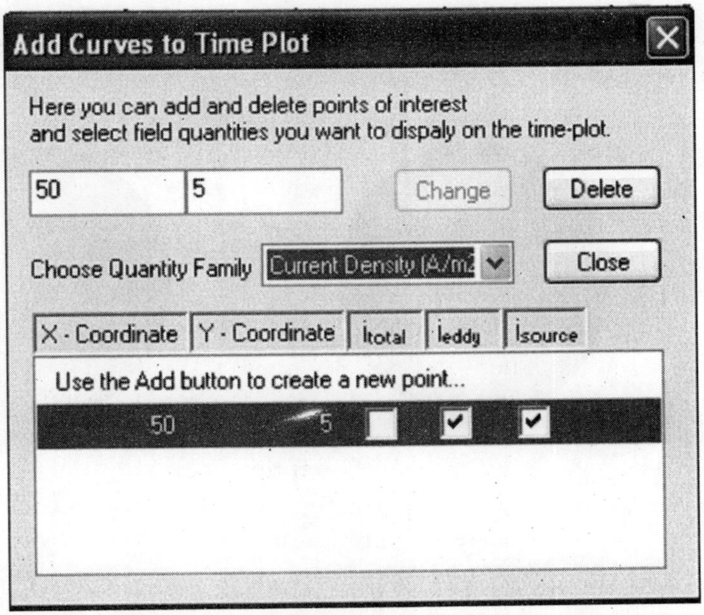

FIGURE 7.5 **Window for recording the physical quantities and the spatial location for time plots.**

Current Sheet Above a Conducting Plane

Next we model a planar sheet inducer with a width of 30 cm, which carries a current density of 1.0 A/m. The inducer is parallel to a conducting plane $\sigma = 10^8$ S/m with a stand-off of 2.0 cm. No explicit time dependence is specified in this model, so the current is treated as if suddenly turned on at $t = 0$ s. Figure 7.8 shows the vector magnetic field and eddy current density plots at two times during the simulation. Contour plots of eddy current density (right of the figure) are made at the two time intervals along the vertical contour shown under the sheet inducer on the left. At $t = 0.016$ s, the eddy current density (top right) is primarily concentrated near the surface and is nearly zero at a depth of 5.0 cm. The eddy current has diffused and attenuated at $t = 0.253$ s (bottom right), decreasing in magnitude by a factor of six at the surface, while penetrating to a depth of 21 cm in the metal.

Current Loop Surrounding a Conducting Sphere

As an example of transient eddy currents induced by a periodic excitation, we model a single current loop surrounding a conducting sphere. The sphere has a radius of 10 cm with $\sigma = 10^8$ S/m. The current loop has a radius of 12 cm and is located 2.0 cm over the equator of the sphere. The excitation current in the

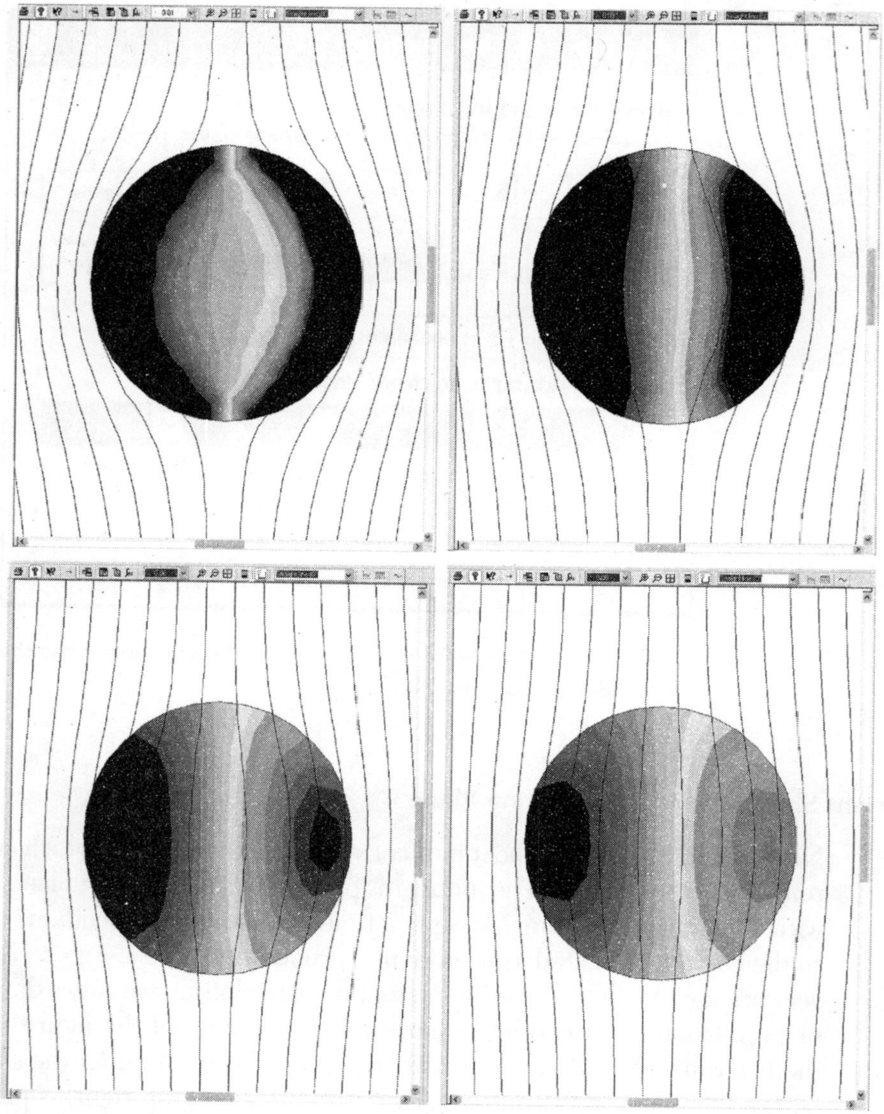

FIGURE 7.6 **Magnetic flux lines and transient eddy current density at times 0.01, 0.03, 0.05, and 0.07 seconds.**

loop is a saw tooth wave defined by the function saw (t,1,1). Figure 7.9 shows the resulting magnetic flux lines and eddy current density plots. A time record of the eddy current density at the location (50,5) inside the sphere is plotted under the flux line plots for the 10 s simulated. In the top plot, recorded at $t = 3.0$ s, the current in the loop is temporarily zero so the flux lines are just due to eddy currents flowing in the sphere. In the bottom plot, recorded at $t = 4.5$ s, the flux lines are mainly due to the excitation current loop.

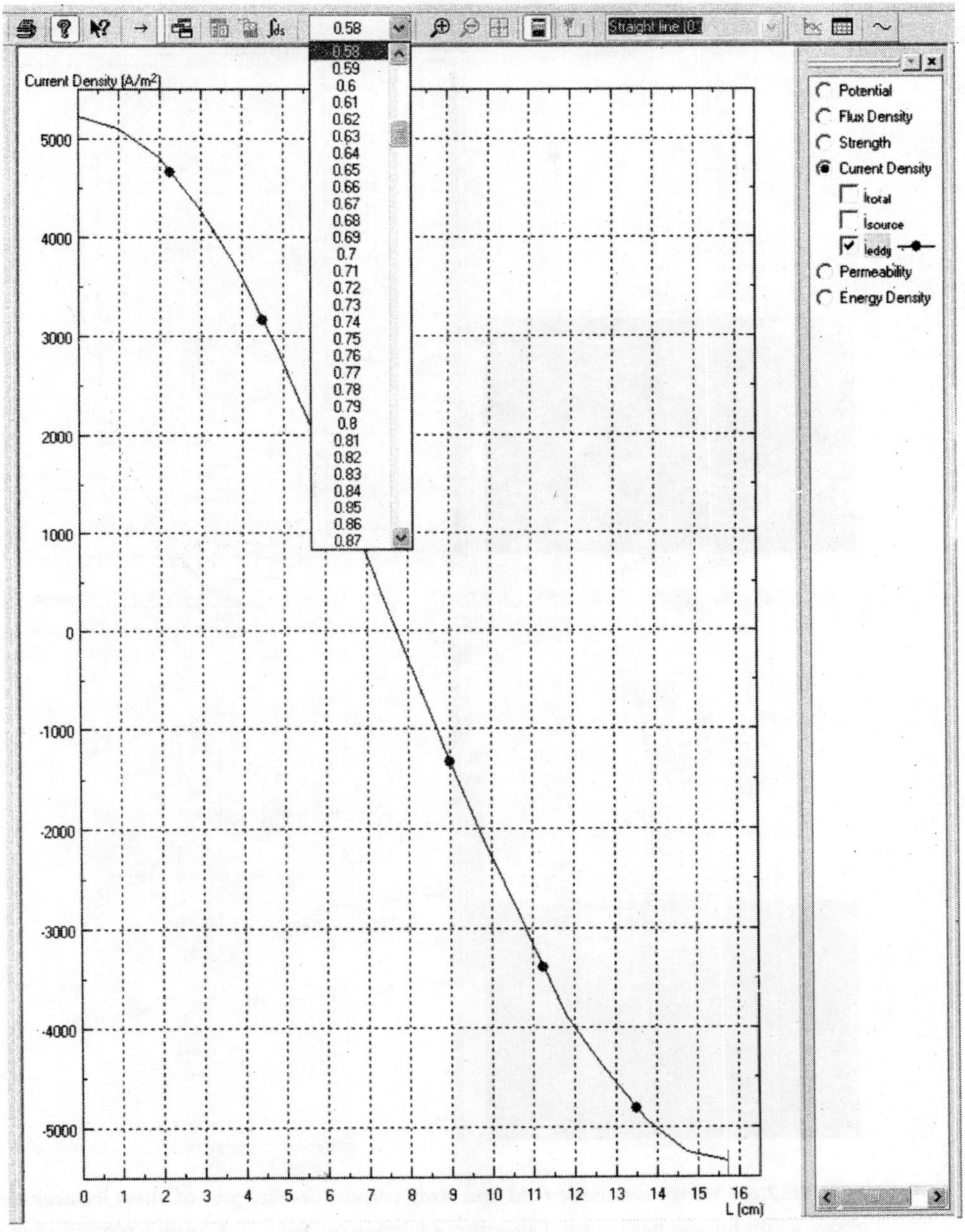

FIGURE 7.7 **Eddy current density around a circular contour surrounding the conducting cylinder at 0.58 seconds. The plot will refresh for each time selected from the pull-down window.**

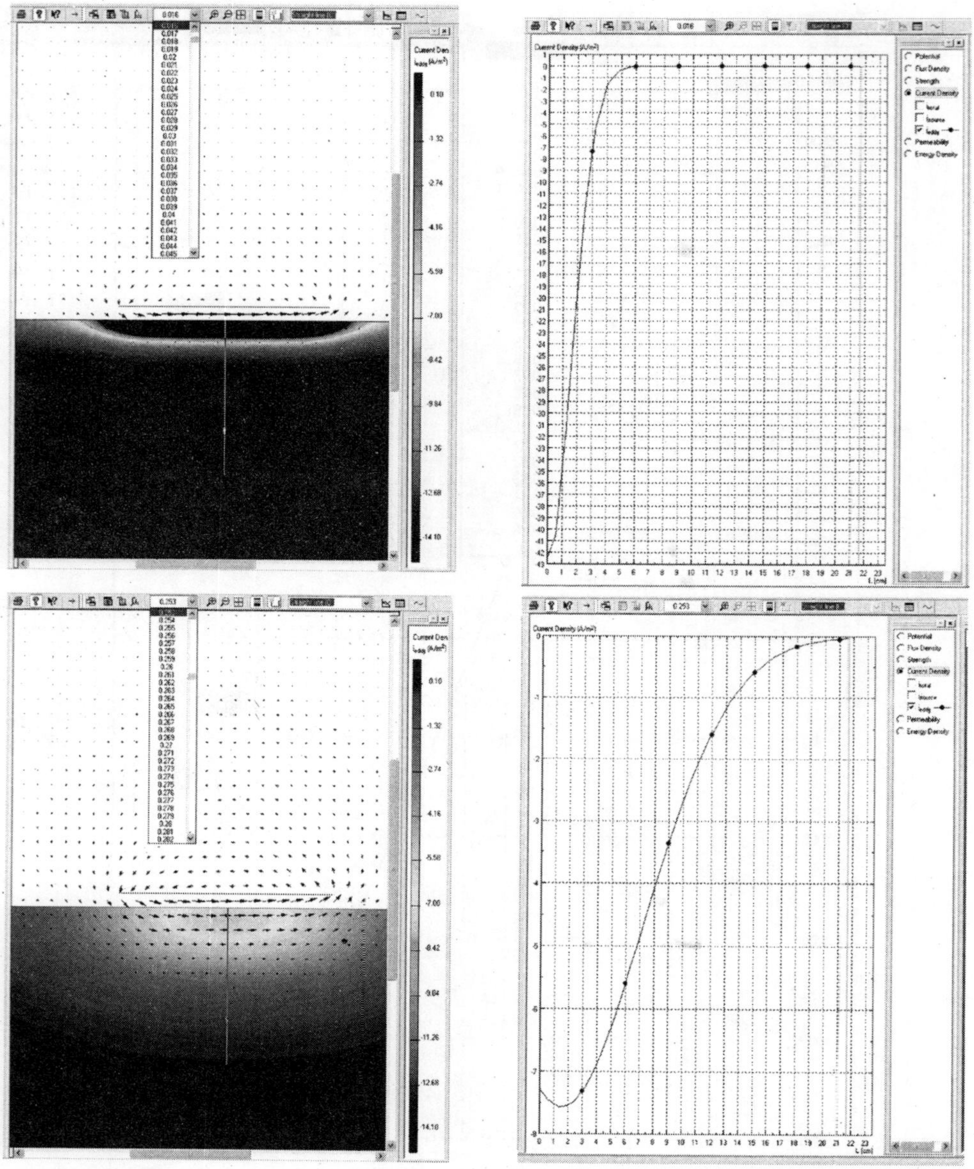

FIGURE 7.8 **Vector magnetic field and eddy current density plot of sheet inducer located above a conducting half-space. Current is switched on at $t = 0$. Contour plot of eddy current density is made at the two time intervals shown along the vertical contour under the sheet inducer.**

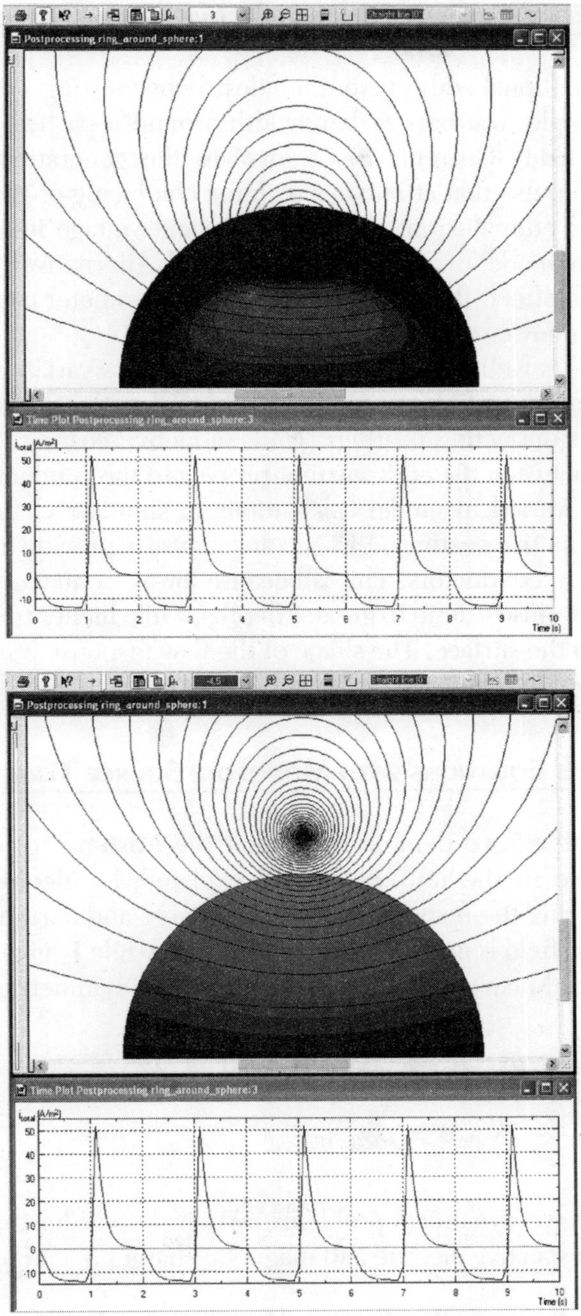

FIGURE 7.9 **Magnetic flux lines and eddy current density plot induced by a current loop surrounding a conducting sphere. The excitation current is a saw tooth wave saw(t,1,1). A time record of the eddy current density at the location (50,5) inside the sphere is plotted over the 10 s simulated.**

7.5 Pulsed Eddy Current Nondestructive Testing Applications

In conventional eddy current nondestructive testing, eddy currents are induced in a sample in a narrow bandwidth around a particular operating frequency. Pulsed Eddy Current (PEC) involves the generation of a square, or other periodic pulse that attenuates within a conducing object under inspection. The resulting eddy current induces a transient voltage in a primary excitation coil (absolute mode) or a separate receiver coil. Alternatively, a magnetic sensor such as a Hall effect, fluxgate, or SQUID magnetometer can detect the field generated by eddy currents directly.

PEC is well suited for measuring thickness variations resulting from corrosion in pipes and planar conductors as well as the thickness of both conducting and nonconducting coatings. Nonconductive coatings essentially increase the lift-off between the eddy current probe and the conducing surface. Lift-off may then be varied in model calculations to simulate changes in the thickness of nonconductive coatings. PEC testing may also provide information on the depth of flaws in conductors. This is because lower frequency signal components of a transient pulse extend to greater depths, while higher frequency components are closer to the surface. The shape of the flaw response waveform will therefore be depth dependent.

7.6 Maxwell's Equations with Monopole Source Term

In Chapter 5 we demonstrated that the existence of even a single magnetic monopole in the universe would explain why electric charge is quantized. Considering the respective magnetic charge and current density ρ_m and $\mathbf{J}_m$, the magnetic field is no longer divergence free while $\mathbf{J}_m$ gives rise to an electric field with curl. Maxwell's equations then have the symmetric form

$$\nabla \times \mathbf{E} = -\mu \mathbf{J}_m - \frac{\partial \mathbf{B}}{\partial t} \quad \nabla \times \mathbf{B} = \mu \mathbf{J}_e + \mu\varepsilon \frac{\partial \mathbf{E}}{\partial t}$$
$$\nabla \cdot \mathbf{E} = \frac{\rho_e}{\varepsilon} \quad \nabla \cdot \mathbf{B} = \mu \rho_m. \tag{7.46}$$

To solve dynamical problems, we must consider the Lorentz forces acting on the respective electric and magnetic charges q_e and q_m

$$\mathbf{F}_e = q_e(\mathbf{E} + \mathbf{v} \times \mathbf{B}) \quad \mathbf{F}_m = q_m(\mathbf{B} - \mu_0\varepsilon_0 \mathbf{v} \times \mathbf{E}) \tag{7.47}$$

in free space. A particle possessing both electric and magnetic charge would experience a force given by the sum $\mathbf{F}_e + \mathbf{F}_m$.

Monopole Moving in Crossed Electric and Magnetic Fields

As an example we consider a magnetic monopole moving with an initial speed $\mathbf{v}_0 = (v_{x0}, v_{y0}, 0)$ that enters a region with crossed electric and magnetic fields $\mathbf{E} = E_x x\mathbf{i}$ $\mathbf{B} = B_y y\mathbf{j}$. From Newton's second law, $\mathbf{F}_m = M\mathbf{a}$, we have the resulting differential equations

$$\begin{Bmatrix} \ddot{x} \\ \ddot{y} \\ \ddot{z} \end{Bmatrix} = \begin{Bmatrix} 0 \\ \frac{q_m}{M}\left(B_y y - \mu_0 \varepsilon_0 E_x x\dot{z}\right) \\ \frac{q_m}{M}\mu_0 \varepsilon_0 E_x x\dot{y} \end{Bmatrix}, \tag{7.48}$$

which may now be integrated numerically in MATLAB.

REFERENCES

M. N. O. Sadiku. (2000). *Numerical Techniques in Electromagnetics 2nd edition*, CRC Press.

QuickField User's Guide: Tera Analysis Ltd. (2007). *Chapter 2: Introductory Guide; Chapter 10 Theoretical Description.*

David J. Griffiths. (1999). *Introduction to Electrodynamics 3rd edition*, Prentice-Hall, Inc., New Jersey. *Chapter 7: Electrodynamics.*

Jack Blitz. (1991). *Electrical and Magnetic Methods of Nondestructive Testing*, Adam Hilger, Bristol England.

Reinhold Ludwig and Xiao-Wei Dai. (1990). "Numerical and Analytical Modeling of Pulsed Eddy Currents in a Conducting Half-Space," *IEEE Transactions on Magnetics* (26) 1, pp. 299–307.

Xiao-Wei Dai. (1990). Reinhold Ludwig and R. Palanisamy, "Numerical Simulation of Pulsed Eddy Current Testing Phenomena," *IEEE Transactions on Magnetics* (26) 6, pp. 3089–3096.

EXERCISES

7.1 From Maxwell's equations (7.1), derive wave equations for the electric and magnetic field in free space

$$\nabla^2 \begin{Bmatrix} \mathbf{E} \\ \mathbf{B} \end{Bmatrix} = \frac{1}{c^2}\frac{\partial^2}{\partial t^2}\begin{Bmatrix} \mathbf{E} \\ \mathbf{B} \end{Bmatrix},$$

where the speed of light $c = 1/\sqrt{\mu_0 \varepsilon_0}$ and taking $\sigma = 0$.

7.2 Write integral forms of Maxwell's equations by applying Stokes' theorem to the curl equations and Gauss' theorem to the divergence equations in (7.1).

7.3 Write a MATLAB program to compare the field and current distributions in equations (7.9) and (7.12) to the finite difference solution for the parabolic field equation.

7.4 Use the Dufort-Frankel discretization

$$\frac{\Psi^{n+1}(i) - \Psi^{n-1}(i)}{2\Delta t} = \frac{\Psi^{n}(i+1) - \Psi^{n+1}(i) - \Psi^{n-1}(i) - \Psi^{n}(i-1)}{k\Delta x^2}$$

of the parabolic differential equation

$$\frac{\partial \Psi}{\partial t} = \frac{1}{k}\frac{\partial^2 \Psi}{\partial x^2}$$

to modify the MATLAB example in Section 7.1 calculating the field distributions inside a conducting half-space resulting from a tangential magnetic field switched on at t = 0.

7.5 Develop a FDTD code in MATLAB to calculate the TM modes inside a microwave cavity

$$\frac{\partial H_\phi}{\partial t} = -\frac{1}{\mu}\left(\frac{\partial E_r}{\partial z} - \frac{\partial E_z}{\partial r}\right)$$

$$\varepsilon \frac{\partial E_r}{\partial t} = -\sigma \frac{\partial E_r}{\partial t} - \frac{\partial H_\phi}{\partial z}$$

$$\varepsilon \frac{\partial E_z}{\partial t} = -\sigma E_z + \frac{1}{r}\frac{\partial}{\partial r} r H_\phi.$$

Give the cavity a square H_ϕ pulse at $t = 0$, and record a time series of one of the field components at a particular grid location. Fourier transform the time series to locate the cavity resonant frequencies. Choose one of the resonant frequencies to harmonically excite the cavity and create mesh plots of the resulting field distributions. Include regions of permittivity, permeability, and conductivity in the cavity.

7.6 Calculate the time-dependent eddy current density in the conducting, permeable sphere in Figure E-7.6 after the external B field is switched on at $t = 0$. Plot the radial and surface eddy current density at several times after the field is switched on.

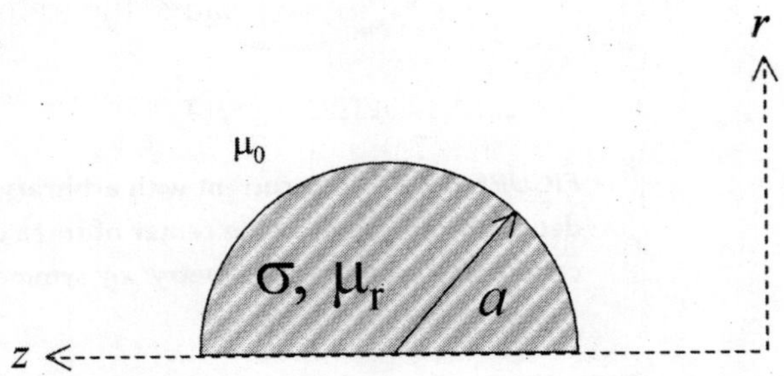

FIGURE E-7.6 **Conducting sphere in an external transient magnetic field Geometry: Axial symmetry.**

7.7 Plot the time-dependent eddy current at several locations and times under the excitation coil carrying current I_0 saw(t) in Figure E-7.7. Make a 3-D plot of the eddy current density along a radial contour for several time intervals.

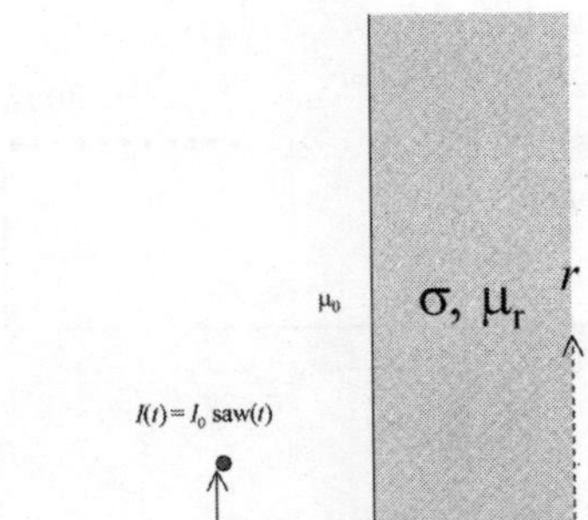

FIGURE E-7.7 **Time-dependent current flowing in a loop above a semi-infinite conducting half-space. Geometry: Axial symmetry.**

7.8 Create time plots of the eddy current density in each conducting cylinder in Figure E-7.8. Also plot the magnetic field in the space between and outside the cylinders. Use a square wave excitation for $I(t)$.

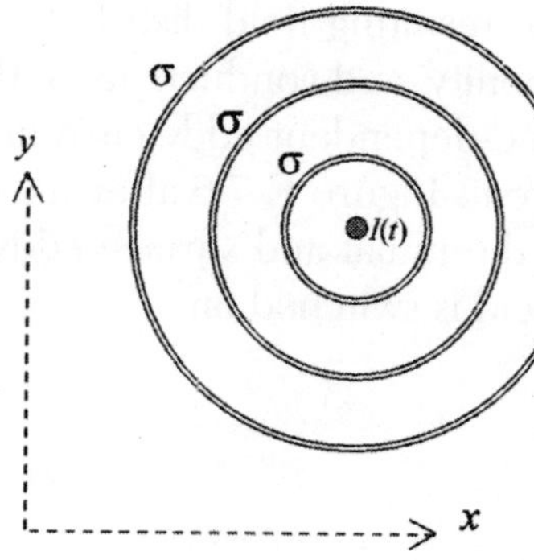

FIGURE E-7.8 **Line current with arbitrary time dependence located at the center of three concentric conducting cylinders. Geometry: *x-y* symmetry.**

7.9 Record the z-component magnetic field directly over the sheet inducer in Figure E-7.9. Note that this field component will be zero in the absence of flaws in the metal from symmetry. Also, the flaw should not be directly below the centerline of the sheet inducer. Using Label Mover, record the flaw signal for various flaw depths inside the metal. Experiment with a saw, dual frequency, and a square wave for $I(t)$.

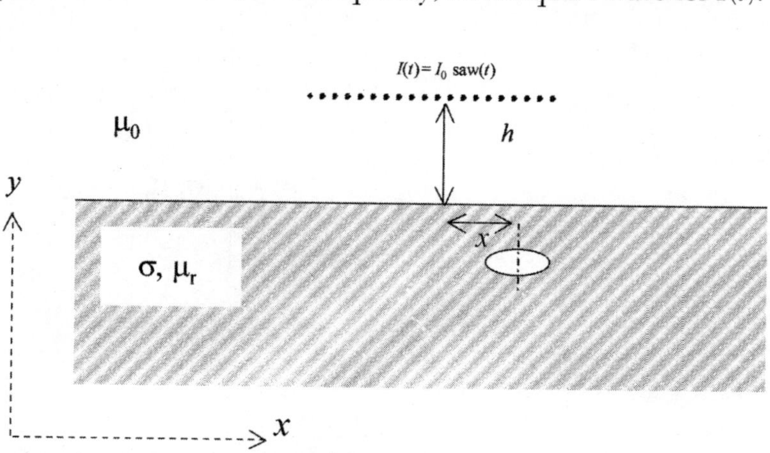

FIGURE E-7.9 **Sheet inducer parallel to a conducting permeable half-space with an elliptical flaw. Geometry: *x-y* symmetry.**

7.10 Plot the time-dependent eddy current at several locations in the conducting, permeable half-space under the cup-core probe eddy current probe shown in Figure E-7.10. Experiment with a single current pulse, a square wave, and a periodic saw current I_0 saw(t). Make a 3-D plot of the eddy current density along a radial contour for at several time intervals. Compare selected results with the same excitation coil without the cup-core.

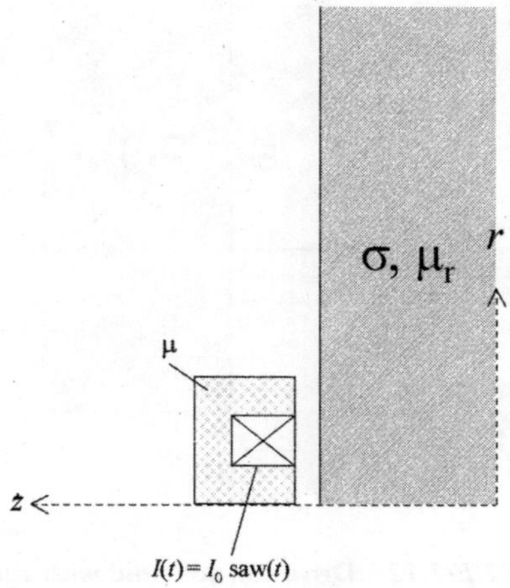

FIGURE E-7.10 **Eddy current probe with ferrite cup-core above a conducting, permeable half-space. Geometry: Axial symmetry.**

7.11 Model a line source carrying a current with arbitrary time dependence above a plate with thickness d. Plot the magnetic field below the plate changing the plate thickness using Label Mover. Experiment with a single current pulse, a square wave, and a periodic saw current I_0 saw(t). Also plot the force acting on the conducing plate.

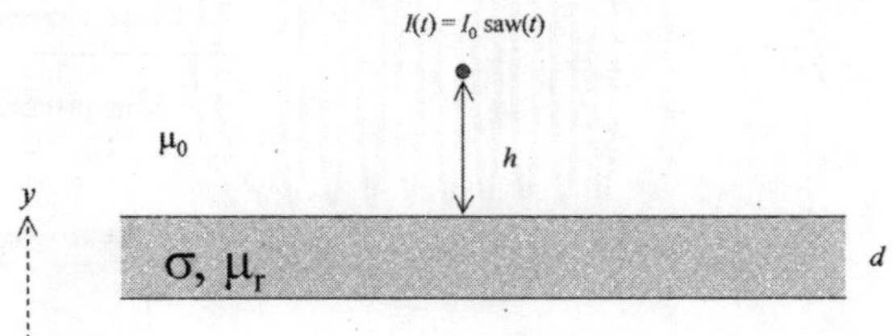

FIGURE E-7.11 **Line source carrying a current with arbitrary time dependence above a plate with thickness *d*. Geometry: *x-y* symmetry.**

7.12 Model a drive coil solenoid surrounded by a highly permeable cylinder located above a semi-infinite conducting half-space with a cylindrical defect. Calculate the perturbation in the z-component magnetic field at the center of the coil for various flaw depths.

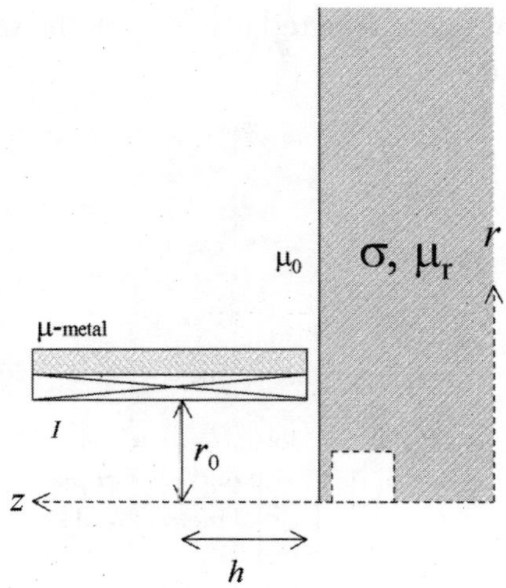

FIGURE E-7.12 **Drive coil solenoid with surrounding high-permeability flux focuser above a semi-infinite conducting half-space with a cylindrical defect. Geometry: Axial symmetry.**

7.13 Model a chloroplast in the nanosecond pulsed magnetic field shown in Figure E-7.13 in x-y symmetry. Calculate the peak fields in the inner and outer chloroplast membrane and the thylakoid membrane.

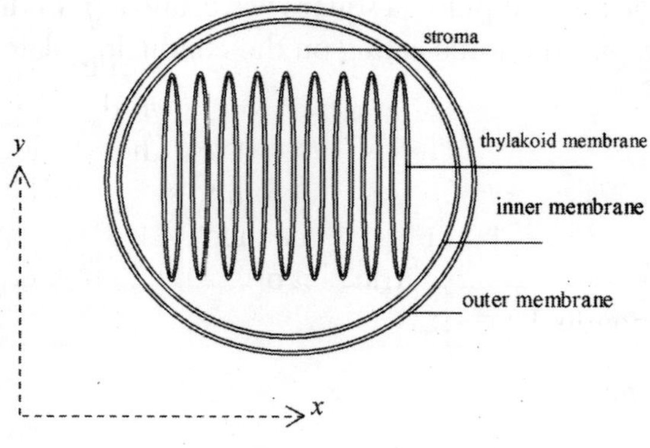

FIGURE E-7.13 **Chloroplast in a nanosecond pulsed magnetic field including inner and outer chloroplast membrane and thylakoid membrane. Geometry: x-y symmetry.**

7.14 Model a chloroplast in the nanosecond pulsed magnetic field shown in Figure E-7.14 in axial symmetry. Calculate the peak fields in the inner and outer chloroplast membrane and the thylakoid membrane.

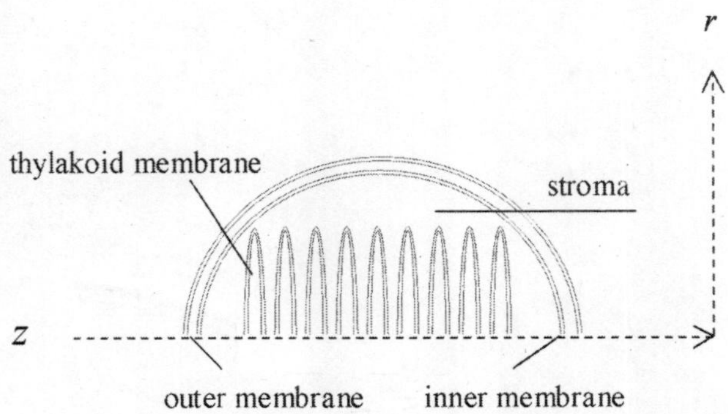

FIGURE E-7.14 **Chloroplast in a nanosecond pulsed magnetic field including inner and outer chloroplast membrane and thylakoid membrane. Geometry: Axial symmetry.**

7.15 Using Gauss' law and Stokes' theorem, write the integral form of Maxwell's equations including the monopole source term.

7.16 Consider an electron moving in the field of a magnetic monopole. Write down the six differential equations in Cartesian coordinates describing the resulting motion of the two particles. Write a MATLAB program to solve for the motion of the particles given the initial conditions of the monopole $\mathbf{r}_m = (0,0,0)$, $\mathbf{v}_m = (0,0,0)$ and the electron $\mathbf{r}_e = (x_0,0,0)$, $\mathbf{v}_e = (0,v_{y0},0)$. Plot 3-D parametric plots of the resulting motion of each particle. Take the mass of the monopole to be $M_m = Qm_e$. (See Appendix 1.)

7.17 Consider a single particle of mass M with both electric and magnetic charges q_e and q_m entering a region with crossed electric and magnetic fields $\mathbf{E} = E_x\mathbf{i}$ and $\mathbf{B} = B_y\mathbf{j}$. Write down the three differential equations in Cartesian coordinates describing the resulting motion of the particles. Write a MATLAB program to solve for the motion of the particles given initial conditions of the electric monopole $\mathbf{r} = (0,0,0)$, $\mathbf{v} = (v_x,0,0)$. Make a 3-D parametric plot of the resulting motion of the particle. (See Appendix 1.)

Chapter 8 SUPERCONDUCTIVITY

In This Chapter

- Theoretical Background
- London's Theory and the Meissner effect
- Flux Quantization
- Type I and Type II Superconductivity
- Superconducting Geometries and Boundary Conditions
- Superconducting Plates
- Hollow Superconducting Shells
- Layered Superconducting and Permeable Shields
- Modeling Type II Superconductors

8.1 THEORETICAL BACKGROUND

In 1911, Dutch physicist H. K. Onnes discovered superconductivity in mercury at a temperature of 4 K soon after he liquefied helium for the first time. Onnes was awarded the Nobel Prize in Physics for his work on the properties of matter at low temperatures in 1913. Later, Fritz London developed a phenomenological theory of superconductivity in 1934 with his brother Heinz. The basic idea is that the electrons in a superconductor condense into a superfluid state characterized by a single rigid wavefunction $\psi(r) = |\psi(r)|e^{i\varphi(r)}$ also known as the order parameter. In 1950 Ginzburg and Landau generalized the London theory by introducing a complex, temperature-dependent order parameter. John Bardeen, Leon Cooper, and J. Robert Schrieffer reported the first microscopic theory of superconductivity in 1957 known as the BCS theory.

The superconducting state is characterized by (1) zero electrical resistance to current below a material dependent transition temperature T_c, (2) the expulsion of magnetic flux from the bulk of the superconductor, and (3) flux quantization. Superconductivity is now known to be a remarkable example of quantum effects acting on a macroscopic scale. Certain metals, including mercury, lead, tin, and more recently discovered copper oxide ceramics, undergo a superconducting phase transition below T_c. Electron spins become oppositely paired resulting in the formation of a spin-zero quasi particle, or Cooper pair, that may pass through the material unimpeded.

Electrons in a superconductor condense into a superfluid-like state described by a single wave function $\psi(r) = |\psi(r)| \exp i\varphi(r)$, where φ is the phase of the superconducting wave function. The resistance free current inside a superconductor is described by

$$\mathbf{J} = \frac{e^* \hbar}{2m^* i}\left(\Psi^* \nabla \Psi - \Psi \nabla \Psi^*\right) - \frac{e^{*2} |\Psi|^2}{m^*} \mathbf{A}, \tag{8.1}$$

where $e^* = 2e$, $m^* = 2m$ and e, m are the electronic charge and mass, respectively. Taking $|\Psi|^2$ to be the density of superconducting electron pairs n_p equal to twice the electron carrier density n_s the supercurrent

$$\mathbf{J} = \frac{n_p e}{m}\left(\hbar \nabla \varphi - 2e\mathbf{A}\right) \tag{8.2}$$

because of rigidity of wave function.

Modeling Superconductors

Analytical modeling techniques can be applied to sufficiently simple superconducting geometries such as a sphere, cylinder, or a disk in uniform external magnetic fields. Modeling complex superconducting shapes in the presence of magnetic sources requires numerical solutions for an accurate determination of the field distribution, especially near the superconductor. In many cases, the problem of a magnetic source in the presence of a superconducting shell can be reduced to a two-dimensional problem having planar or axial symmetry. In carrying out simulations in this chapter, the magnetostatics problem type is initially specified along with the designated x-y or r-z symmetry. Superconductors can also be modeled in Transient and Time-harmonic Magnetics in QuickField using boundary conditions on the vector potential or by assigning zero permeability.

8.2 London Theory and the Meissner Effect

The Meissner effect refers to the expulsion of magnetic flux from the interior of a superconductor, or perfect diamagnetism. In a normal metal with conductivity σ, Ohm's law describes the current density $\mathbf{J} = \sigma\mathbf{E}$, where the current density is proportional to electric field.

To explain the Meissner effect, London's theory proposes, in part due to the assumed rigidity of the wave function, that the current density $\vec{J}_s$ in a superconductor is proportional to the vector potential

$$\mathbf{J}_s = -\frac{1}{\mu_0 \lambda_L^2}\mathbf{A}, \tag{8.3}$$

where λ_L is the London penetration depth

$$\lambda_L = \left(\frac{m}{\mu_0 n_s e^2}\right)^{1/2} \tag{8.4}$$

Because $\mathbf{E} = -\frac{\partial \mathbf{A}}{\partial t}$ according to equation (8.1) the electric field is then proportional to the time derivative of the supercurrent density

$$\mathbf{E} = \mu_0 \lambda_L^2 \frac{\partial \mathbf{J}_s}{\partial t}. \tag{8.5}$$

London's first equation also obtained from Newton's second law $\mathbf{F} = -e\mathbf{E} = m\mathbf{a} = m\dot{\mathbf{v}}_s$, where $\mathbf{J}_s = n_s e\mathbf{v}$. The magnetic field $\mathbf{B} = \nabla \times \mathbf{A}$ is proportional to the curl of $\mathbf{J}_s$

$$\mathbf{B} = -\mu_0 \lambda_L^2 \nabla \times \mathbf{J}_s. \tag{8.6}$$

From the Maxwell relations

$$\nabla \times \mathbf{E} = -\frac{\partial \mathbf{B}}{\partial t} \tag{8.7}$$

$$\nabla \times \mathbf{B} = \mu_0 \sigma \mathbf{E} + \mu_0 \varepsilon_0 \frac{\partial \mathbf{E}}{\partial t} \tag{8.8}$$

and substituting the London equations (8.5) and (8.6) we obtain

$$\nabla^2 \mathbf{J} = \frac{1}{\lambda_L^2}\mathbf{J} + \mu_0\sigma\frac{\partial \mathbf{J}}{\partial t} + \mu_0\varepsilon\frac{\partial^2 \mathbf{J}}{\partial t^2} \tag{8.9}$$

$$\nabla^2 \mathbf{A} = \frac{1}{\lambda_L^2}\mathbf{A} + \mu_0\sigma\frac{\partial \mathbf{A}}{\partial t} + \mu_0\varepsilon\frac{\partial^2 \mathbf{A}}{\partial t^2} \tag{8.10}$$

$$\nabla^2 \mathbf{B} = \frac{1}{\lambda_L^2}\mathbf{B} + \mu_0\sigma\frac{\partial \mathbf{B}}{\partial t} + \mu_0\varepsilon\frac{\partial^2 \mathbf{B}}{\partial t^2} \tag{8.11}$$

$$\nabla^2 \mathbf{E} = \frac{1}{\lambda_L^2}\mathbf{E} + \mu_0\sigma\frac{\partial \mathbf{E}}{\partial t} + \mu_0\varepsilon\frac{\partial^2 \mathbf{E}}{\partial t^2}. \tag{8.12}$$

For time-harmonic fields that vary as $\exp(i\omega t)$, the above equations become

$$\nabla^2 \begin{Bmatrix} \mathbf{J} \\ \mathbf{A} \\ \mathbf{B} \\ \mathbf{E} \end{Bmatrix} = \left(\frac{1}{\lambda_L^2} + i\omega\mu_0\sigma - \omega^2\mu_0\varepsilon\right)\begin{Bmatrix} \mathbf{J} \\ \mathbf{A} \\ \mathbf{B} \\ \mathbf{E} \end{Bmatrix}. \tag{8.13}$$

These equations reduce to

$$\nabla^2 \begin{Bmatrix} \mathbf{J} \\ \mathbf{A} \\ \mathbf{B} \end{Bmatrix} = \left(\frac{1}{\lambda_L^2}\right)\begin{Bmatrix} \mathbf{J} \\ \mathbf{A} \\ \mathbf{B} \end{Bmatrix} \tag{8.14}$$

for static applied magnetic fields. For static electric fields

$$\nabla^2 \mathbf{E} = \frac{1}{\lambda_L^2}\mathbf{E}. \tag{8.15}$$

For a uniform magnetic field with flux density $\mathbf{B} = B_z\mathbf{k}$ tangential to a superconducting half-space located at $x = 0$, the magnetic field, vector potential, and current density attenuate in the superconductor according to

$$\begin{Bmatrix} J_y \\ A_y \\ B_z \end{Bmatrix} = \begin{Bmatrix} J_y \\ A_y \\ B_z \end{Bmatrix}_0 \exp\left(-\frac{x}{\lambda_L}\right) \tag{8.16}$$

so that these field components fall off exponentially inside the superconductor.

8.3 FLUX QUANTIZATION

Flux quantization can be derived considering a superconducting ring with a thickness much larger than the London penetration depth. The current inside the ring will be zero so that equation (8.2) gives

$$\hbar \nabla \varphi = 2e\mathbf{A}. \tag{8.17}$$

Integrating this expression around a closed path around the superconducting ring gives

$$\hbar \oint_\Gamma \nabla \varphi \cdot dl = 2e \oint_\Gamma \mathbf{A} \cdot dl, \tag{8.18}$$

where

$$\oint_\Gamma \nabla \varphi \cdot d\ell = 2n\pi \tag{8.19}$$

due to the single-valuedness of the wave function. The total flux contained inside the superconducting ring is given by

$$\oint_\Gamma \mathbf{A} \cdot d\ell = \iint_{\text{surf}} \nabla \times \mathbf{A} \cdot d\mathbf{a} = \iint_{\text{surf}} \mathbf{B} \cdot d\mathbf{a} = \Phi, \tag{8.20}$$

which gives the total flux contained inside the superconducting ring. This flux is quantized in integral multiples of the flux quantum Φ_0 or,

$$\Phi = n\frac{h}{2e} = n\Phi_0 = n \cdot 2.07 \cdot 10^{-15}\,Wb. \tag{8.21}$$

This result may be obtained by using the Bohr-Sommerfield quantization condition

$$\oint_\Gamma \mathbf{p} \cdot d\ell = nh \tag{8.22}$$

and substituting the momentum of an electron pair $p = 2(m\mathbf{v} + e\mathrm{A})$.

This results in an expression for the flux density threading the superconducting ring

$$\Phi = n\frac{h}{2e} - \frac{m}{n_s e^2}\oint_{\Gamma} \mathbf{J}_s \cdot d\ell, \tag{8.23}$$

where the supercurrent $\mathbf{J}_s = n_s e\mathbf{v}$ vanishes within a few penetration depths so that the quantization condition is obtained.

8.4 TYPE-I AND TYPE-II SUPERCONDUCTORS

A key quantity that distinguishes different types of superconductors is the Ginzburg-Landau parameter

$$\kappa \equiv \frac{\lambda}{\xi}, \tag{8.24}$$

where the coherence length ξ is a measure of transition distance between superconducting and normal regions and λ is the London penetration depth. Superconductors with smaller values of $\kappa < 1/\sqrt{2}$ are referred to as Type-I and are characterized by a positive surface energy. Type-I superconductors tend to minimize the normal-superconductor interfacial area if both normal and superconducting regions exist in the sample. Type-II superconductors, on the other hand, have values of $\kappa > 1/\sqrt{2}$ with a negative surface energy and a maximization of the normal-superconductor interface area. The Ginzburg-Landau theory gives the lower critical field h_{C1} of a Type-II superconductor below which no flux will penetrate the superconductor

$$h_{C1} \cong \frac{\Phi_0}{4\pi\lambda^2}\ln\kappa. \tag{8.25}$$

The upper critical field h_{C2} of a Type-II superconductor is given by

$$h_{C2} \cong \frac{\Phi_0}{2\pi\xi^2}. \tag{8.26}$$

If a transport supercurrent $\mathbf{J}_s$ also flows in addition to the vortex current, there will be a Lorentz force $F = \mathbf{J}_s \times \mathbf{\Phi}_0$ acting on the vortices, where $\mathbf{\Phi}_0$ points in the direction of flux penetration into the superconductor.

QuickField calculations of superconductors in external fields will be handled with the application of boundary conditions and material properties discussed in the following sections.

8.5 SUPERCONDUCTING GEOMETRIES AND BOUNDARY CONDITIONS

Geometric Effects

Type-I superconductors, such as lead, become normal in magnetic fields greater than the thermodynamic critical field h_c, which decreases with increasing temperature. For fields sufficiently weaker than h_c, flux is completely expelled from the bulk of the sample, except for a thin surface layer λ_L of about 50 nm. The expelled field results in larger flux densities near the edges of the superconductor. Demagnetization occurs in nearly critical fields, where the local field strength exceeds h_c resulting in an intermediate state with coexisting superconducting and normal regions. The interface between these superconducting and normal regions will have a complex geometry that tends to minimize the total energy of the system.

Nonlinear Superconductors

Type-II superconductors such as Nb_3Sn are characterized by two critical fields h_{c1} and h_{c2}. Flux is expelled from the superconductor below h_{c1} and the sample becomes normal above h_{c2}. Between these field values, the superconductor is penetrated by flux vortices each with a normal core threaded by one flux quantum $\Phi_0 = 2.07 \cdot 10^{-15}\ Wb$. For magnetic fields strengths greater than h_{c1}, the nonlinear $B(H)$ characteristics need to be accounted for. For high field applications such as magnetic resonance imaging, accelerator magnets and fault current limiters, the nonlinear $B(H)$ curve is entered in as a data file that is used by the solver during the calculation of the resulting field distribution.

Boundary Conditions

Calculations in this chapter will neglect field penetrations into the superconductor entirely. This approximation is valid if the dimensions of the superconductor are much larger than λ_L. In modeling superconductors, where demagnetization effects do not occur, the appropriate boundary condition is zero normal flux density on the superconducting surface. This boundary condition can be applied explicitly by forcing $\mathbf{B} \cdot n = 0$ on all superconducting surfaces, or implicitly by choosing the relatively permeability of the superconductor to be nearly zero.

Calculation of Supercurrent Density

Once the field is calculated, the supercurrent density at the surface of the superconductor may be determined by the discontinuity in the tangential component of the field strength $\boldsymbol{H}_t$.

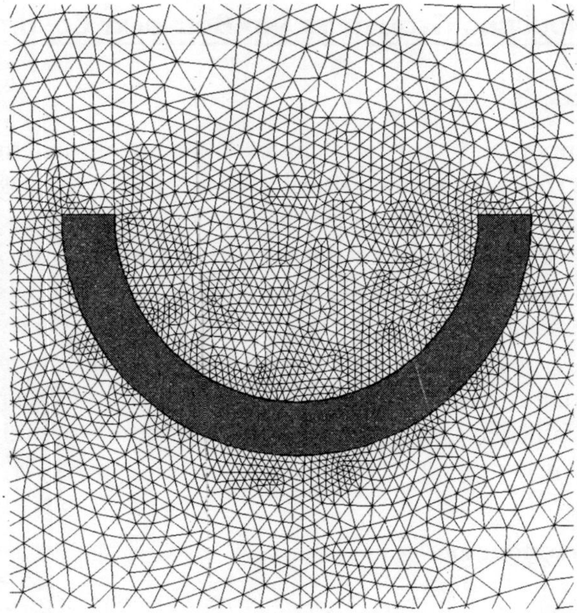

FIGURE 8.1 **Finite element mesh of a superconducting bowl modeled in *x-y* coordinates.**

Color images for many of the figures may be found on the CD-ROM.

Mesh Requirements

Greater computational accuracy is desired in regions with large spatial field variations near the superconductor than at remote parts of the solution region, where the field is slowly varying. This is accomplished with a finer finite element mesh, where an accurate determination of the field distribution is required and a coarser mesh in remote regions. Adaptive mesh refinement may be used to adjust the node spacing depending on the variation of energy density in a particular region. Figure 8.1 is a FEM mesh exterior to a superconducting bowl. The mesh grading is apparent in this figure with a finer mesh near the bowl. Roughly 1/10 of the solution region is shown. Note that, in the linear case, it is not necessary to mesh interior regions of the superconductor, where the fields are taken to be zero.

8.6 SUPERCONDUCTING PLATES

Superconducting Strip

A superconducting strip can be modeled analytically in two dimensions by the conformal mapping method. The deviation of flux lines by a superconducting

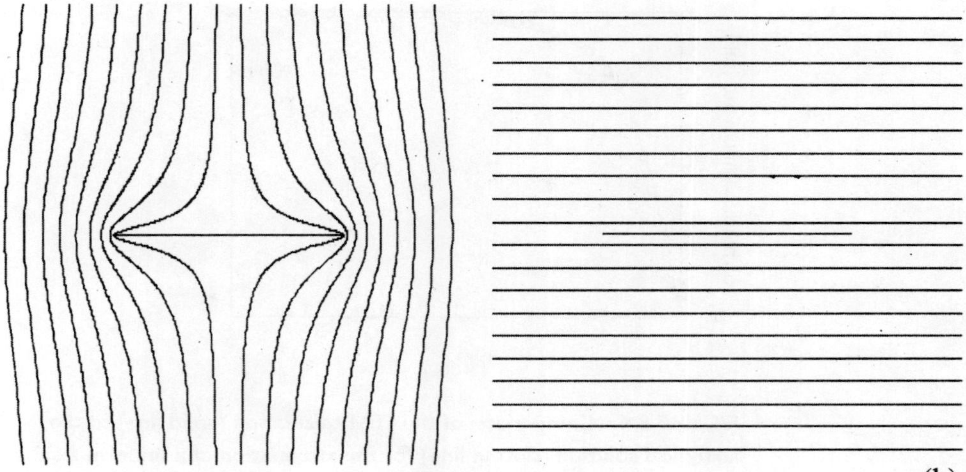

(a) (b)

FIGURE 8.2 **Thin superconducting disk in (a) a normally incident and (b) tangentially incident magnetic field modeled in *x-y* coordinates.**

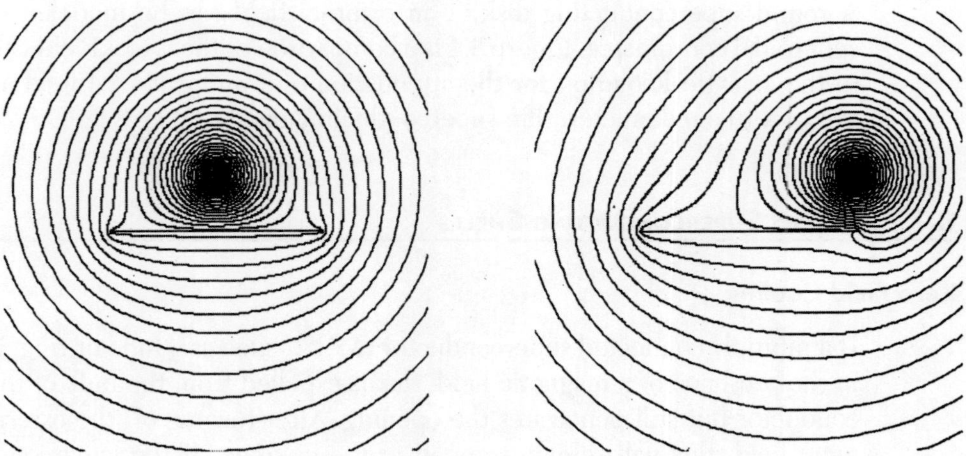

(a) (b)

FIGURE 8.3 **Superconducting strip in the presence of (a) a centered and (b) an off-center line current modeled in *x-y* coordinates.**

strip, where the field is normal to the plane of the strip, is shown in Figure 8.2 (a). The 10 cm width strip is modeled as a single boundary with $\mathbf{B}\cdot\hat{\mathbf{n}}=0$ in x-y coordinates. When the external field is tangential to the superconducting disk there is no distortion of the field lines in Figure 8.2 (b). Figure 8.3 (a) shows the flux lines of an infinite line source located directly above the center of the superconducting disk. Figure 8.3 (b) shows the flux line distribution with the dipole located over the edge of the disk.

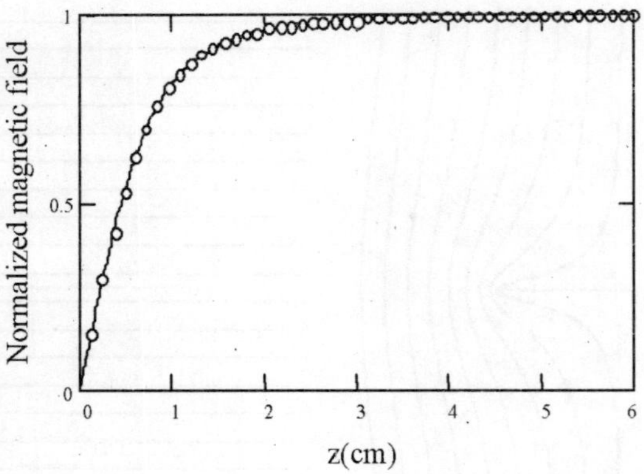

FIGURE 8.4 **Comparison of the FEM calculation (solid line) to the analytical solution (dotted line) for the attenuation of a uniform axial field along the axis of a 5 cm radius thin superconducting disk. The magnetic field is normalized to unit magnitude far away from the disk.**

Superconducting Disk

A round superconducting disk in an external field can be modeled in oblate spheroidal coordinates. Figure 8.4 is a comparison of the QuickField calculation to the analytical solution for the attenuation of a uniform axial field along the axis of a 10 cm diameter, thin superconducting disk.

8.7 HOLLOW SUPERCONDUCTING SHELLS

Zero Field Cooling

If a multiply connected superconductor (MCS), such as a tube or ring, is cooled in the presence of a magnetic field, flux is expelled from the bulk of the superconductor but still penetrates the opening. After turning off the external magnetic field, flux will remain trapped in the opening. If the superconductor is cooled in zero field (ZFC) and the field is then switched on, screening currents will flow that cancel the applied magnetic field in such a way that the total flux through the MCS remains zero. The appropriate MCS boundary condition depends on the field present at the time that the superconducting state is established. In two dimensions, constraining the vector potential to be zero on the surface of the superconductor satisfies the zero flux condition. The field-cooled case is modeled with zero normal field, or permeability.

Flux Trapping

A multi-connected superconducting body such as a ring or a tube will trap any flux present in the hole of the superconductor upon transition to the superconducting state. If the field is then changed after the superconducting state is established, screening currents will flow so as to keep the amount of trapped flux a constant. In this way a ZFC superconductor will maintain the zero flux condition upon the application of an external magnetic field.

Superconducting Ring

A field-cooled superconducting washer is simulated in an axial background field in Figure 8.5 (a). The cross-section of the washer appears as two strips in *x-y* coordinates. Here the background field is seen to be diverted by the sides of the washer but still penetrates the hole of the superconductor. The boundary condition applied here is $\mathbf{B}\cdot\hat{n}=0$ on the superconductor. This boundary condition is also applicable modeling a superconducting washer with a thin slit so that the superconducting path is not complete. In this case the flux density is enhanced in the opening of the washer as is evident by the increased density of flux lines in that region. The washer then acts as a superconducting *flux-focuser*. In Figure 8.5 (b) the boundary condition $A_z = 0$ is applied to the strip washer simulating the ZFC condition, where flux lines are now expelled from the interior.

A ZFC superconducting ring in a uniform axial magnetic field is modeled in r-z coordinates with the boundary condition $A_\phi = 0$ applied to the superconductor so that, by Stokes' theorem, the total flux

$$\Phi_M = \int B_z da = \int A_\phi dl \tag{8.27}$$

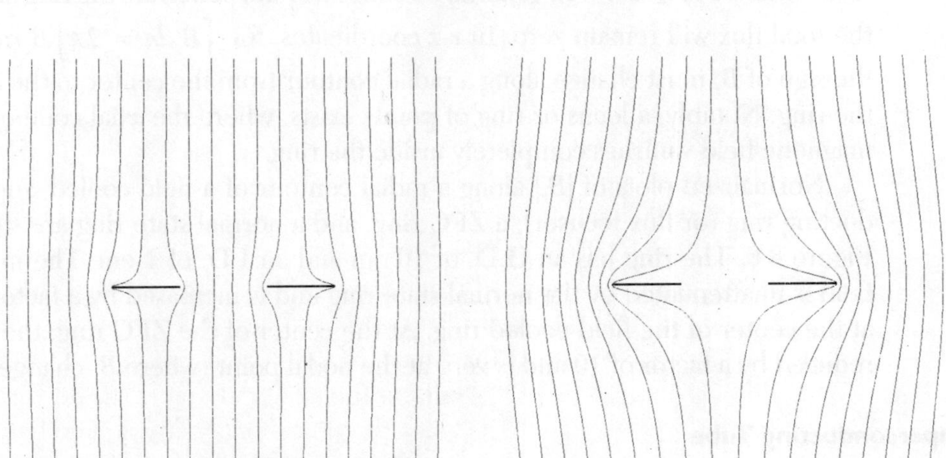

FIGURE 8.5 **(a) Field cooled and (b) zero field cooled superconducting ring in an axial magnetic field modeled in *x-y* coordinates.**

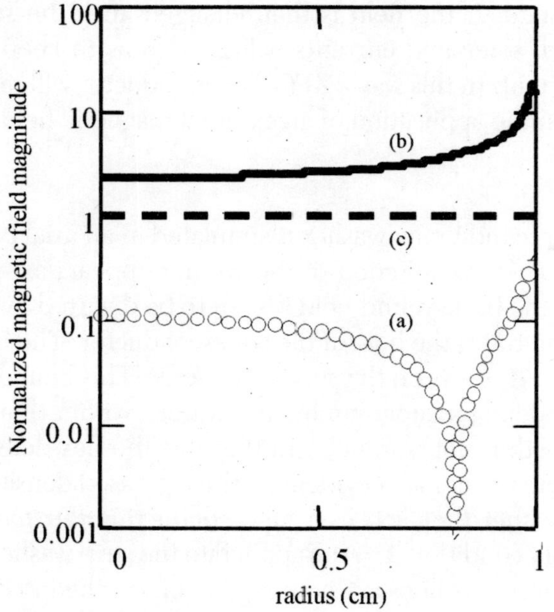

FIGURE 8.6 **Normalized B_z along a radial contour of a superconducting ring with a 5 cm O.R. and a 1 cm I.R. modeled in *r-z* coordinates (a) zero field-cooled ring (b) field-cooled ring and (c) ring in normal state.**

is constrained to be zero. In general, flux lines will still penetrate the ring although the total flux will remain zero. In r-z coordinates, for $\int B_z da = 2\pi \int B_z r dr = 0$, the sign of B_z must change along a radial contour from the center to the edge of the ring. Notably, a locus or ring of points exists, where the axial component of magnetic field vanishes completely inside the ring.

Normalized plots of $|B_z|$ along a radial contour of a field cooled, superconducting ring (or flux focuser), a ZFC ring, and a normal state ring are shown in Figure 8.6. The ring has an O.D. of 10 cm and an I.D. of 1 cm. The magnetic field is unattenuated by the normal state ring and is increased by a factor of 2.3 at the center of the field-cooled ring. At the center of the ZFC ring, the field is reduced by a factor of 10 and is zero at the nodal point (where B_z changes sign).

Superconducting Tube

Figure 8.7 (a) illustrates the flux line distribution of a superconducting tube in an axial uniform field. The tube is modeled as two parallel strips in x-y coordinates with boundary condition $A_z = 0$. As a comparison Figure 8.7 (b) shows the calculated flux density with the same tube in a transverse x-directed field. Note the greater field penetration into the tube by the transverse field compared to

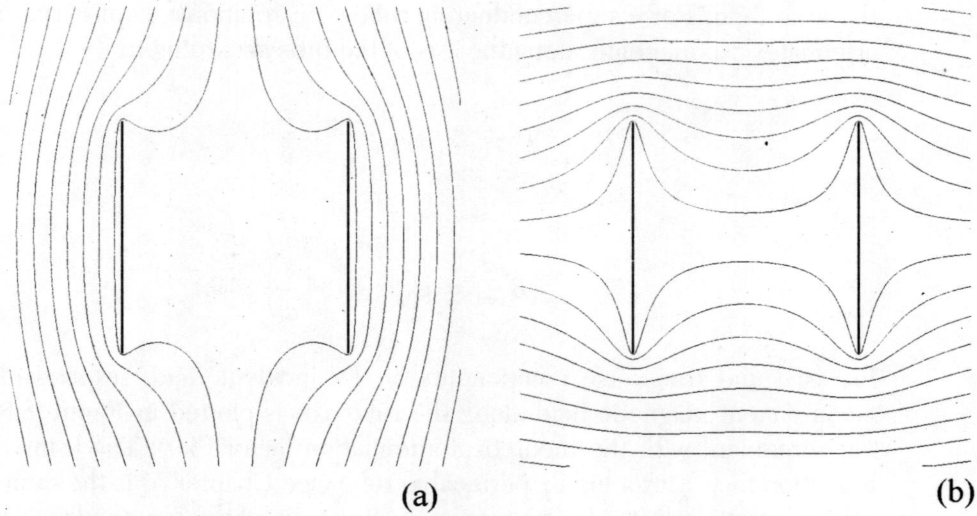

FIGURE 8.7 **Superconducting tube in the presence of (a) axial and (b) transverse fields modeled in *x-y* coordinates.**

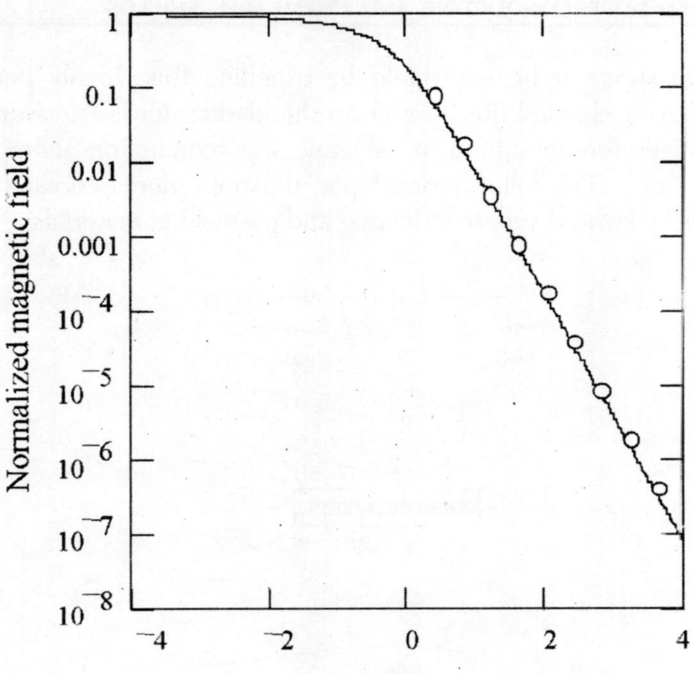

FIGURE 8.8 **FEM calculation of the field attenuation along the axis of a 1.0-cm radius superconducting tube modeled in *r-z* coordinates (solid line) compared to the analytical solution (dotted line).**

the axial field. For a superconducting tube with rotational symmetry, the field attenuates exponentially along the axis of the tube according to

$$B_{\text{axial}} \propto \exp\left(-3.832\frac{z}{a}\right) \tag{8.28}$$

and

$$B_{\text{trans}} \propto \exp\left(-1.84\frac{z}{a}\right) \tag{8.29}$$

for axial and transverse components of the incident field, respectively. The z-component magnetic field along the tube axis is plotted in Figure 8.8 showing agreement with the theoretical attenuation factor (8.9). The transverse attenuation factor for a highly permeable tube (see Chapter 5) is the same as the axial attenuation factor for a superconducting tube of the same radius. This suggests combining superconducting and highly permeable materials for shielding applications.

8.8 LAYERED SUPERCONDUCTING AND PERMEABLE SHIELDS

While superconductors shield by expelling flux, highly permeable μ-metals effectively channel flux away from shielded regions. Transverse fields are preferentially screened by μ-metal while superconductors more effectively screen axial fields. The following examples illustrate increased shielding of transverse fields by layered superconducting and permeable materials.

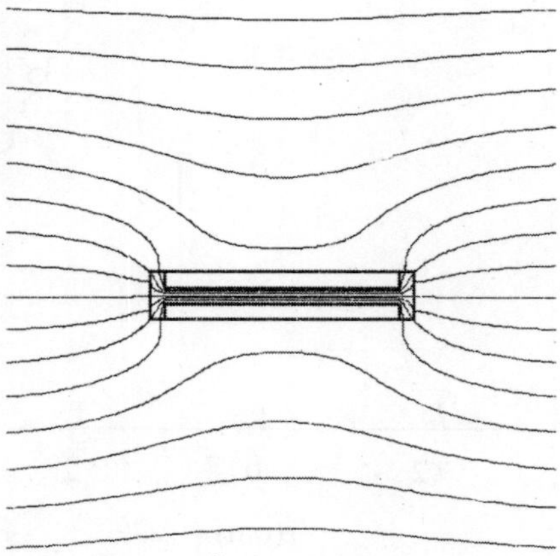

FIGURE 8.9 **Tri-layered superconducting and μ-metal disk in a transverse field. The μ-metal side tabs help channel flux in between the superconducting disks (modeled in *x-y* coordinates).**

Superconducting & Highly Permeable Metal Sandwich

A tri-layered superconducting and μ-metal disk in a transverse field is illustrated in Figure 8.9. Additional μ-metal side tabs help channel flux in between the superconducting disks (modeled in x-y coordinates). The tangential field directly above the superconductor is reduced by roughly a factor of two compared to the superconducting disk without the μ-metal layer.

8.9 MODELING TYPE-II SUPERCONDUCTORS

The partial penetration of magnetic flux into Type-II superconductors may be simulated in QuickField using the nonlinear magnetics curve editor shown in Figure 8.10. Values are entered in the table at the right in this window and a continuous *B-H* curve is generated to fit to the tabulated values using a cubic spline fitting.

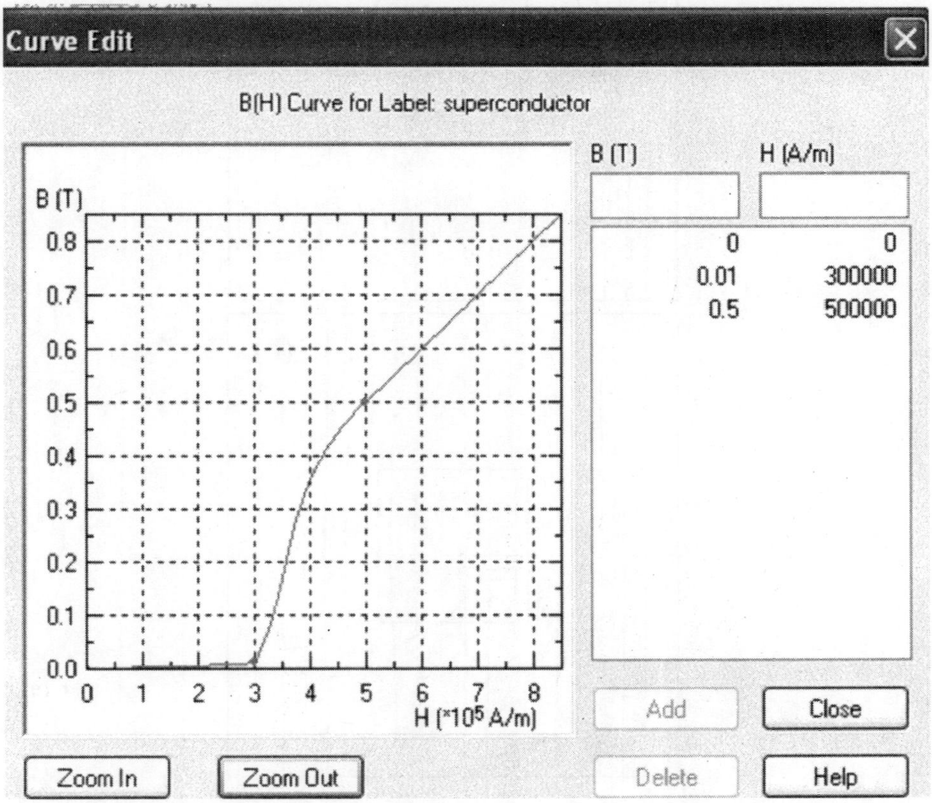

FIGURE 8.10 **Nonlinear *B-H* curve of a Type-II superconductor specified in the nonlinear magnetics curve editor.**

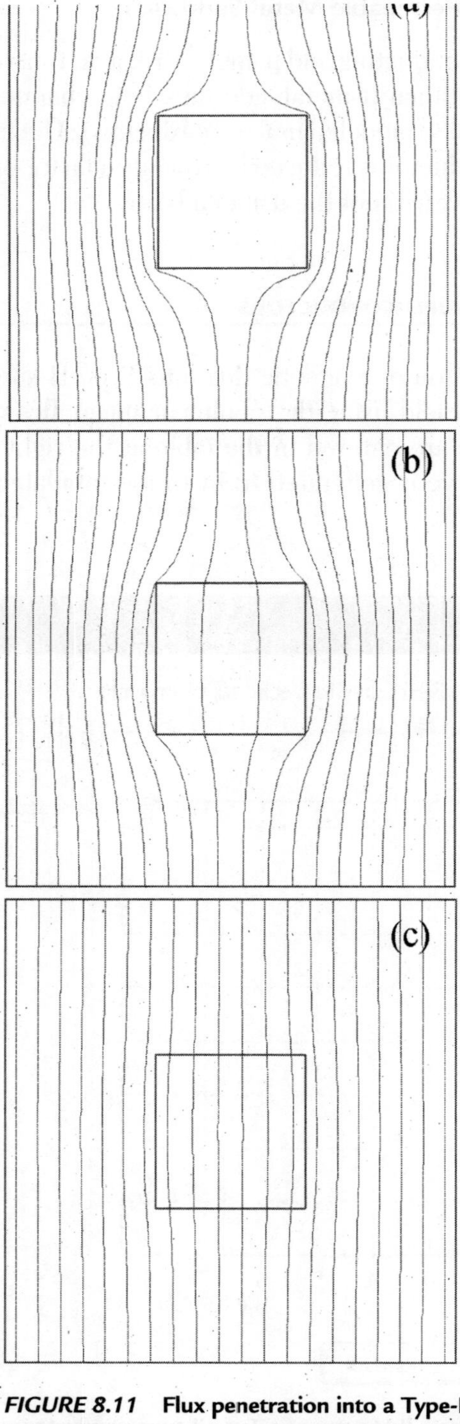

FIGURE 8.11 **Flux penetration into a Type-II superconductor with the nonlinear *B-H* curve given in Figure 8.13 for (a) $B = 0.07\ T$ (b) $B = 0.2\ T$ and (c) $B = 0.7\ T$.**

According to the *B-H* curve, the internal flux density is zero for magnetic field strengths below about $3 \cdot 10^5$ A/m. Figure 8.11 shows the penetration of flux into a Type-II superconductor with a square cross-section and the nonlinear *B-H* curve given in Figure 8.10. For an external flux density $B = 0.07\ T$ (a) there is no penetration of flux into the superconductor. Flux lines begin to penetrate the superconductor for $B = 0.2\ T$ (b) with greater flux penetration at $B = 0.7\ T$ (c).

REFERENCES

R. P. Feynman, R. B. Leighton, and M. Sands. (1966). *Volume III The Feynman Lectures on Physics,* Addison-Wesley Publishing Company. *Chapter 21: The Schrödinger Equation in a Classical Context: A Seminar on Superconductivity.*

Charles Kittel. (1986). *Introduction to Solid State Physics*, 6th edition, John Wiley & Sons, Inc. *Chapter 12: Superconductivity.*

A. C. Rose-Innes and E. H. Rhoderick. (1994). *Introduction to Superconductivity*, Pergamon.

J. R. Reitz, F. J. Milford, R. W. Christy. (1992). *Foundations of Electromagnetic Theory 4th edition.* Addison-Wesley Publishing Company. (Current edition includes discussions of numerical methods (finite difference) and superconductivity.)

EXERCISES

8.1 Model a solid superconducting sphere in a uniform magnetic field in axial symmetry; (a) plot the attenuation of magnetic field along the z-axis outside of the sphere and (b) plot the tangential H-field equal to the supercurrent density along the boundary contour of the sphere.

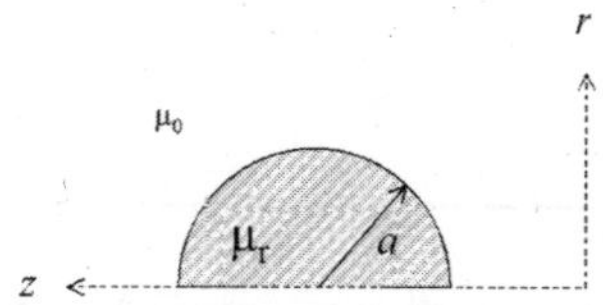

FIGURE E-8.1 **Superconducting sphere in uniform field. Geometry: Axial symmetry.**

8.2 Simulate a solid superconducting sphere near a current loop in axial symmetry; (a) compare contour plots of the magnetic field along the z-axis and also between the current loop and the sphere when the sphere is in the superconducting and the normal state and (b) calculate the force between the current loop and the sphere for several locations above the sphere.

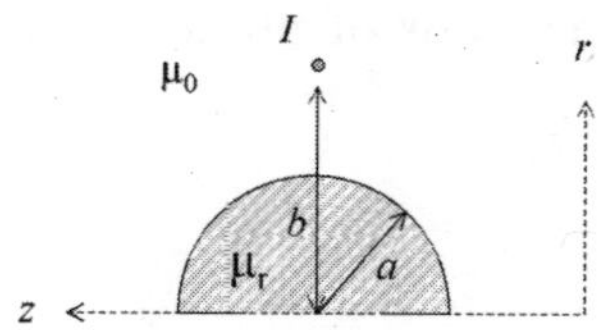

FIGURE E-8.2 **Superconducting sphere near a current loop. Geometry: Axial symmetry.**

8.3 Model a current loop inside a hollow superconducting sphere in axial symmetry. Treat the superconducting shell as a single boundary with $\mathbf{B}\cdot\hat{n}=0$; (a) calculate the total magnetic field energy inside the sphere for several sphere radii and (b) plot the induced supercurrent density (tangential $\boldsymbol{H}$) on the inner surface of the sphere.

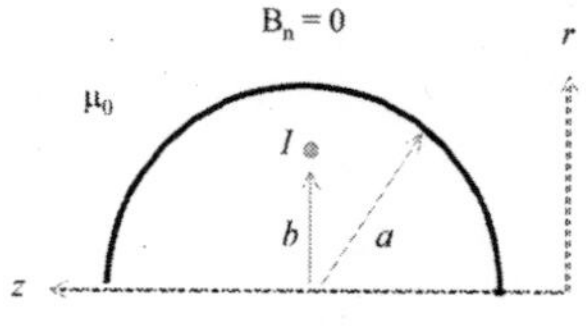

FIGURE E-8.3 **Current loop inside a hollow superconducting sphere. Geometry: Axial symmetry.**

8.4 Magnetic levitation: Model a permanent magnet above a superconducting plane in x-y symmetry; (a) compare the force between the magnet and the superconducting plane when the coercive force vector is oriented normal and tangential to the superconducting plane and (b) compare the supercurrent density (tangential $\boldsymbol{H}$) at the surface of the superconductor for each case.

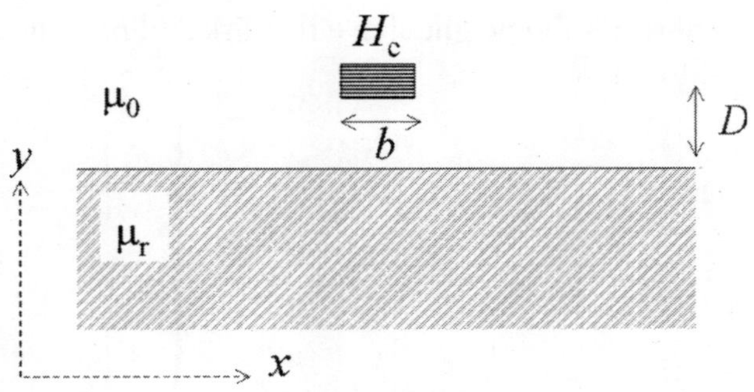

FIGURE E-8.4 **Permanent magnet above a superconducting plane. Geometry: *x-y* symmetry.**

8.5 Simulate a line current; (a) outside a superconducting and (b) a permeable cylinder in x-y symmetry. Calculate the force between a line current and the cylinder when the cylinder is superconducting (take $\mu_r = 10^{-6}$) and compare to the force between the permeable cylinder (take $\mu_r = 10^4$) with the same geometry.

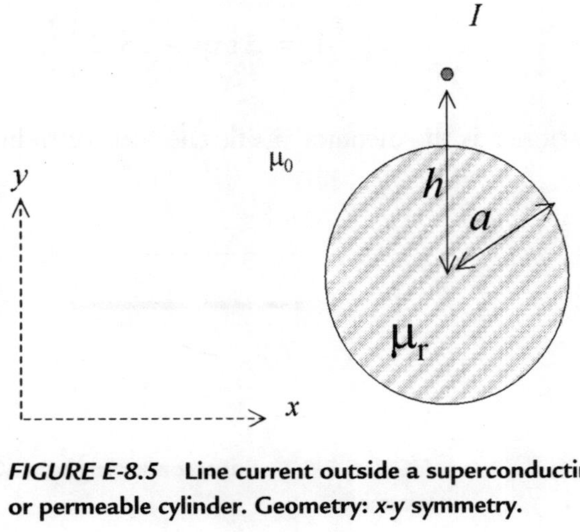

FIGURE E-8.5 **Line current outside a superconducting or permeable cylinder. Geometry: *x-y* symmetry.**

8.6 Plot the z-component magnetic field along the axis of a superconducting disk of radius a in a uniform field modeled in r-z coordinates. Compare to the analytical solution

$$B_z = B_0\left[1 - \frac{2}{\pi}\left(\tan^{-1}\frac{a}{z} - \frac{az}{a^2 + z^2}\right)\right],$$

where z is the height above the disk and B_0 is the flux density in absence of the disk.

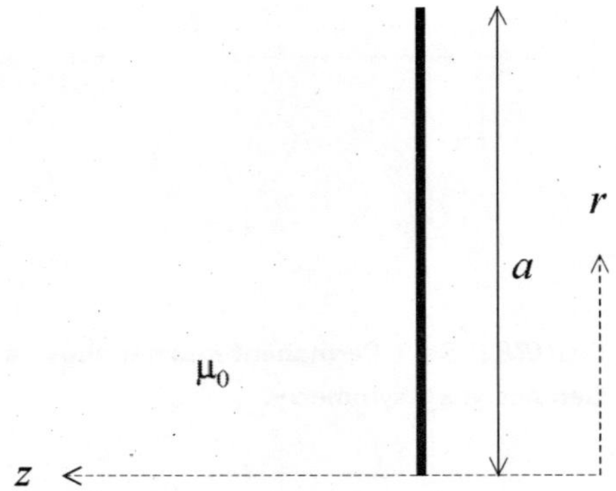

FIGURE E-8.6 **Superconducting disk. Geometry: Axial symmetry.**

8.7 Model a superconducting tube in r-z coordinates. Treat the tube as a single boundary with zero vector potential. Plot the z-component magnetic field along the axis of the superconducting tube and compare to the asymptotic value

$$B_z = A\exp\left(-3.832\frac{z}{a}\right),$$

where z is the distance inside the tube of radius a and A is a constant.

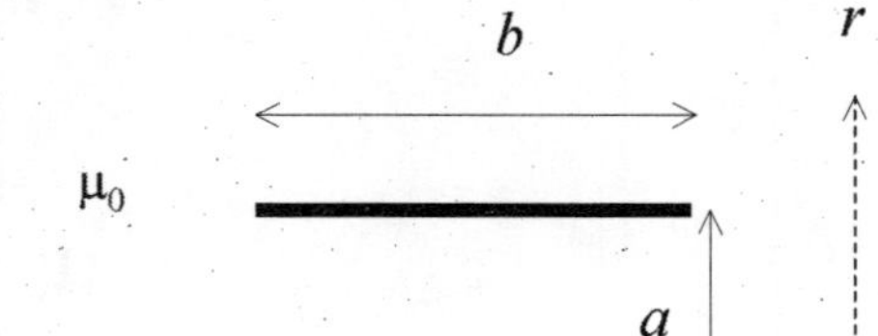

FIGURE E-8.7 **Superconducting tube. Geometry: Axial symmetry.**

8.8 Model a superconducting ring in r-z coordinates with a specified inner and outer radius. Treat the ring as a single boundary. Compare the z-component magnetic field along a radial contour inside the ring for both field-cooled and zero field-cooled boundary conditions. Show that the total flux through the zero field-cooled ring is zero.

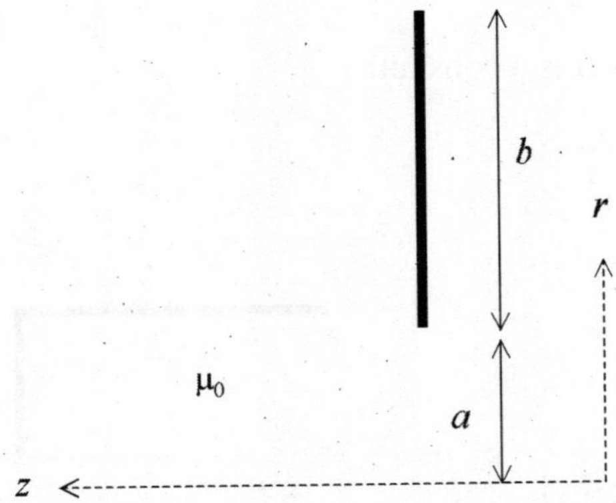

FIGURE E-8.8 **Superconducting ring. Geometry: Axial symmetry.**

8.9 Magnetic levitation: Model a permanent magnet above a superconducting plane in r-z symmetry; (a) calculate the force between the magnet and the superconducting plane with the coercive force vector oriented normal to the superconducting plane and (b) calculate the supercurrent density (tangential $\boldsymbol{H}$) at the surface of the superconductor.

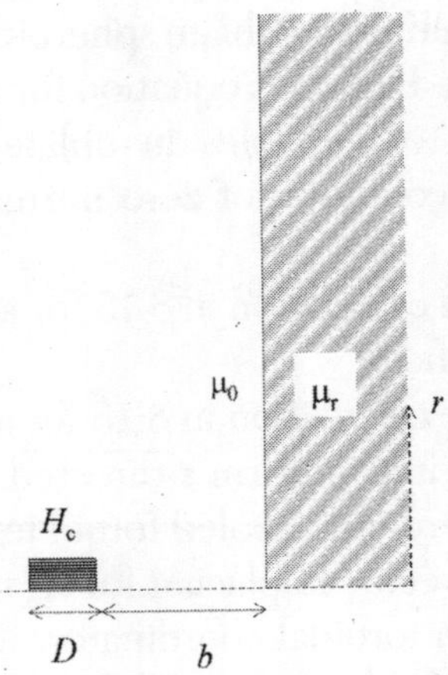

FIGURE E-8.9 **Permanent magnet above a superconducting plane.**

8.10 Model a superconducting can of radius a in an axial magnetic field. Make a contour plot of B_z along the axis of the cup. Show that the value of B_z inside the cup measured a distance z from the bottom is given by

$$B_z = A\sinh\left(3.832\frac{z}{a}\right),$$

where A is a constant.

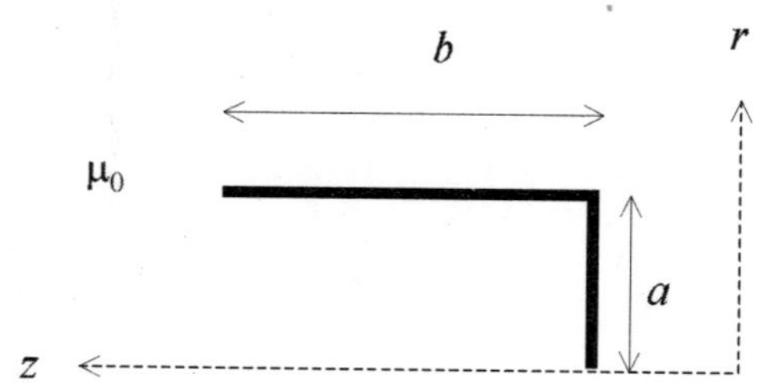

FIGURE E-8.10 **Superconducting can. Geometry: Axial symmetry.**

8.11 Model the dual layer superconducting μ-metal bowl in a uniform magnetic field in axial symmetry. Compare contour plots of the tangential field along the inside surface of the bowl with the outer μ-metal layer to contour plots without the outer μ-metal layer.

8.12 Model a Type-II hollow superconducting sphere in a uniform field. Plot the penetration of flux into the sphere for several external field values.

8.13 A superconducting oblate spheroid is positioned in a uniform z-directed field. Solve Laplace's equation for the scalar magnetic potential Ω using separation of variables in oblate spheroidal coordinates. Apply the boundary condition of zero normal field on the surface of the oblate spheroid.

8.14 Repeat the calculation in 8.13 for a superconducting prolate spheroid in a uniform field.

8.15 Repeat the calculation in 8.13 for a field cooled superconducting toroid positioned in a uniform z-directed field.

8.16 Model a zero field cooled toroid in a uniform z-directed magnetic field. Solve the vector Laplacian for A_ϕ in axial symmetry using separation of variables in toroidal coordinates. Apply the zero field-cooled condition $A_\phi = 0$ on the boundary of the toroid. Find the radial position of the nodal points of vanishing B_z in the center of the ring.

Chapter 9

DC CONDUCTION ANALYSIS

In This Chapter

- DC Current Flow Theory
- DC Conduction Analysis in QuickField
- Alternative Energy Applications
- Resistivity Measurements
- Printed Circuit Boards

9.1 DC CURRENT FLOW THEORY

The basic equation of steady state current flow is $\nabla \cdot \mathbf{J} = 0$ or $\nabla \cdot (\mathbf{E}/\rho) = 0$, where $\mathbf{E} = -\nabla U$ so that

$$\nabla \cdot \left(\frac{1}{\rho} \nabla U \right) = 0, \tag{9.1}$$

where ρ is resistivity with units of $\Omega \cdot$ m. Note that $\rho = 1/\sigma$, where σ is the electrical conductivity in S/m. Once U is determined, the current density may be calculated from Ohm's law $\mathbf{J} = \sigma \mathbf{E}$ or

$$\mathbf{J} = -\frac{1}{\rho} \nabla U. \tag{9.2}$$

From equation (9.1) the scalar potential will satisfy Laplace's equation $\nabla^2 U = 0$ in regions of constant resistivity.

9.2 DC CONDUCTION ANALYSIS IN QUICKFIELD

In Cartesian coordinates, equation (9.1) for steady state current flow is

$$\frac{\partial}{\partial x}\left(\frac{1}{\rho_x}\frac{\partial U}{\partial x}\right)+\frac{\partial}{\partial y}\left(\frac{1}{\rho_y}\frac{\partial U}{\partial y}\right)=0. \tag{9.3}$$

In axial symmetric cylindrical coordinates

$$\frac{1}{r}\frac{\partial}{\partial r}\left(r\frac{1}{\rho_r}\frac{\partial U}{\partial r}\right)+\frac{\partial}{\partial z}\left(\frac{1}{\rho_z}\frac{\partial U}{\partial z}\right)=0. \tag{9.4}$$

The electric field may be calculated from the gradient of the scalar potential as in Chapter 4. The total energy may be calculated from the electrical current density

$$W = \frac{1}{2}\iiint_{\text{vol.}}(\mathbf{E}\cdot\mathbf{J})dV. \tag{9.5}$$

Boundary Conditions

Dirichlet Boundary Conditions

Assign specific electric potential U values on exterior boundaries or edges within the solution region. This condition must be applied to at least one point in the solution region. Position dependent Dirichlet conditions may be entered using the formula editor.

Neumann Boundary Conditions

Specify the normal DC current density on either exterior boundaries as $J_n = \hat{\mathbf{n}}\cdot\mathbf{J}$ or interior boundaries as $\Delta J_n = (\mathbf{J}^+ - \mathbf{J}^-)\cdot\hat{\mathbf{n}}$ The Neumann condition $J_n = 0$ is the default condition on all outward boundaries unless otherwise specified.

Periodic Boundary Bonditions

Can be specified on opposite edges of the model. The potential is given by $U_1 = U_2$ on even periodic boundaries and $U_1 = -U_2$ on odd periodic boundaries.

Constant Potential or "floating" Boundary Conditions

Are used for isolated conductors whose surfaces are at constant but unknown potentials. Edges assigned a constant "floating" potential boundary condition should not touch fixed potential Dirichlet boundaries. Figure 9.1 displays the window for editing edge label data for DC current flow problems: To open the edge label window click on the corresponding edge label in the tree diagram on the left of the screen.

Material Properties

Block labels are associated with resistivity values in $\Omega \cdot m$ that may be isotropic or anisotropic. Block labels with no specified resistivity are excluded from the solution region. Figure 9.2 displays the window for editing block label data for DC current flow problems: To open the block label window click on the corresponding block label in the tree diagram on the left of the screen.

Vertex Conditions

Local potential or current values may be assigned to individual vertices. The potential and charge values may be specified by a numerical value or as a

Color images for many of the figures may be found on the CD-ROM.

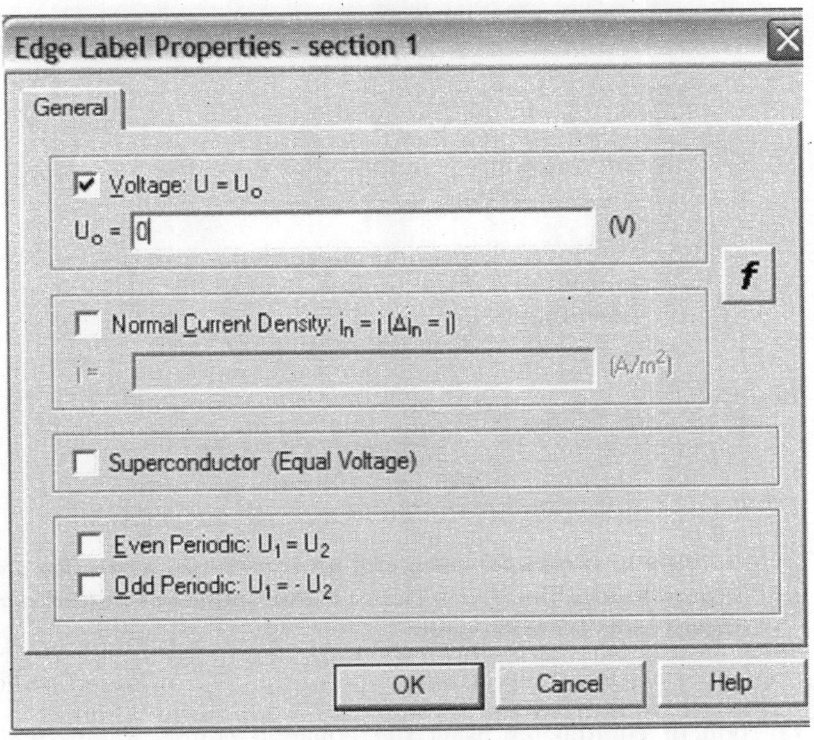

FIGURE 9.1 **Window for editing block label data for DC current flow problems: To open the block label window click on the corresponding block label in the tree diagram on the left of the screen.**

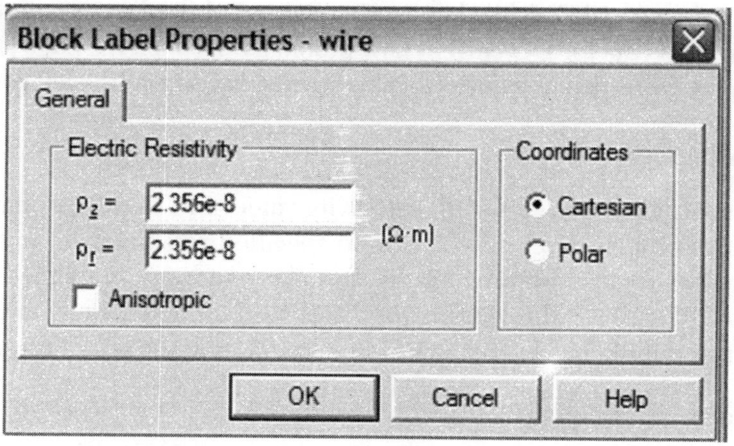

FIGURE 9.2 **Window for editing vertex label data for DC current flow problems: To open the vertex label window click on the corresponding vertex label in the tree diagram on the left of the screen.**

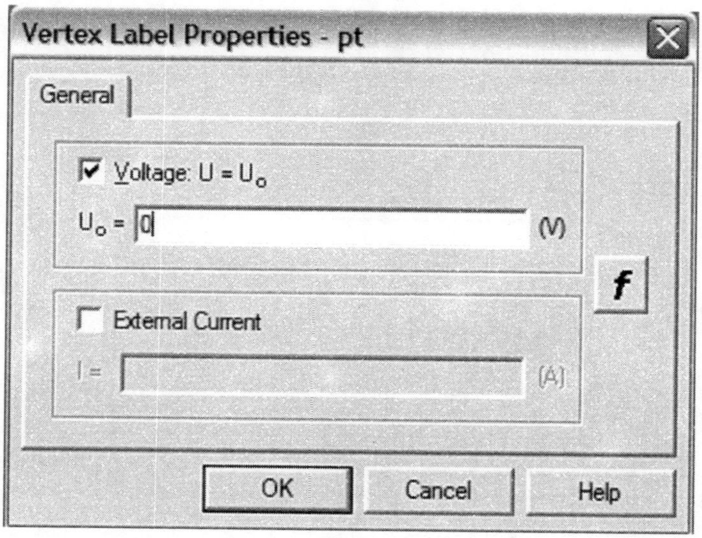

FIGURE 9.3 **Window for editing edge label data for DC current flow problems: To open the edge label window click on the corresponding edge label in the tree diagram on the left of the screen.**

function of coordinates using the equation editor. Figure 9.3 displays the window for editing vertex label data for DC current flow problems: To open the vertex label window click on the corresponding vertex label in the tree diagram on the left of the screen.

Ellipsoidal Crack in a Current Carrying Plate

Current flowing through a thin lamina with no component in the z-direction may be modeled in x-y symmetry. Figure 9.4 shows the DC current flow around an ellipsoidal crack in a metal plate with ρ equal to $1.0 \cdot 10^{-8}\ \Omega \cdot m$. The crack

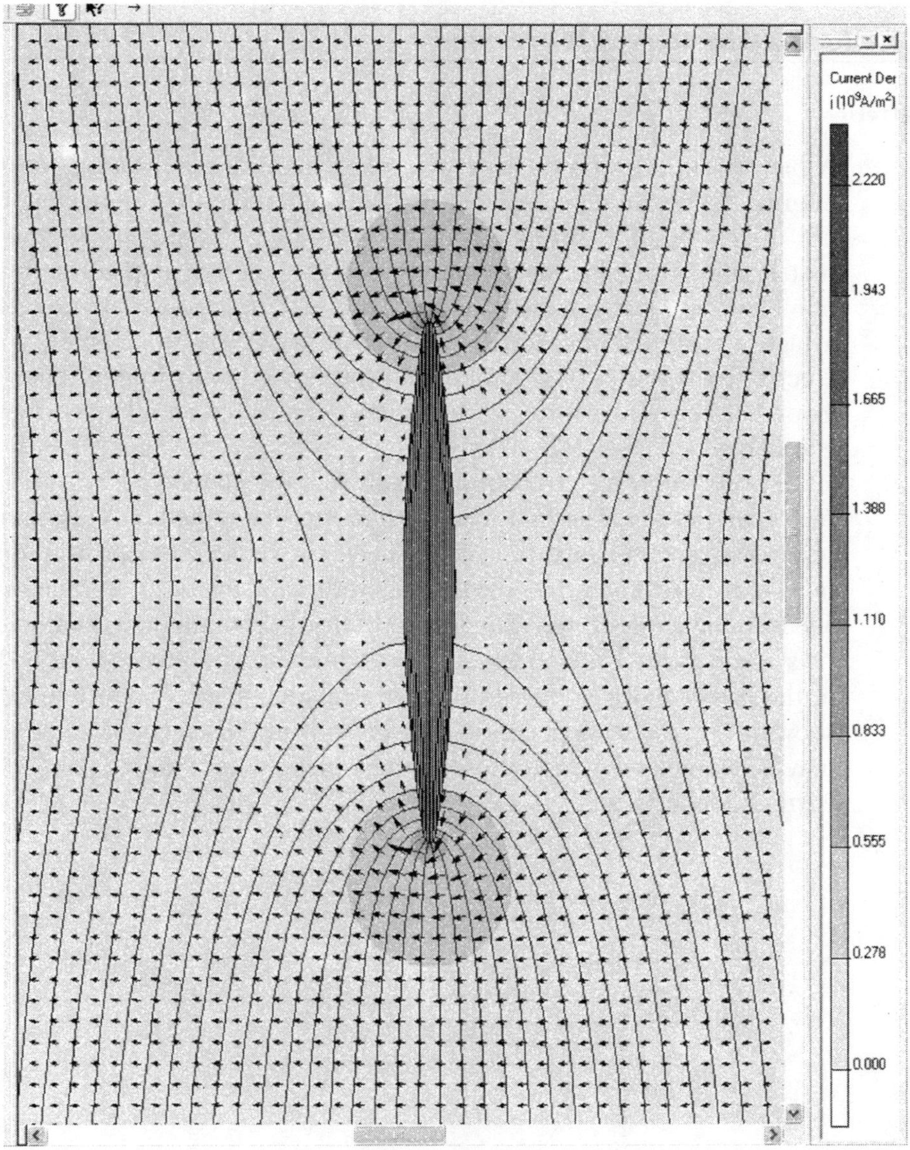

FIGURE 9.4 **DC current surrounding an ellipsoidal crack with resistivity $1.0 \cdot 10^{6}\ \Omega \cdot m$ in a metal plate with resistivity $1.0 \cdot 10^{-8}\ \Omega \cdot m$ with a potential difference of 2 volts across the 100-cm plate. Equipotentials are shown with an arrow plot of DC current with a greyscale map of the current density.**

resistivity is taken to be $1.0 \cdot 10^6\ \Omega \cdot$ m. The current is driven by a 2-volt potential difference across a 100-cm square plate. This plot shows equipotentials, an arrow plot of DC current and an intensity map of current density.

9.3 ALTERNATIVE ENERGY APPLICATIONS

Batteries

A battery supplies power by converting chemical bond energy into electrical energy. An electrochemical cell may be constructed by immersing two different metals in an electrolyte such as brine or a dilute acid. As the dissimilar metals slowly dissolve, metal ions go into the solution at different rates leaving one electrode less negative than the other. The least negative electrode is at a higher potential and is therefore taken to be positive. Table 9.1 shows the electromotive series of several metals referenced to hydrogen. The voltage of a given battery depends on its constituent metals independent of their size and geometry.

Two or more electrochemical cells may be connected in series to form a multi-cell battery. The first working battery constructed by Volta consisted of a brine-soaked cardboard disk sandwiched between silver and zinc discs. These metal-cardboard layers were stacked on top of each other to increase the voltage forming a voltaic pile. The weight of the disks limited the number that could be stacked as the brine in lower layers tended to be squeezed out by the weight of the layers above. Figure 9.5 shows a schematic of a modern dry cell battery showing the zinc casing, central carbon rod, and silver chloride paste. Batteries may be connected in series for greater voltage output and in parallel for higher current applications.

TABLE 9.1 Electromotive series of several metals.

Lithium	- 3.0 V
Aluminum	- 1.7 V
Zinc	- 0.76 V
Lead	- 0.13 V
Copper	+ 0.34 V
Silver	+ 0.80 V

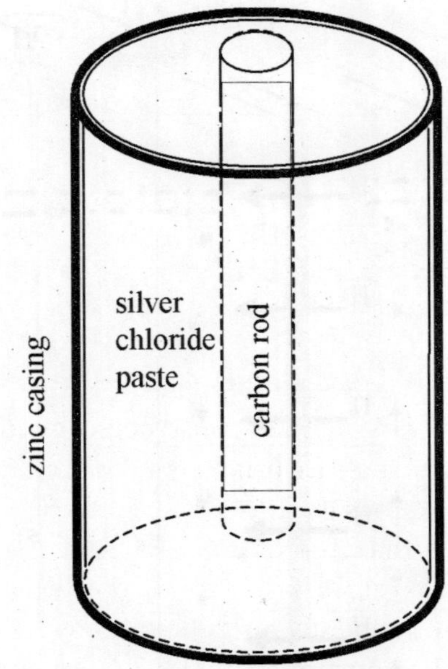

FIGURE 9.5 **Dry cell battery showing the zinc casing, central carbon rod, and silver chloride paste.**

Fuel Cells

Unlike batteries, where electrochemical reactants are stored within the cell, fuel cells are supplied with fuel sources contained outside the cell that are continuously supplied during cell operation.

Various types of fuel cells include solid-oxide, metal hydride, and hydrogen fuel cells to name a few. Proton Exchange Membrane (PEM) hydrogen fuel cells operate by combining hydrogen and oxygen to produce electricity with water and heat as byproducts. Hydrogen-oxygen PEM fuel cells are environmentally friendly and do not emit carbon dioxide or toxic pollutants. Because fuel cells have no moving parts, they operate at a very high efficiency compared to internal combustion engines.

Figure 9.6 shows a schematic of a single PEM fuel cell consisting of an anode and cathode with the proton exchange membrane located at the center. Hydrogen gas is supplied to the fuel cell anode, where a platinum coating facilitates separation of the hydrogen into protons and electrons. The PEM at the center of the fuel cell only allows the passage of protons through the membrane towards the cathode. Only protons are able to pass through the PEM while the electrons pass through an external circuit supplying current to a resistive load. Having supplied electrical power through the circuit, the electrons flow into the cathode region of the fuel cell, where another platinum catalyst facilitates the combination of oxygen, protons, and electrons to form water

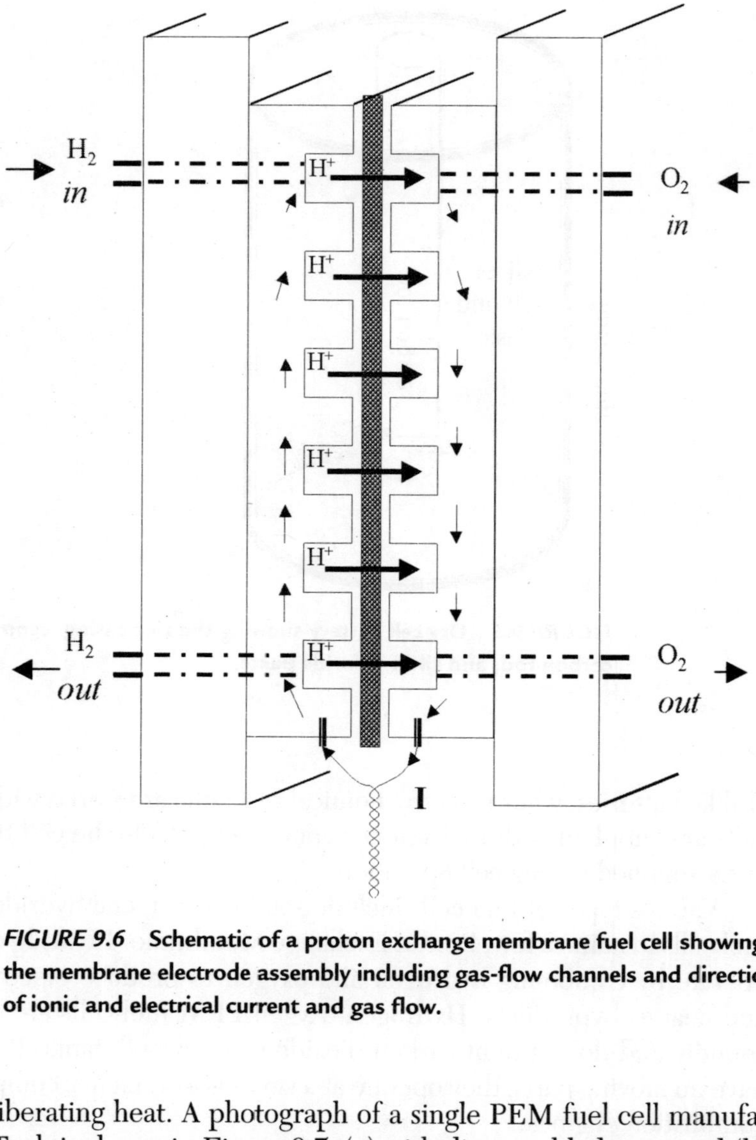

FIGURE 9.6 **Schematic of a proton exchange membrane fuel cell showing the membrane electrode assembly including gas-flow channels and directions of ionic and electrical current and gas flow.**

while liberating heat. A photograph of a single PEM fuel cell manufactured by GlobeTech is shown in Figure 9.7: (a) with disassembled parts and (b) showing gas-flow channels and gasket.

Individual fuel cells can be stacked in series with the total number of cells proportional to the stack voltage. The surface area of each cell determines the total current. Fuel cells may be arranged in parallel for higher current and in series for higher voltage. The total power generated by a battery, fuel cell, or fuel cell stack is equal to the current voltage product $P = IV$.

A fuel cell stack consists of multiple structural elements that have the same geometry. In creating a fuel cell model in QuickField one structural element may be created and then duplicated as needed. Figure 9.8 shows the duplication

FIGURE 9.7 **Photo of a proton exchange membrane fuel cell manufactured by GlobeTech showing (a) the membrane electrode assembly and (b) gas-flow channels surrounded by a Teflon gasket.**

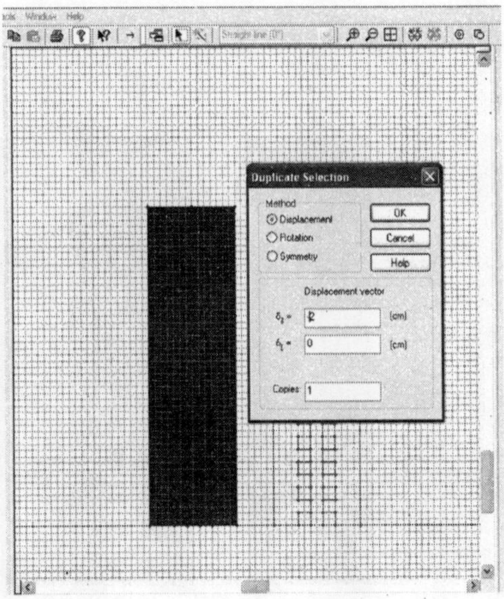

FIGURE 9.8 **Duplicating and translating periodic fuel cell structural elements. The highlighted membrane electrode assembly consists of a graphite block with the gas-flow channel surrounding the PEM. The Duplicate Selection window shows that the copied element is being translated 2.0 cm along the horizontal *z*-axis.**

and stacking of a circular fuel cell component with axial symmetry. The highlighted membrane electrode assembly consists of a graphite block with gas-flow channel surrounding the PEM. The Duplicate Selection window shows that the copied element is being translated 2.0 cm along the horizontal z-axis.

Figure 9.9 illustrates labeling and assigning material properties to fuel cell components. The graphite region surrounding the PEM and gas-flow channels is highlighted in this figure. The gas-flow channels consist of circular grooves that flank each side of the PEM.

The finite element mesh in each fuel cell stack component is shown in Figure 9.10. This figure illustrates the greater mesh density near the edges of the PEM.

A vector plot of current flow and with an intensity plot of electric potential near one membrane electrode assembly is shown in Figure 9.11. Figure 9.12

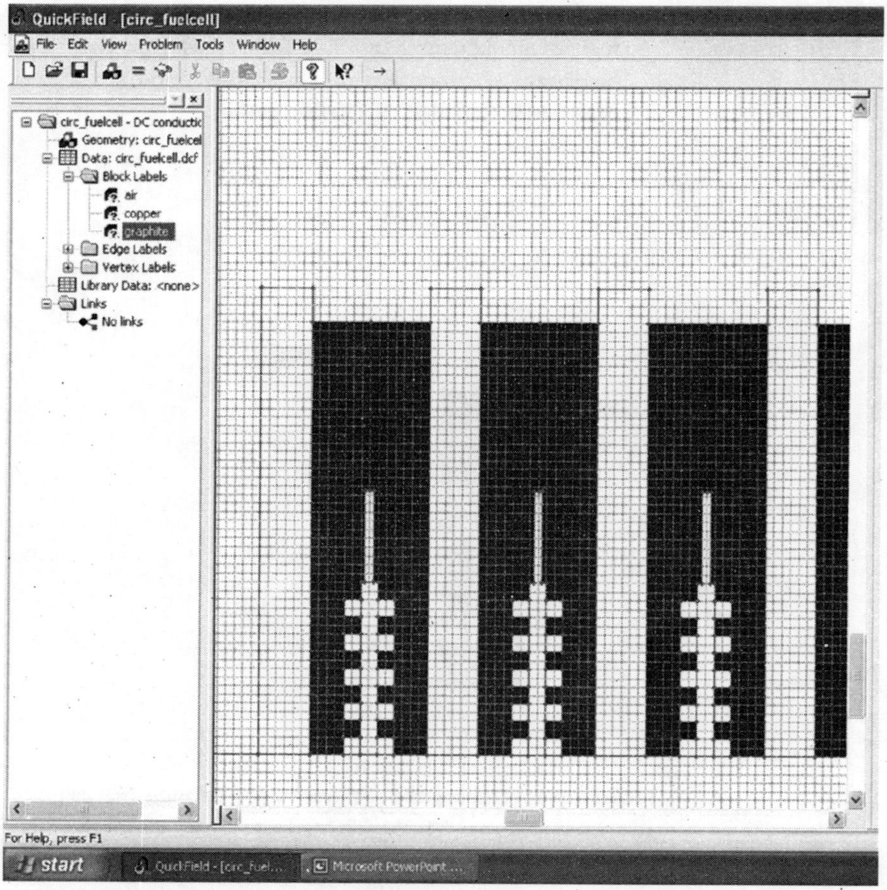

FIGURE 9.9 **Labeling and assigning material properties to fuel cell components. The graphite region surrounding the PEM and gas-flow channels is highlighted above.**

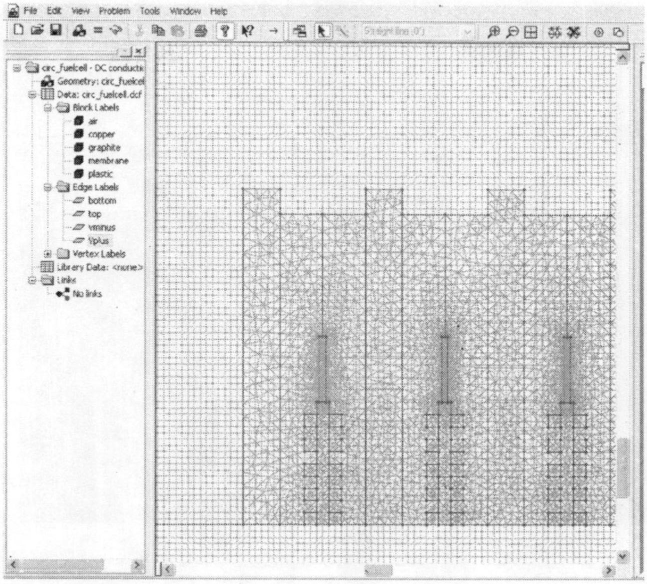

FIGURE 9.10 **Building the finite element mesh in each region of the fuel cell stack.**

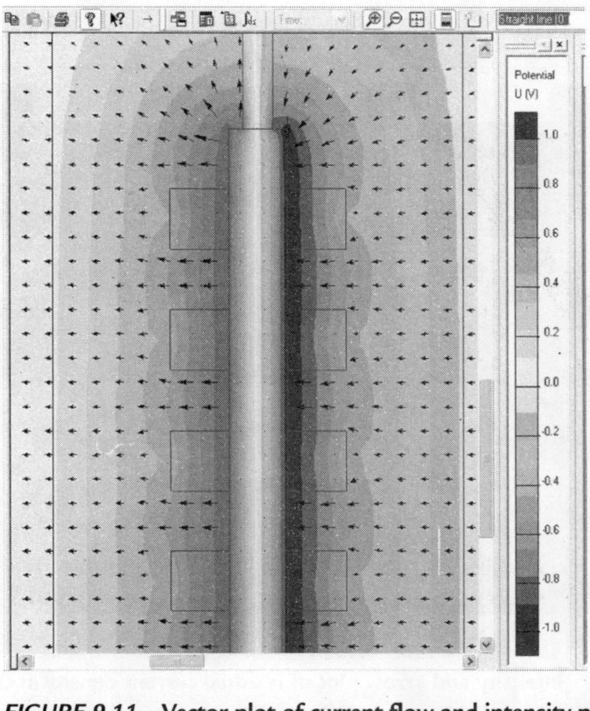

FIGURE 9.11 **Vector plot of current flow and intensity plot of electric potential near the membrane electrode assembly.**

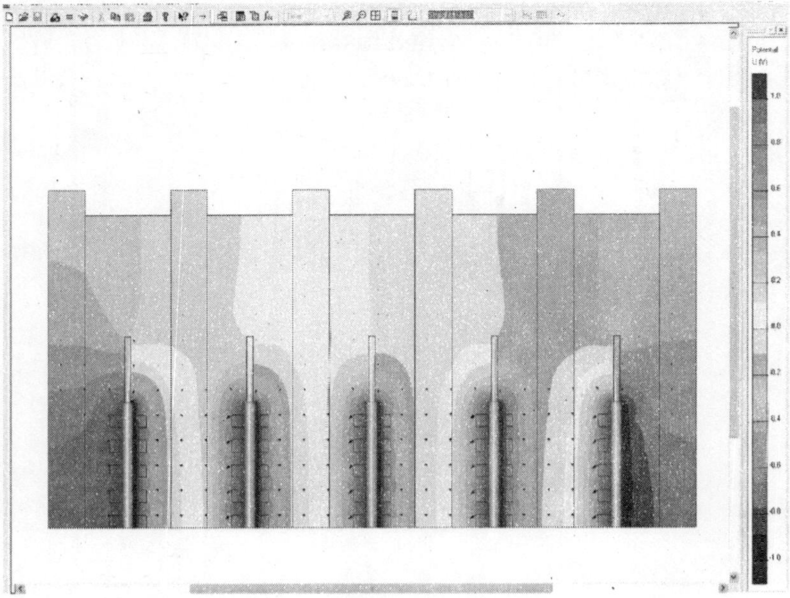

FIGURE 9.12 **Open-circuit voltages and current flow within the fuel cell stack. Zero current density is assigned to the ends of the fuel cell.**

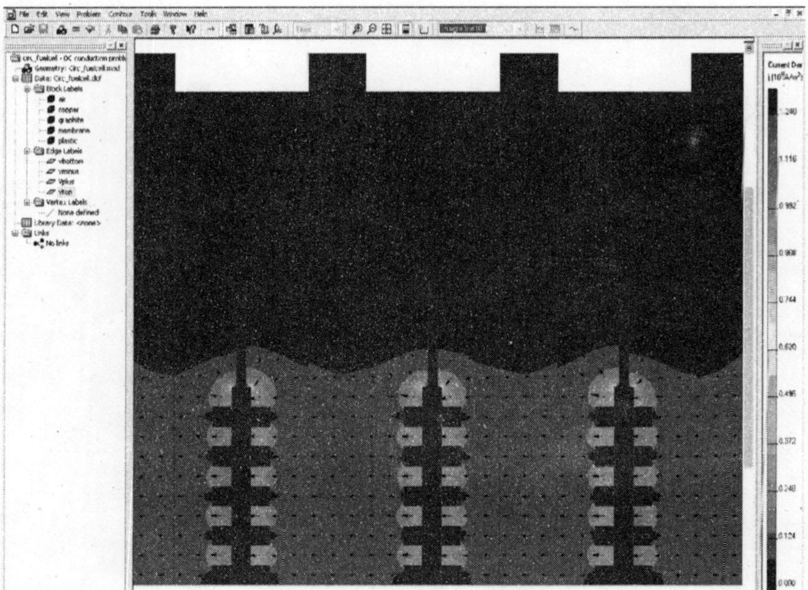

FIGURE 9.13 **Intensity and arrow plot of residual current density in the fuel cell stack without external load. Odd periodic voltage boundary conditions are applied to opposite ends of the fuel cell stack. The current density is highest near the corners of the PEM.**

shows the open-circuit potentials and residual current flow within the fuel cell stack. Zero current density is assigned to the ends of the fuel cell. When the fuel cell is not connected to a load resistance, an open-circuit voltage still exists across each membrane.

Figure 9.13 shows a grayscale arrow plot of current density in the fuel cell stack with zero resistance load. Odd periodic voltage boundary conditions are applied to opposite ends of the fuel cell stack. In this plot the current density is highest near the corners of the PEM. This higher current results from the greater potential variation near the corners of the PEM. Note that a very localized higher current density may result in a burn through the membrane resulting in catastrophic failure of the fuel cell. The temperature distribution in the fuel cell will be investigated in Chapter 11.

9.4 RESISTIVITY MEASUREMENTS

Figure 9.14 shows the current density vectors and equipotentials in a resistive plate resulting from surface electrodes held at opposite potential. The circuit for measuring the effective resistivity is shown at the top of this figure. Given the distances between the electrodes, the apparent resistivity of the plate may be measured as

$$\rho_a = \frac{2\pi V}{I}\left(\frac{1}{r_1} - \frac{1}{r_2} - \frac{1}{r_3} + \frac{1}{r_4}\right)^{-1}, \tag{9.6}$$

where the separations are r_1 between the first two electrodes r_2 between the second and fourth, and r_3 between the first and third and r_4 between the third and fourth electrodes. Equation (9.6) gives the true resistivity for a homogenous plate. This measurement technique is used for geophysical prospecting,, where the effective ground resistivity is determined by the measurement setup in Figure 9.14. For a layered inhomogeneous medium, the apparent resistivity will depend on the electrode separation. An estimation of the thickness of surface layers can then be obtained by plotting ρ_a for different electrode separations.

9.5 PRINTED CIRCUIT BOARDS

The DC current flow through circuit board elements may be calculated in x-y symmetry, where there is no current flowing in the z-direction. Figure 9.15 shows an intensity map of the current density in a printed circuit board with twelve parallel conducting rails.

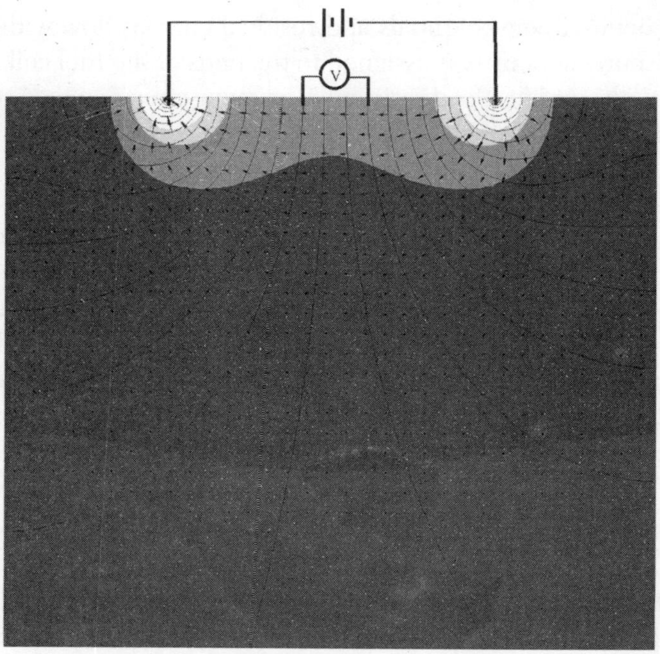

FIGURE 9.14 Density map of current density in the printed circuit board.

FIGURE 9.15 Current density in a resistive plate resulting from surface electrodes held at opposite potential. The circuit for measuring the effective resistivity is shown at the top of this figure.

A voltage difference is applied to single nodes at opposite ends of the board. Figure 9.16 shows a contour plot of the current density along a line crossing through the center of the board and perpendicular to the rails. Current is not equal in each leg but decreases by roughly a factor of two between the center of the board and its edge. This is because of the longer current path and hence greater resistance of the outer rails as measured from the current source points.

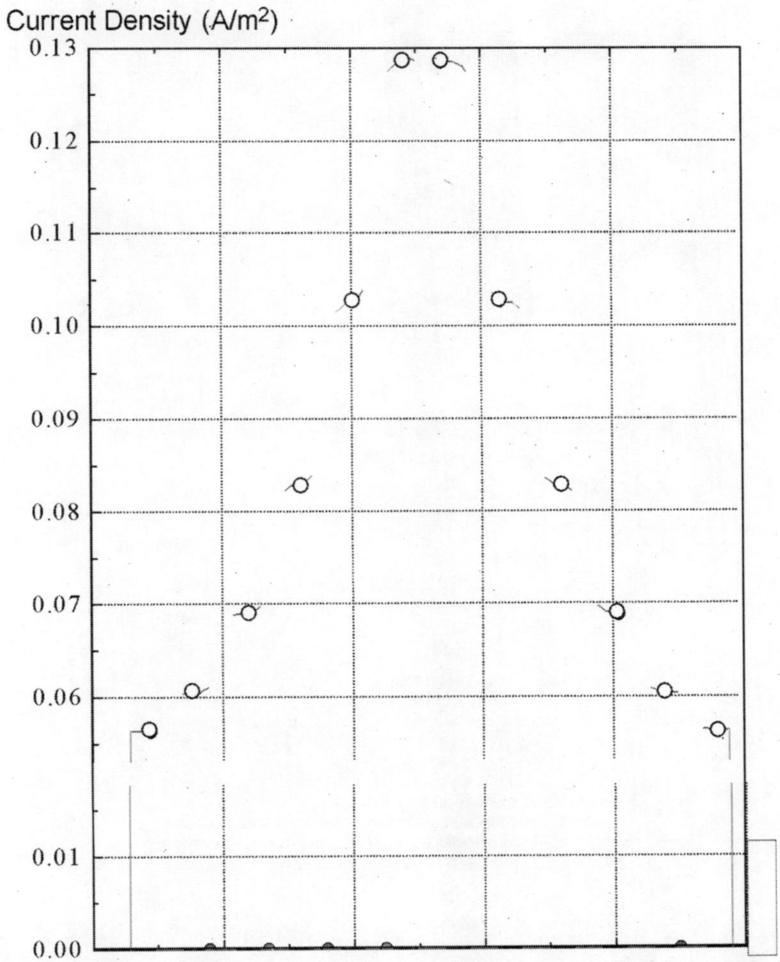

FIGURE 9.16 **Current density plotted across the center of the printed circuit board. Current is not equal in each leg but varies by a factor of two from the center of the board to its edges.**

REFERENCES

QuickField User's Guide: Tera Analysis Ltd. (2007). *Chapter 2: Introductory Guide; Chapter 10. Theoretical Description.*

William R. Smythe. (1989). *Static and Dynamic Electricity*, Taylor & Francis. *Chapter VI: Electric Current.*

Rhett Herman. (2001). "An introduction to electrical resistivity in geophysics," *American Journal of Physics*, **69** (9) pp. 943–952.

EXERCISES

9.1 Calculate the current flow in a plate with a decreasing cross-section as shown in Figure E-9.1.

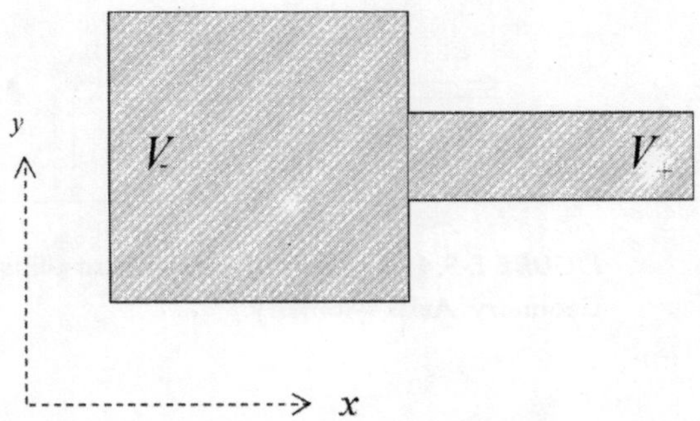

FIGURE E-9.1 **Current carrying plate with decreasing cross-section. Geometry: *x-y* symmetry.**

9.2 Plot the current density inside a dry cell battery with internal resistances as shown in Figure E-9.2.

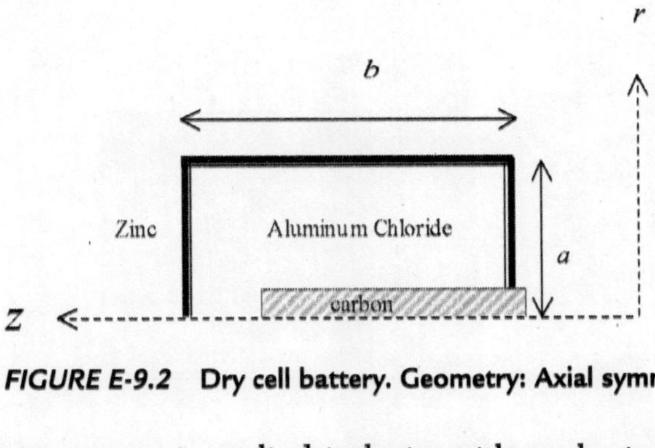

FIGURE E-9.2 **Dry cell battery. Geometry: Axial symmetry.**

9.3 Model a current carrying cylindrical wire with a spherical void as shown in Figure E-9.3. Display the current density as a vector field arrow plot.

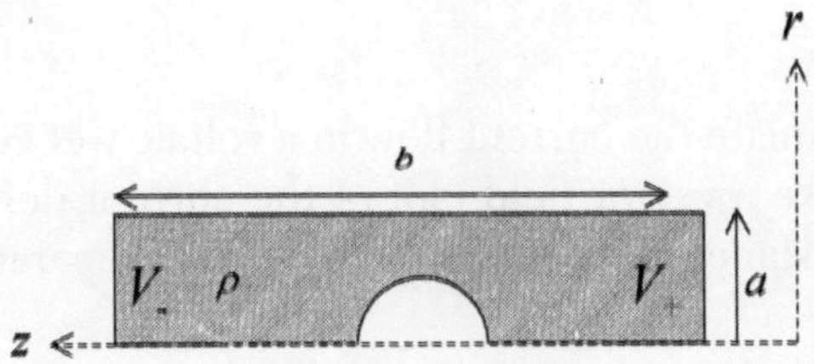

FIGURE E-9.3 **Cylindrical wire with a spherical void. Geometry: Axial symmetry.**

9.4 Model a current carrying wire with an ellipsoidal void as shown in Figure E-9.4. Display the current density as a vector field arrow plot.

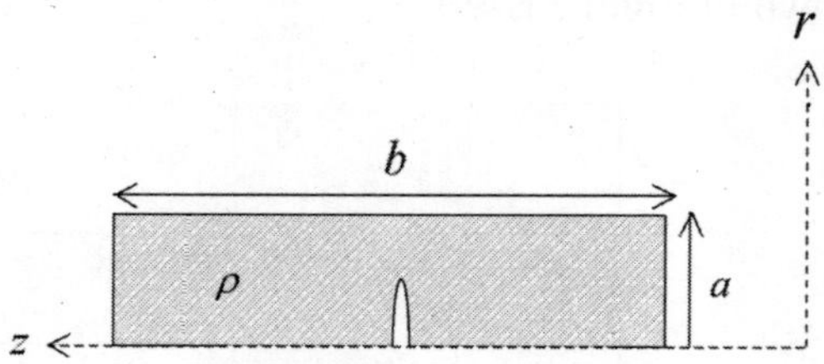

FIGURE E-9.4 **Cylindrical wire with an ellipsoidal void. Geometry: Axial symmetry.**

9.5 Model the current flow through a layered conducting medium applying a potential difference to the outer two electrodes as shown in Figure E-9.5. Plot the equipotentials and the current density as a vector plot. Calculate the effective resistance as a function of inner electrode separation using Label Mover for several values of surface layer thickness.

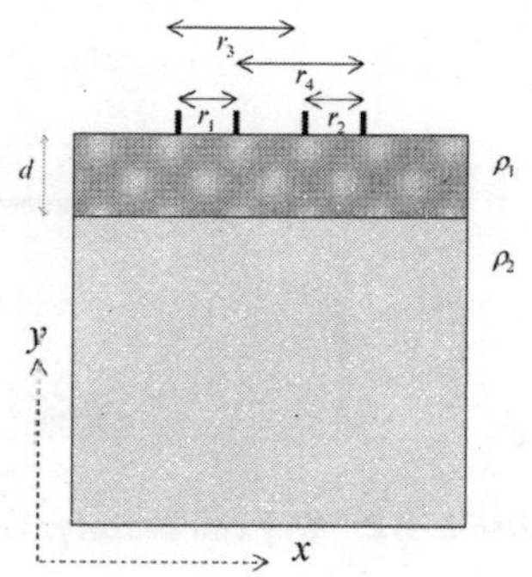

FIGURE E-9.5 **Layered conducting medium. Geometry: *x-y* symmetry.**

9.6 Simulate the current flow in a voltaic wet cell as shown in Figure E-9.6. Make a vector field plot of the current density. Calculate the effective resistance as a function of electrode separation.

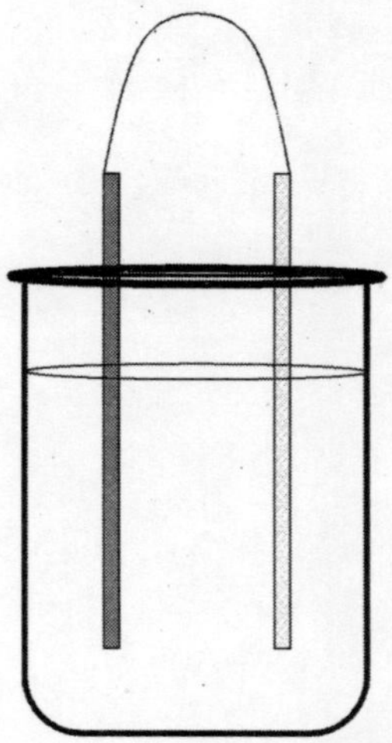

FIGURE E-9.6 **Wet electrochemical cell. Geometry: *x-y* symmetry.**

9.7 Simulate the current flow through a layered resistive sphere with opposite potentials applied at the poles of the sphere. Make a radial plot of the current density from the center of the sphere to its equator.

9.8 Model two cylindrical electrodes held at opposite potential in bicylindrical coordinates. The electrodes are embedded in an infinite resistive medium with resistivity. Solve Laplace's equation for the scalar potential using separation of variables in bicylindrical coordinates. Calculate the current density near the electrodes.

Chapter 10 AC Current Flow

In This Chapter

- AC Conduction Theory
- AC Conduction Analysis in QuickField
- Application to Biophysical Measurements

10.1 AC Conduction Theory

The continuity equation equates the divergence of the current density to the time variation of volume charge density $\nabla \cdot \mathbf{J} + \partial\rho/\partial t = 0$. For fields that vary harmonically as $\exp(i\omega t)$, the continuity equation can be written

$$\nabla \cdot \mathbf{J} = -i\omega\rho. \tag{10.1}$$

Substituting Ohm's law $\mathbf{J} = \sigma\mathbf{E}$ and $\nabla \cdot \varepsilon\mathbf{E} = \rho$ into equation (10.1) gives

$$\nabla \cdot \sigma\mathbf{E} = -i\omega\nabla \cdot \varepsilon\mathbf{E}, \tag{10.2}$$

where $\mathbf{E} = -\nabla U$ so that

$$\nabla \cdot \left(\varepsilon - \frac{i\sigma}{\omega}\right)\nabla U = 0. \tag{10.3}$$

This equation may be used to solve for the AC electric potential, field, and current flow in regions with varying dielectric permittivity and conductivity.

10.2 AC CONDUCTION ANALYSIS IN QUICKFIELD

AC conduction analysis is performed in QuickField solving equation (10.3) that may be written in x-y symmetry as

$$\frac{\partial}{\partial x}\left[\left(\varepsilon_x - \frac{i\sigma_x}{\omega}\right)\frac{\partial U}{\partial x}\right] + \frac{\partial}{\partial y}\left[\left(\varepsilon_y - \frac{i\sigma_y}{\omega}\right)\frac{\partial U}{\partial y}\right] = 0, \tag{10.4}$$

where $(\varepsilon_x, \varepsilon_y)$ and (σ_x, σ_y) are the permittivity and conductivity components in the x- and y-directions, respectively. In axial symmetry we have that.

$$\frac{1}{r}\frac{\partial}{\partial r}\left[r\left(\varepsilon_r - \frac{i\sigma_r}{\omega}\right)\frac{\partial U}{\partial r}\right] + \frac{\partial}{\partial z}\left[\left(\varepsilon_z - \frac{i\sigma_z}{\omega}\right)\frac{\partial U}{\partial z}\right] = 0, \tag{10.5}$$

where $(\varepsilon_r, \varepsilon_z)$ and (σ_r, σ_z) are the permittivity and conductivity components in the r- and z-directions, respectively.

Once the potential is solved for, the electric field is calculated from $\mathbf{E} = -\nabla U$ as in Electrostatics. The active and reactive current density components may then be obtained from $\mathbf{J}_{\text{active}} = \sigma\mathbf{E}$ and $\mathbf{J}_{\text{reactive}} = i\omega\varepsilon\mathbf{E}$. The apparent current density is the sum of active and reactive current densities $\mathbf{J}_{\text{apparent}} = \mathbf{J}_{\text{active}} + \mathbf{J}_{\text{reactive}}$. The local time average and peak power loss densities are evaluated from $Q = \mathbf{J} \cdot \mathbf{E}$ with apparent, active, or reactive current densities corresponding to Q_{active}, Q_{reactive}, and Q_{apparent}.

Integral Quantities

The total active, reactive, or apparent current through a given surface is given by

$$I = \iint_{\text{surf}} \mathbf{J} \cdot \mathbf{n} da. \tag{10.6}$$

The active, reactive, or apparent power dissipated over a volume V is

$$P = \iiint_{\text{vol}} \mathbf{E} \cdot \mathbf{J} dV \tag{10.7}$$

evaluated as either time average and peak values. The time average and peak electric force and torques are calculated as in Electrostatics.

Figure 10.1 shows the window for editing problem properties for AC current flow problems. Once the AC Current type is selected, the frequency may then be specified.

Material Properties

Block labels are associated with dielectric permittivity components and electrical conductivity (in S/m) that may be isotropic or anisotropic. Figure 10.2 shows

ON THE CD: Color images for many of the figures may be found on the CD-ROM.

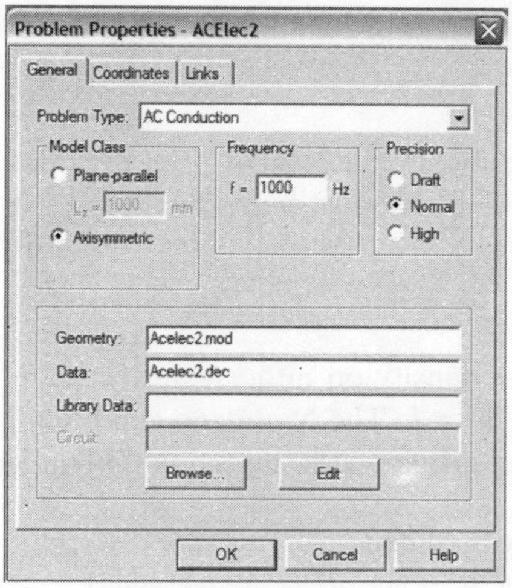

FIGURE 10.1 **Window for editing problem properties data for AC current flow problems: To open the problem properties window click on problem properties under the problem tab on the top menu bar. Choices include Model Class (Plane-parallel or Axisymmetric), Frequency, and Precision.**

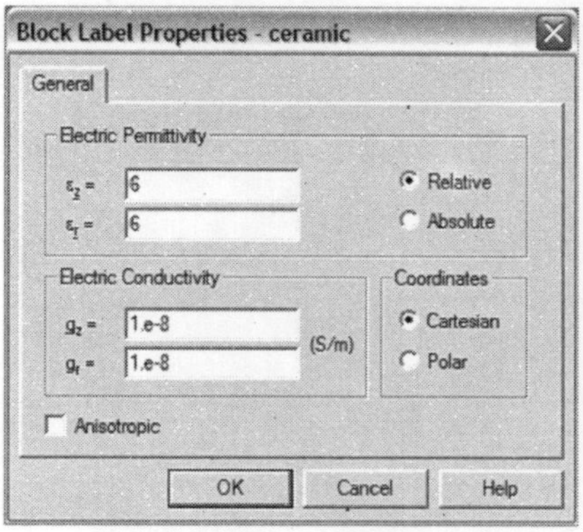

FIGURE 10.2 **Window for editing block label data for AC current flow problems: To open the block label window click on the corresponding block label in the tree diagram on the left of the screen. Enter electric permittivity and electric conductivity values. Check the Anisotropic box to separately specify *z*- and *r*-component values.**

the window for editing block label data for AC current flow problems including permittivity and conductivity components that may be isotropic or anisotropic.

Boundary Conditions

Figure 10.3 shows the window for editing edge label data for AC current flow problems. Boundary conditions are the same as in DC Magnetics. Choices include Voltage, Normal Current Density, Floating Conductor (Equal Voltage), and Even and Odd periodic boundary conditions.

Dirichlet boundary conditions

Assign specific electric potential U values on exterior boundaries or edges within the solution region. This condition must be applied to at least one point in the solution region.

Neumann boundary conditions

Specify the normal current density on either exterior boundaries as $J_n = J$ or interior boundaries as $J_n^+ - J_n^- = J$. The Neumann condition $J_n = 0$ is the default condition on all outward boundaries with unspecified boundary conditions.

Constant potential boundary conditions

Are used for isolated conductors whose surfaces are at constant but unknown potential values. Edges assigned a constant potential boundary condition should not contain points assigned a Dirichlet fixed potential.

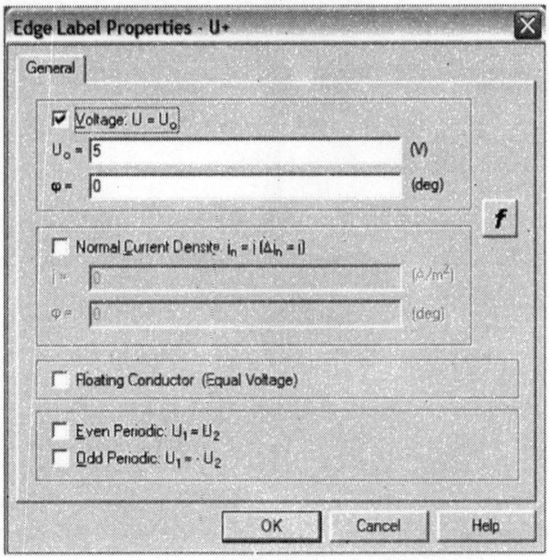

FIGURE 10.3 Window for editing edge label data for AC current flow problems: To open the edge label window click on the corresponding edge label in the tree diagram on the left of the screen.

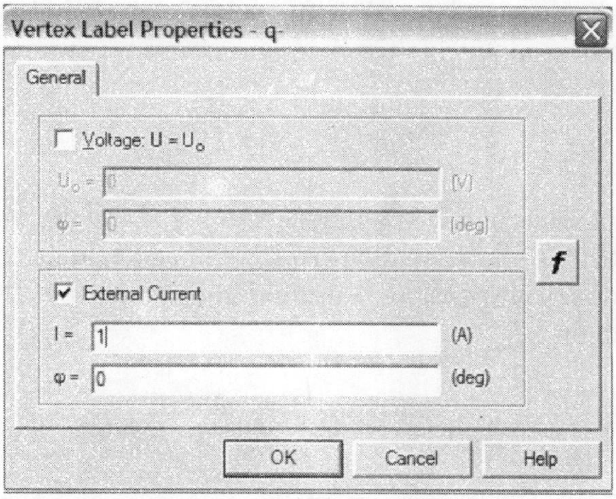

FIGURE 10.4 Window for editing vertex label data for AC current flow problems: To open the vertex label window click on the corresponding vertex label in the tree diagram on the left of the screen.

Periodic boundary conditions

Can be specified on opposite edges of the model. The potential is given by $U_1 = U_2$ on even periodic boundaries while $U_1 = -U_2$ on odd periodic boundaries.

Vertex Conditions

Figure 10.4 shows the window for editing vertex label data for AC current flow problems. Local potential or current values may be assigned to individual vertices. The potential and charge values may be specified by a numerical value or as a function of coordinates using the equation editor.

10.3 APPLICATION TO BIOPHYSICAL MEASUREMENTS

Measurement of Effective Conductivity and Permittivity

Biological tissue is composed of mostly water, is weekly conducting, and behaves as a lossy dielectric in AC electromagnetic fields. In general, AC currents flowing in tissue under electrical stimulation will consist of a sum of conduction and displacement current densities

$$J = \sigma E + i\omega\varepsilon E. \tag{10.8}$$

The active and reactive components of the current density J are given by

$$J_{\text{active}} = \sigma E \quad J_{\text{reactive}} = \omega\varepsilon E \tag{10.9}$$

so that the conductivity and permittivity can be determined from experimentally measured current components

$$\sigma = \frac{J_{\text{active}}}{E} \quad \varepsilon = \frac{J_{\text{reactive}}}{E\omega}. \tag{10.10}$$

Biological tissue is highly inhomogeneous with local variations in conductivity and permittivity so that equation (10.10) will give effective values for a given tissue sample or cell suspension. A useful figure of merit is the loss tangent (or loss factor) defined as

$$\tan\delta = \frac{J_{\text{active}}}{J_{\text{reactive}}} = \frac{\sigma}{\omega\varepsilon}. \tag{10.11}$$

The loss tangent is a measure of the rms or peak energy lost per cycle divided by the energy stored per cycle. The loss angle

$$\delta = \tan^{-1}\frac{\sigma}{\omega\varepsilon} \tag{10.12}$$

will be zero for a simple capacitor with no resistive losses. Given a uniform current density $J = I/A$, the total current may be written in terms of Ohm's law $V = IZ$. The current may be written as a product of admittance (or inverse impedance) times the voltage $I = Y^*V$ or from equation (10.8)

$$I = \left(\frac{\sigma A}{d} + i\omega\frac{\varepsilon A}{d}\right)V, \tag{10.13}$$

where $E = V/d$ and Y^* is the complex admittance

$$Y^* = G + i\omega C. \tag{10.14}$$

The conductance G and the capacitance C are proportional to the conductivity and permittivity, respectively. For inhomogeneous tissue regions, C and G will each be dependent on the conductivity and permittivity distributions in a complicated way. Equation (10.8) is sometimes written as $J = \sigma^*E$, where the complex conductivity is $\sigma^* = \sigma' + i\sigma''$ so that $\sigma = \sigma' = \text{Re}\,\sigma^*$ and $\varepsilon = \text{Im}\,\sigma^*/\omega$. Other conventions exist involving complex permittivity or complex resistivity. As well, there are slight notational variations in the literature. Comparison with equation (10.8) can help sort out the meaning of a given notational convention.

Dielectric Spectroscopy of Cells in Suspension and Tissue

The electrical response of living cells to AC electric fields exhibits similar behavior for cells in suspension and tissue. In general, the effective dielectric constant

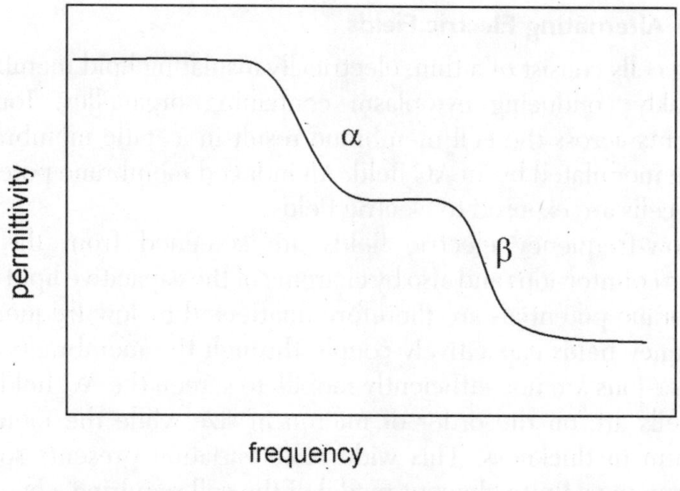

FIGURE 10.5 **Sketch of the frequency-dependent dielectric permittivity of living cells in suspension or tissue.**

of biological material decreases while the conductivity increases with increasing frequency.

Three rather broad frequency ranges have been identified, where different processes affect the dielectric permittivity and conductivity in biological material. These frequency ranges correspond to the α-, β-, and γ-dispersions.

α-dispersion

(mHz-kHz) Results mostly from the motion of charges, or counter ions next to the cell's plasma membrane in suspension, where the relative dielectric permittivity can exceed 10^7. It will not be possible to simulate the motion counter ions in QuickField directly.

The mechanism for α-dispersion is different in tissue than in cell suspensions. The AC response of gap junctions connecting cells contributes to the enormous low-frequency dielectric permittivity of tissue. The gap junction may be simulated in QuickField as a weakly conducting junction between connected cells.

β-dispersion

(0.1–100 MHz) Results from the passive cell capacitance, Maxwell-Wagner effects with organelle membrane contributions. The β-dispersion may be simulated in QuickField.

γ-dispersion

Dispersion (0.1–100 GHz) range results mainly from relaxations of molecular dipole moments such as water, salts and proteins. QuickField models of cells in eternal fields will not include the γ-dispersion.

Living Cells in Alternating Electric Fields

Living cells consist of a thin, electrically insulating lipid membrane surrounding a weakly conducing cytoplasm containing organelles. Ionic concentration gradients across the cell membrane result in a static membrane potential that can be modulated by an AC field. An induced membrane potential is developed when cells are exposed to electric fields.

Low-frequency electric fields are screened from the cell interior by mobile counter-ions and also by charging of the capacitive lipid bilayer. Organelle membrane potentials are therefore unaffected by low-frequency fields. Higher frequency fields capacitively couple through the membrane while the massive counter-ions are not sufficiently mobile to screen the AC field.

Cells are on the order of micron in size while the membrane is roughly 5–10 nm in thickness. This wide scale variation presents some challenges in constructing a finite element model of the cell requiring a hyper-fine mesh with thousands of nodes near the membrane. The easi est way to construct a membrane with such a small thickness compared to the spatial extent of the model is to use the 'Insert Shape' feature on the Model Toolbar to create two circles differing in radius by 5–10 nm. Unwanted bits can then be cropped forming the desired arcs.

Exaggerated membrane thickness will be required using nodefe-limited Student Versions of QuickField, where meaningful qualitative results may be obtained for visualizing the field and current distribution near the cell and plotting the frequency-dependent induced transmembrane potential.

Figure 10.6 (a) shows the finite element mesh near a single membrane consisting of about $0.5 \cdot 10^6$ nodes. In this figure one quarter of the cell (radius 0.25 µm, thickness 5 nm) is modeled making use of symmetry about the equatorial plane. For a z-directed field the equatorial plane is an equipotential.

A much finer mesh is required near the membrane because of the 5-nm width compared to the near micron size of the cell. Values of relative permeability and conductivity for the intracellular space are ($\varepsilon_r = 87.7$, $\sigma = 0.95$ S/m), the plasma membrane ($\varepsilon_r = 16.8$, $\sigma = 5.0 \cdot 10^8$ S/m), and extracellular region ($\varepsilon_r = 80$, $\sigma = 2.0$ S/m). Figure 10.7 shows the equipotentials and current flow vectors resulting from a (a) 50 Hz and (b) 1.0 MHz excitations. The current density is nearly zero inside the cell at low frequencies. At high frequencies, the AC current readily penetrates the cell. It turns out that the current density is uniform inside the cell at both high and low frequencies. Figure 10.8 shows the tabulated total active and reactive current components recorded while sweeping the excitation frequency from 50 Hz to 50 MHz using Label Mover. These values represent the total current flowing through the equatorial plane, or the left-hand edge of the model.

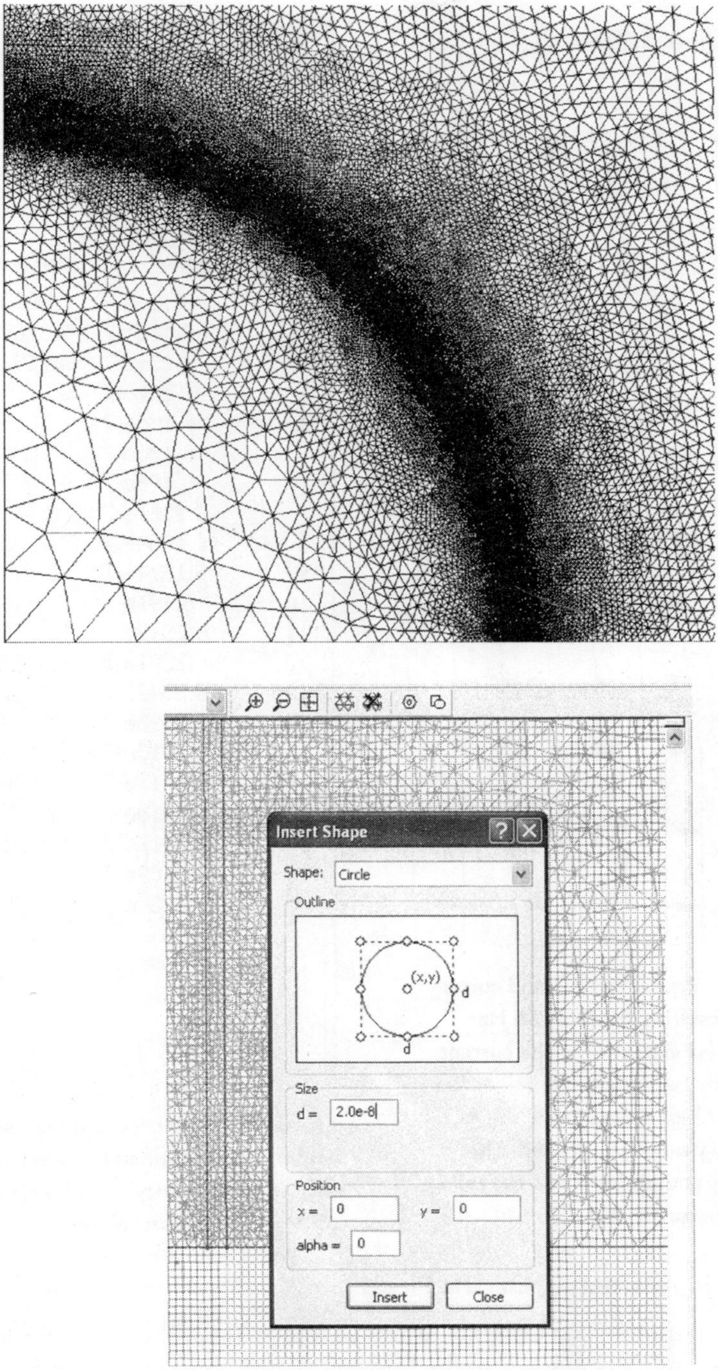

FIGURE 10.6 **Finite element mesh of a spherical cell showing the extremely fine mesh near biological membrane. Because of symmetry only one quarter of organelle is simulated.**

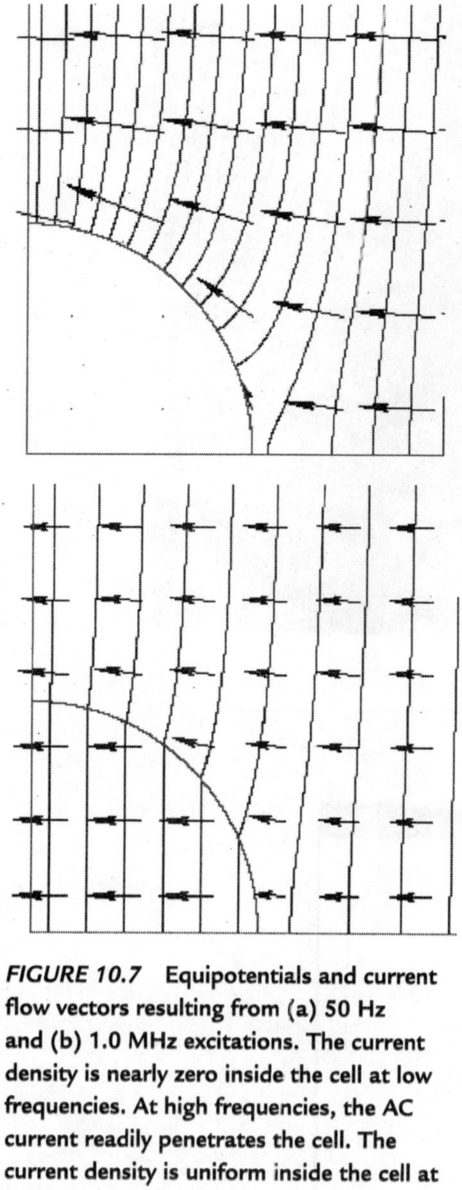

FIGURE 10.7 **Equipotentials and current flow vectors resulting from (a) 50 Hz and (b) 1.0 MHz excitations. The current density is nearly zero inside the cell at low frequencies. At high frequencies, the AC current readily penetrates the cell. The current density is uniform inside the cell at high and low frequencies.**

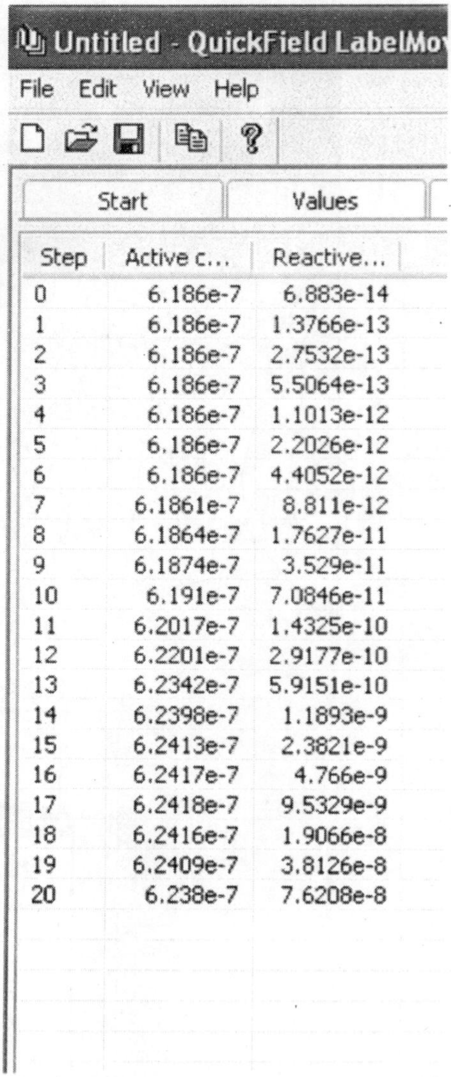

Step	Active c...	Reactive...
0	6.186e-7	6.883e-14
1	6.186e-7	1.3766e-13
2	6.186e-7	2.7532e-13
3	6.186e-7	5.5064e-13
4	6.186e-7	1.1013e-12
5	6.186e-7	2.2026e-12
6	6.186e-7	4.4052e-12
7	6.1861e-7	8.811e-12
8	6.1864e-7	1.7627e-11
9	6.1874e-7	3.529e-11
10	6.191e-7	7.0846e-11
11	6.2017e-7	1.4325e-10
12	6.2201e-7	2.9177e-10
13	6.2342e-7	5.9151e-10
14	6.2398e-7	1.1893e-9
15	6.2413e-7	2.3821e-9
16	6.2417e-7	4.766e-9
17	6.2418e-7	9.5329e-9
18	6.2416e-7	1.9066e-8
19	6.2409e-7	3.8126e-8
20	6.238e-7	7.6208e-8

FIGURE 10.8 **Active and reactive current components obtained by sweeping the excitation frequency from 50 Hz to 50 MHz using Label Mover.**

Cellular Organelles in Alternating Electric Fields

The outer cell membrane effectively acts as a high-pass filter so that membrane potentials are only developed across internal membranes at higher frequencies. Isolated organelles extracted from cells may be electrically stimulated, or micro-electrodes may be inserted inside the cell to electrically excite organelles.

Spherical Mitochondrion Model

Mitochondria are two-membrane organelles consisting of an outer membrane and a convoluted inner membrane separating the central matrix, inner membrane space, and the outside intracellular region. The simplest mitochondrion model consists of two nested dielectric insulating spheres in close proximity. Figure 10.9 shows the equipotentials and RMS current density resulting from a two-shelled spherical mitochondrion in an external field.

At low frequencies the current density is negligible inside the membranes. Conductivity and permittivity in the medium outside the mitochondrion ($\varepsilon_r = 40$, $\sigma = 5.0\,\text{S/m}$), the outer and inner mitochondrial membranes ($\varepsilon_r = 7$, $\sigma = 2.0\,\text{S/m}$), the inner membrane space ($\varepsilon_r = 5.0$, $\sigma = 0.001\,\text{S/m}$), and the matrix ($\varepsilon_r = 54$, $\sigma = 0.001\,\text{S/m}$). The mitochondrion model has an outer radius 0.25 μm; each membrane and the inner membrane space is 5 nm in thickness.

FIGURE 10.9 Equipotentials and RMS current density resulting from a two-shelled spherical mitochondrion in an external field.

Mitochondrion with Complex Geometry

Figure 10.10 shows a more complex finite element mitochondrion modeled in axial symmetry. This model is about ten times the size of an actual mitochondrion. The outer and inner mitochondrial membranes are highlighted in this figure.

Conductivity and permittivity in the cellular medium outside the mitochondria is set equal to ($\varepsilon_r = 40$, $\sigma = 5.0\,\text{S/m}$), the membranes ($\varepsilon_r = 7$, $\sigma = 2.0\,\text{S/m}$), and the inner membrane space ($\varepsilon_r = 5.0$, $\sigma = 0.001\,\text{S/m}$).

The right corner section of the mitochondrion with meshed intracellular and inner mitochondrial membrane regions is illustrated in Figure 10.11. The rounded corners are formed using the 90-degree arc element to connect the nodes on each side.

Equipotentials and electric field magnitude are plotted in a section of the mitochondrion at (a) 500 kHz and (b) 5 MHz in Figure 10.12. A zoomed in region near the outer mitochondrial membrane in Figure 10.13 plots the active current, equipotentials and electric field strength at 500 kHz.

Mitochondrion with Periodic Structure

Much time can be saved in building models with periodic elements by copying and pasting repetitive sections instead of constructing each element individually.

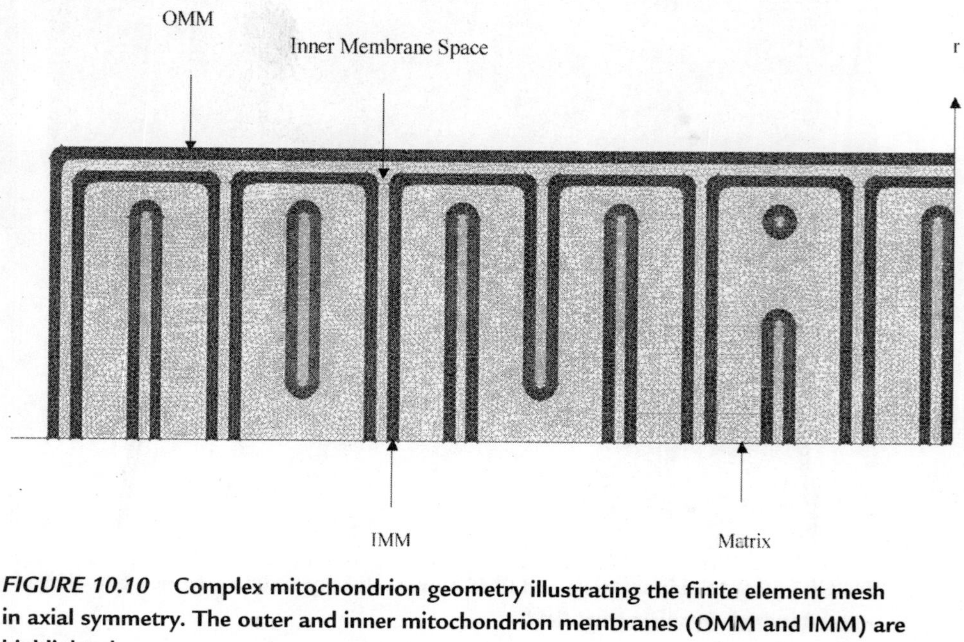

FIGURE 10.10 **Complex mitochondrion geometry illustrating the finite element mesh in axial symmetry. The outer and inner mitochondrion membranes (OMM and IMM) are highlighted.**

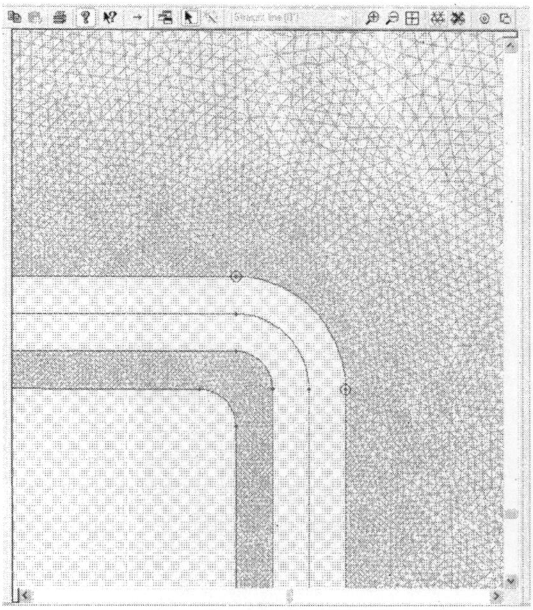

FIGURE 10.11 **Corner section of mitochondrion with partial FEM mesh. So far the intracellular and IMM regions are meshed. The corners are formed using 90-degree arc elements.**

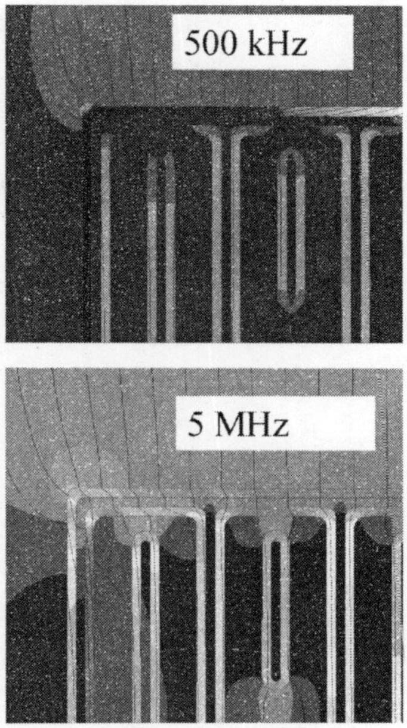

FIGURE 10.12 **Equipotentials and electric field magnitude for AC current flow in the region of a mitochondrion at (a) 500 kHz and (b) 5 MHz.**

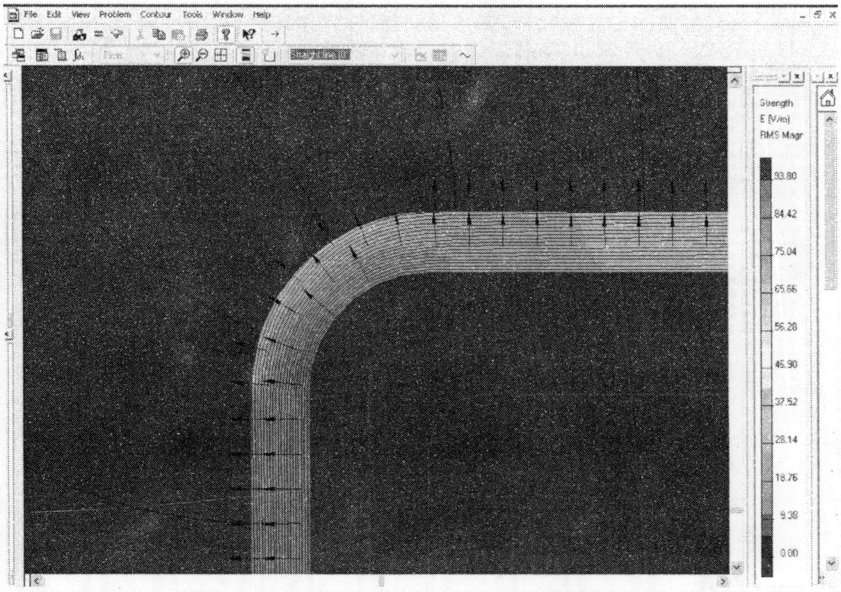

FIGURE 10.13 **Close up of outer mitochondrial membrane showing active current, equipotentials, and electric field strength at 500 kHz.**

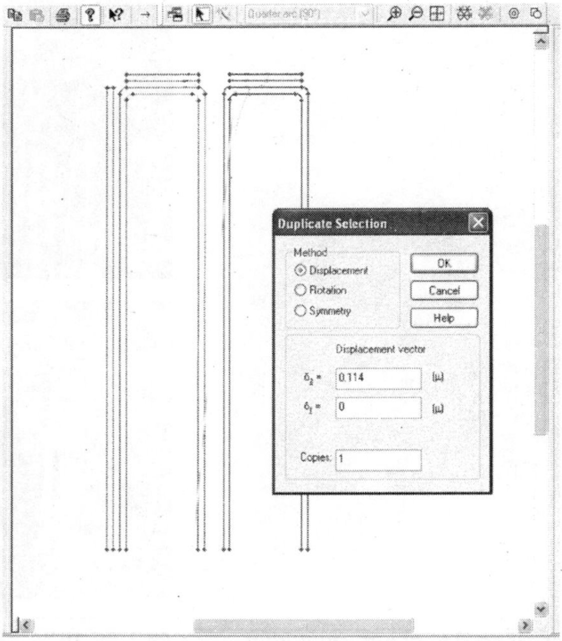

FIGURE 10.14 **Duplicating and translating periodic structures. The duplicated element is translated 0.114 μm to the right along the *z*-axis.**

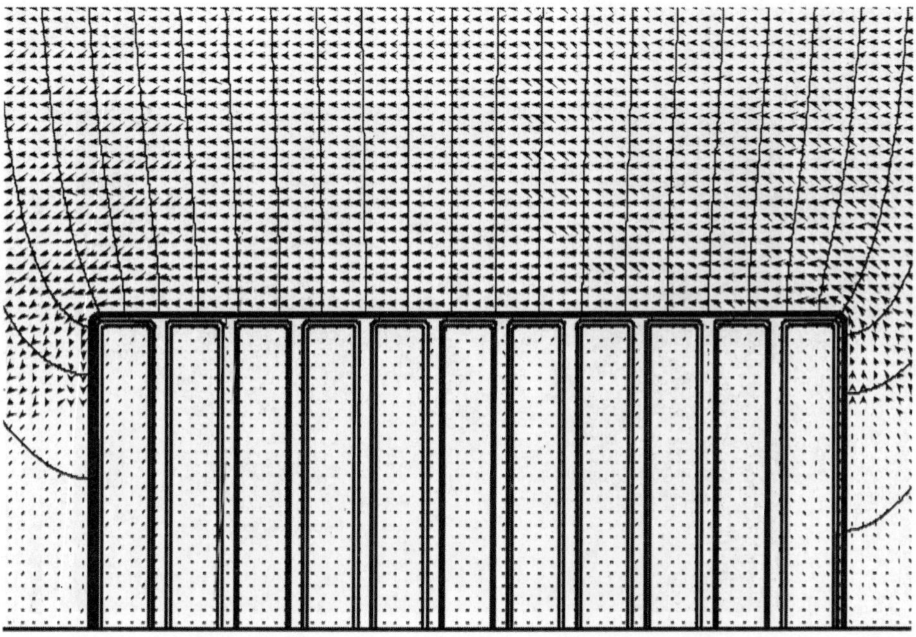

FIGURE 10.15 **Mitochondrion in a 100 kHz external field showing the vector plot of active current density and equipotentials.**

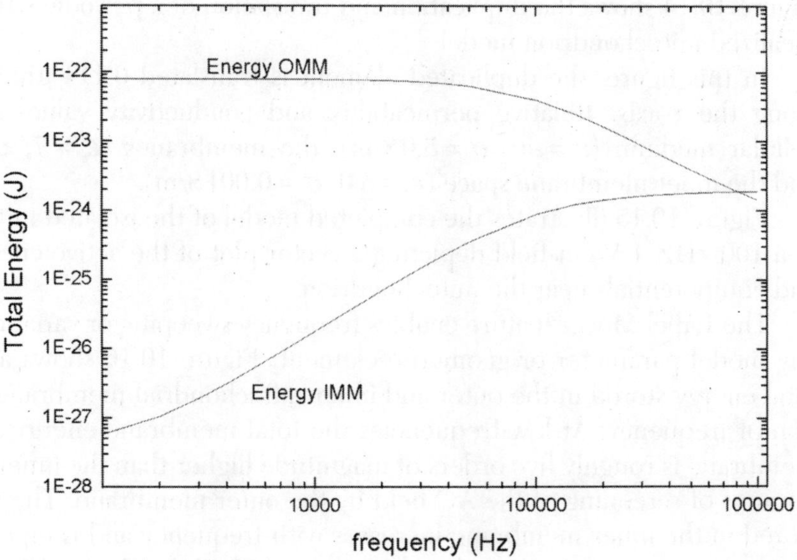

FIGURE 10.16 **Total electrical energy stored in the outer and inner mitochondrial membranes as a function of frequency of a 1 V/cm excitation field.**

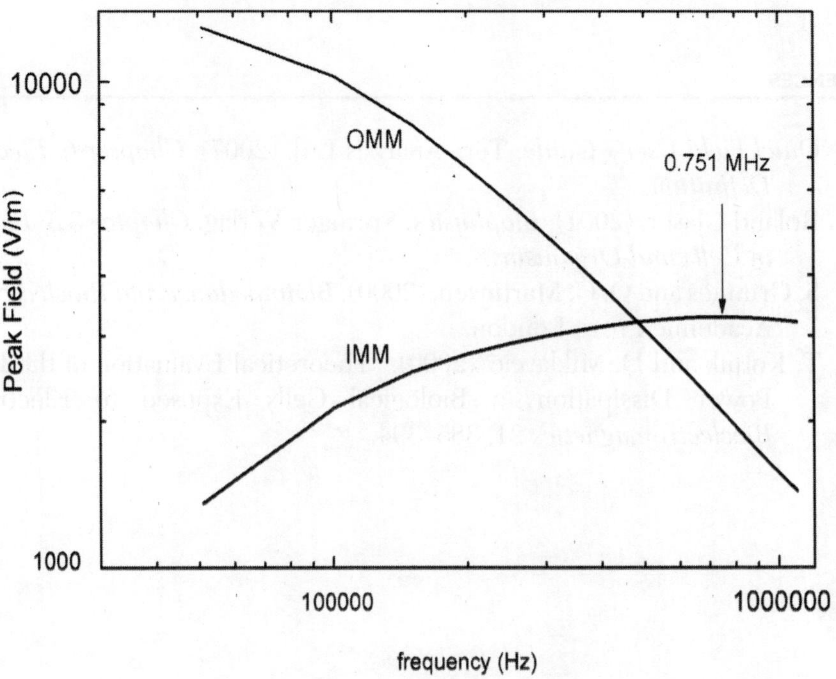

FIGURE 10.17 **Peak electric field in the outer and inner mitochondrial membrane as a function of frequency of a 1 V/cm excitation field.**

Figure 10.14 shows the duplication and translation of a periodic structure in an idealized mitochondrion model.

In this figure, the duplicated element is translated 0.114 μm to the right along the z-axis. Relative permeability and conductivity values are for the cellular medium ($\varepsilon_r = 40$, $\sigma = 5.0\,\text{S/m}$), the membranes ($\varepsilon_r = 7$, $\sigma = 2.0\,\text{S/m}$), and the inner membrane space ($\varepsilon_r = 5.0$, $\sigma = 0.001\,\text{S/m}$).

Figure 10.15 illustrates the completed model of the isolated mitochondrion in a 100 kHz, 1 V/cm-field depicting a vector plot of the active current density and equipotentials near the mitochondrion.

The Label Mover feature enables frequency sweeping or variation of almost any model parameter or geometric element. Figure 10.16 shows a plot of the total energy stored in the outer and inner mitochondrial membranes as a function of frequency. At low frequencies the total membrane energy of the outer membrane is roughly five orders of magnitude higher than the inner membrane because of screening of the AC field by the outer membrane. The total energy stored in the inner membrane increases with frequency and is equal to the energy stored in the outer membrane at about 0.75 MHz. Figure 10.17 shows the peak electric field as a function of frequency in the outer and inner membranes for the 1 V/cm excitation. To obtain the peak-induced membrane potential these values are multiplied by the membrane thickness of $\delta = 7.0\,\text{nm}$.

REFERENCES

QuickField User's Guide: Tera Analysis Ltd. (2007). *Chapter 6: Electric Circuit Definition.*

Roland Glaser. (2001). *Biophysics*, Springer-Verlag. *Chapter 3.5: Electric Fields in Cells and Organisms.*

S. Grimnes and Ø. G. Martinsen. (2000). *Bioimpedance and Bioelectricity Basics*, Academic Press, London.

T. Kotnik and D. Miklavčič. (2000). "Theoretical Evaluation of the Distribution Power Dissipation in Biological Cells Exposed to Electric Fields," *Bioelectromagnetics* 21, 385–394.

EXERCISES

10.1 Using Label Mover, calculate the frequency-dependent total active and reactive current components flowing through the inhomogeneous capacitor shown in Figure E-10.1.

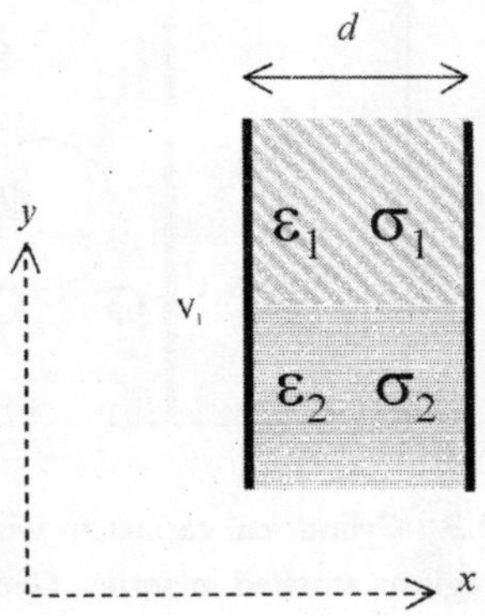

FIGURE E-10.1 **Capacitor with parallel regions of different conductivity and permittivity. Geometry: *x-y* symmetry.**

10.2 Using Label Mover, calculate the frequency-dependent total active and reactive current components flowing through the inhomogeneous capacitor shown in Figure E-10.2.

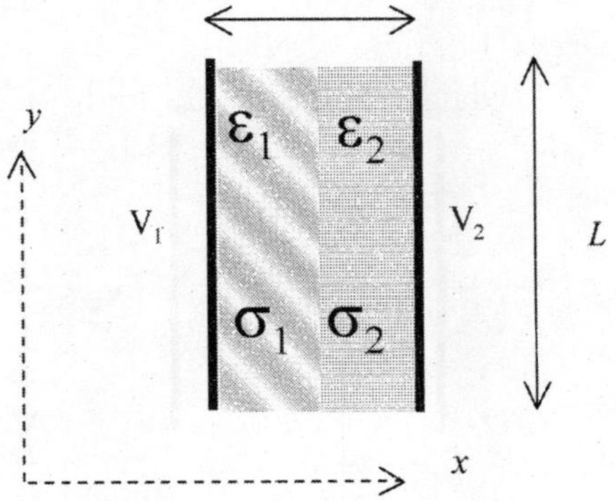

FIGURE E-10.2 **Capacitor regions with different conductivity and permittivity regions stacked in series. Geometry: *x-y* symmetry.**

10.3 Using Label Mover, calculate the frequency-dependent total active and reactive current components flowing through the inhomogeneous capacitor shown in Figure E-10.3.

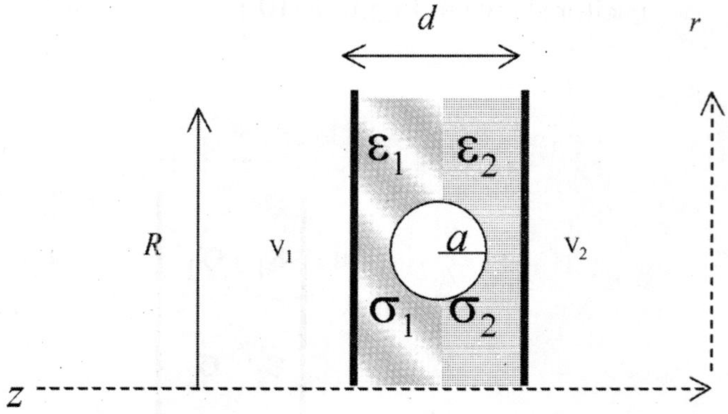

FIGURE E-10.3 **Cylindrical capacitor with different conductivity and permittivity regions stacked in series. Geometry: axial symmetry.**

10.4 Using Label Mover, calculate the peak induced transmembrane potential as a function of excitation frequency for two cells connected by a gap junction. Model the gap as a weekly conducting segment with a size equal to about one membrane thick. Compare your results with simulations involving several gap junctions. You may wish to duplicate and translate periodic elements of your model. Perform the simulation in axial symmetry. Qualitatively, similar results may be obtained using x-y symmetry as well.

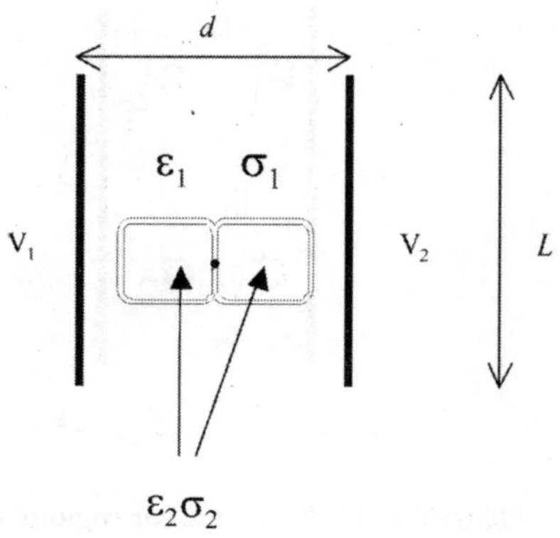

FIGURE E-10.4 **Two cells connected by a gap junction.**

10.5 Model the electrical response of a cell undergoing mitosis in axial symmetry. Calculate the frequency-dependent peak transmembrane potential for several stages shown in Figure E-10.5 (full view shown for clarity).

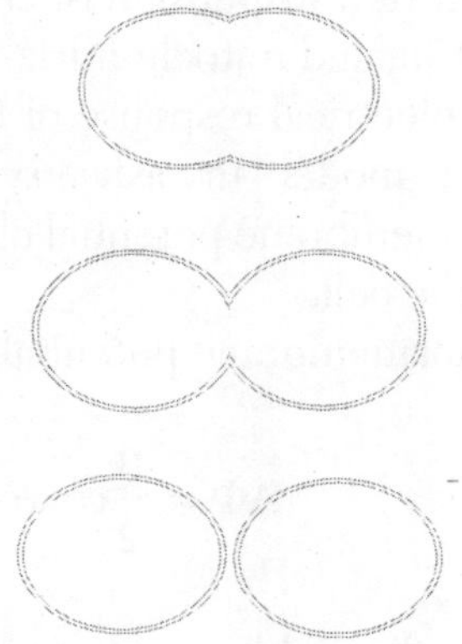

FIGURE E-10.5 **Deformation of plasma membrane during mitosis. Geometry: axial symmetry.**

10.6 Model a red blood cell in suspension under AC electrical stimulation in axial symmetry. Plot the frequency-dependent membrane potential of the blood cell using Label Mover.

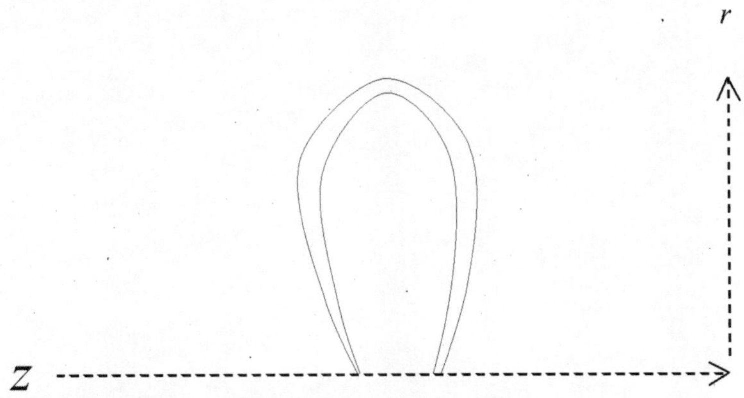

FIGURE E-10.6 **Red blood cell. Geometry: axial symmetry**

10.7 Simulate a cell containing a nucleus in suspension under AC electrical stimulation in axial symmetry. Plot the frequency-dependent membrane potential of the outer membrane and the nuclear membrane using Label Mover.

10.8 Simulate a 4-probe measurement connecting microelectrodes outside the cell and measuring potential difference between locations in the cell.

10.9 Calculate the individual organelle contributions to the effective dielectric constant of a suspension of cells containing cell nucleus, endoplasmic reticulum, and mitochondria. Model the cell in axial symmetry.

10.10 Model the electrical response of two spherical cells in close proximity. Create the model in axial symmetry. Compare the frequency-dependent membrane potential close to the neighboring cell and on the far side of the cell.

10.11 The AC transmembrane potential for a spherical cell of radius R is

$$\Delta\Phi = \frac{3}{2}\cos\theta\frac{E_0 R}{1 + i\omega\tau},$$

where E_0 is the field amplitude, ω is the angular frequency, τ is the membrane charging time constant, and θ is the polar angle measured from the direction of the incident field. Plot the $\mathrm{Re}\Delta\Phi$ vs. $\mathrm{Im}\Delta\Phi$ in MATLAB as a function of frequency for different values of τ.

Chapter 11

THERMAL ANALYSIS

In This Chapter

- The Heat Equation
- Steady State Heat Flow
- Transient Heat Flow
- Thermal Analysis in QuickField
- Coupled AC Magnetic and Heat Transfer Problems
- Coupled Current Flow and Heat transfer Problems

11.1 THE HEAT EQUATION

Thermal conduction acts to equalize temperature differences between regions of higher and lower temperatures. The rate of thermal energy Q transferred between two reservoirs at temperatures T_1 and T_2 separated by an insulator of thickness Δx is given by

$$\frac{\Delta Q}{\Delta t} = \lambda A \frac{T_2 - T_1}{\Delta x}, \tag{11.1}$$

where A is the cross-sectional area and λ is the thermal conductivity of the insulating barrier. The rate $\Delta Q/\Delta t$ is positive in the lower temperature reservoir and vice versa as energy is transferred from the high temperature to the low temperature reservoir. Equation (11.1) may be written as a partial differential equation describing local heat flow in a material body with a one-dimensional temperature gradient

$$\frac{\partial Q}{\partial t} = \lambda \frac{\partial T}{\partial x}. \tag{11.2}$$

We seek to write equation (11.2) as a partial differential equation over a single scalar field $T(x,t)$. The thermal energy stored in a body of volume V with constant temperature is given by $Q = mcT$, or with variable temperature and mass density

$$Q = c\iiint_{\text{vol}} T\rho dV, \tag{11.3}$$

where c is the specific heat in J/kg · K, T is the absolute temperature in Kelvins, and ρ is the mass density in kg/m^3. The power loss due to the dissipation of thermal energy through a surface bounding the volume V is given by

$$\frac{\partial Q}{\partial t} = \lambda \iint_{\text{surf}} \nabla T \cdot d\mathbf{a}. \tag{11.4}$$

Setting the time derivate of equation (11.3) equal to equation (11.4) and applying Gauss' divergence theorem gives

$$c\iiint_{\text{vol}} \frac{\partial T}{\partial t} \rho dV = \lambda \iiint_{\text{vol}} \nabla^2 T dV. \tag{11.5}$$

Note that equation (11.5) holds over arbitrary volumes so that the integrands can be equated giving the heat equation

$$\frac{\partial T}{\partial t} = \alpha \nabla^2 T, \tag{11.6}$$

where $\alpha = \lambda / c\rho$.

11.2 STEADY STATE HEAT FLOW

In steady state $\partial T/\partial t = 0$, temperature distribution is described by Laplace's equation

$$\nabla^2 T = 0 \tag{11.7}$$

in regions without heat source and constant thermal conductivity. In regions with heat source q, the temperature satisfies Poisson's equation

$$\lambda\nabla^2 T = -q \tag{11.8}$$

and in regions with both heat source and inhomogeneous and /or anisotropic λ

$$\nabla\cdot(\lambda\nabla T) = -q. \tag{11.9}$$

Note that λ and q may also depend on the temperature resulting in a non-linear heat flow equation. In general, a heat source will give rise to a heat flux with divergence

$$\nabla\cdot\mathbf{F} = q, \tag{11.10}$$

where the flux vector $\mathbf{F} = -\nabla T$. Both the heat flux and temperature must be continuous at the interface between regions 1 and 2 with differing thermal conductivities so that

$$\left\{\begin{array}{c} T_1 = T_2 \\ \lambda_1\dfrac{\partial T_1}{\partial n} = \lambda_2\dfrac{\partial T_2}{\partial n}\end{array}\right\} \tag{11.11}$$

in the absence of a heat source on the boundary. The analysis of steady state heat flow is similar to electrostatics with the potential V and electric field $\mathbf{E}$ analogous to T and $\mathbf{F}$, respectively. Techniques for solving Laplace's and Poisson's in Chapter 2 may also be used to solve for steady state temperature distributions.

Temperature Distribution in a Semi-Infinite Plate

If a constant temperature is specified along the edge of a semi-infinite half-space or a thin plate insulated so that are no convective or radiative losses in the z-direction, the steady state temperature distribution may be obtained from Poisson's integral formulation for a half-space

$$T(x,y) = \frac{1}{\pi}\int_{-\infty}^{\infty}\frac{yG(x')dx'}{y^2 + (x - x')^2}, \tag{11.12}$$

where the boundary temperature $G(x')$ is specified along the x-axis. As an example, consider the boundary temperature distribution

$$G(x') = \begin{cases} T_0 & x' \le -1 \\ 0 & -1 < x' < 1, \\ T_0 & x' \ge 1 \end{cases} \tag{11.13}$$

where the temperature in the half-space $y \ge 0$ is given by

$$T(x,y) = \frac{T_0}{\pi}\int_{-\infty}^{-1}\frac{y\,dx'}{y^2 + (x - x')^2} + \frac{T_0}{\pi}\int_{1}^{\infty}\frac{y\,dx'}{y^2 + (x - x')^2} \tag{11.14}$$

$$T(x,y) = T_0\left(1 + \tan^{-1}\left(\frac{-x-1}{y}\right) - \tan^{-1}\left(\frac{1 - x}{y}\right)\right). \tag{11.15}$$

Inspection of this formula reveals that $T(x,y) \to T_0$ for large values of x and y.

Temperature Distribution in a Circular Plate

The steady state temperature distribution inside a circle of radius R with a specified temperature on the boundary may be obtained from Poisson's integral formulation for a circle

$$T(r,\theta) = \frac{1}{2\pi}\int_{0}^{2\pi}\frac{(R^2 - r^2)F(\theta')}{R^2 - 2Rr\cos(\theta - \theta') + r^2}d\theta', \tag{11.16}$$

where $F(\theta')$ is the boundary temperature specified on $r = R$.

Conformal Transformation

The steady state temperature distribution inside a plate with temperature T_0 specified at one boundary shown in Figure 11.1 (a) is given by

$$T(x,y) = \frac{4}{\pi}\sum_{n=1,3,5\ldots}^{\infty}\frac{1}{n}\frac{\sinh\left(n\pi\frac{y}{a}\right)}{\sinh\left(n\pi\frac{b}{a}\right)}\sin\left(\frac{n\pi x}{a}\right). \tag{11.17}$$

The temperature steady state distribution in the half ring illustrated in Figure 11.1 (b) may be obtained by conformal mapping $w = e^z$

$$T(r,\theta) = \frac{4}{\pi}\sum_{n=1,3,5\ldots}^{\infty}\frac{1}{n}\frac{\sinh\left(\frac{n\pi\theta}{\ln r_0}\right)}{\sinh\left(\frac{n\pi^2}{\ln r_0}\right)}\sin\left(\frac{n\pi\ln r}{\ln r_0}\right). \tag{11.18}$$

ON THE CD

Color images for many of the figures may be found on the CD-ROM.

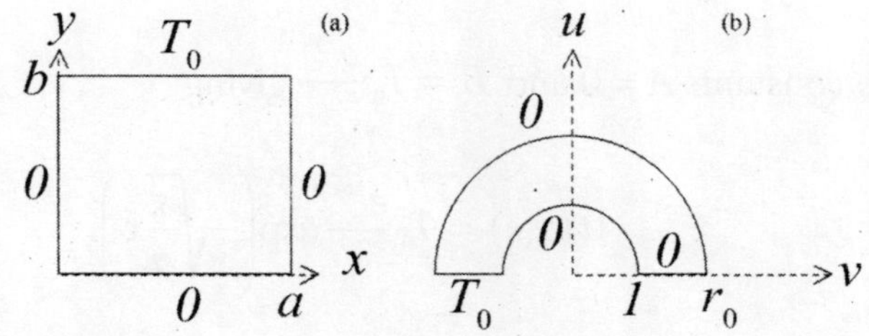

FIGURE 11.1 **Conformal mapping of a rectangular region (a) to a half-ring (b).**

Conformal mapping may also be used to transform a more complicated region to a half-plane or a circle, where the appropriate Poisson integral formula is then used to calculate the temperature distribution.

11.3 TRANSIENT HEAT TRANSFER

Consider the one-dimensional problem, where the $x = 0$ plane of a semi-infinite heat conductor is held at a constant temperature T_0 beginning at time $t = a$. The heat equation with boundary conditions is given by

$$\frac{\partial T}{\partial t} = \alpha \frac{\partial^2 T}{\partial x^2} \qquad \begin{Bmatrix} T(x,0) \\ T(0,t) \\ T(\infty,t) \end{Bmatrix} = \begin{Bmatrix} 0 \\ T_0 \Theta(t-a) \\ 0 \end{Bmatrix}. \tag{11.19}$$

Applying the Laplace transform operator to both sides of the heat equation and the boundary conditions gives

$$\frac{s}{\alpha} t(x,s) = \frac{\partial^2 t(x,s)}{\partial x^2} \qquad \begin{Bmatrix} t(x,0) \\ t(0,s) \\ t(\infty,s) \end{Bmatrix} = \begin{Bmatrix} 0 \\ T_0 \dfrac{e^{-as}}{s} \\ 0 \end{Bmatrix}, \tag{11.20}$$

where $t(x,s) = LT(x,t)$. Solving the ordinary differential equation for $t(x,s)$ we have

$$t(x,s) = A \exp\left(\sqrt{\frac{s}{\alpha}}x\right) + B \exp\left(-\sqrt{\frac{s}{\alpha}}x\right), \tag{11.21}$$

where the constants $A = 0$ and $B = T_0 \dfrac{e^{-as}}{s}$ giving

$$t(x,s) = T_0 \frac{e^{-as}}{s} \exp\left(-\sqrt{\frac{s}{\alpha}}x\right). \tag{11.22}$$

Note that MATLAB will not return the inverse Laplace transform of equation (11.22) directly. We first need to use the result

$$L^{-1}\{e^{-as}F(s)\} = \begin{pmatrix} f(t-a) & t \geq a \\ 0 & t < a \end{pmatrix}, \tag{11.23}$$

where $L^{-1}F(s) = f(t)$ so that

$$T(x,t) = T_0 \cdot \operatorname{erfc}\left(\frac{x}{2\sqrt{\alpha(t-a)}}\right) \quad t \geq 0, \tag{11.24}$$

where erfc(z) is the complementary error function defined by equation (7.10). The temperature distribution resulting from a finite duration temperature pulse at $x = 0$ may also be constructed as a superposition of solutions of the form of equation (11.24).

11.4 Thermal Analysis in QuickField

Thermal analysis in Quickfield is based on the heat equation with convective, radiative, heat flux, and temperature boundary conditions. Quantities of interest include temperature distributions and heat losses. Transient analysis may be performed imposing time-dependent boundary conditions and also simulating bodies at different temperature in thermal contact.

In two-dimensional x-y symmetry, the heat equation is given by

$$\frac{\partial}{\partial x}\left(\lambda_x \frac{\partial T}{\partial x}\right) + \frac{\partial}{\partial y}\left(\lambda_y \frac{\partial T}{\partial y}\right) = -q + c\rho\frac{\partial T}{\partial t} \tag{11.25}$$

with thermal conductivity components λ_x, λ_y, and heat source volume power q. In axial symmetry the heat equation is given by

$$\frac{1}{r}\frac{\partial}{\partial r}\left(\lambda_r r\frac{\partial T}{\partial r}\right)+\frac{\partial}{\partial z}\left(\lambda_z\frac{\partial T}{\partial z}\right)=-q+c\rho\frac{\partial T}{\partial t} \tag{11.26}$$

with thermal conductivity components λ_r and λ_z. For a temperature-dependent thermal conductivity, heat source and specific heat in x-y symmetry, the heat equation is calculated according to

$$\frac{\partial}{\partial x}\left(\lambda(T)\frac{\partial T}{\partial x}\right)+\frac{\partial}{\partial y}\left(\lambda(T)\frac{\partial T}{\partial y}\right)=-q(T)+c(T)\rho\frac{\partial T}{\partial t}. \tag{11.27}$$

In axial symmetry, the heat equation with nonlinear coefficients is calculated as

$$\frac{1}{r}\frac{\partial}{\partial r}\left(\lambda(T)r\frac{\partial T}{\partial r}\right)+\frac{\partial}{\partial z}\left(\lambda(T)\frac{\partial T}{\partial z}\right)=-q(T)+c(T)\rho\frac{\partial T}{\partial t}, \tag{11.28}$$

where only isotropic thermal conductivities are supported in the nonlinear case. In steady state heat transfer problems, the last term in equations (11.25) through (11.28) are zero. Temperature-dependent coefficients equations in (11.27) and (11.28) are determined by a cubic spline approximation to data entered in their corresponding material properties curve editors discussed below.

Adaptive Time Step

In QuickField the time step is adjusted depending on properties of the mesh and potential variation. The initial time step Δt_0 is chosen $\Delta t_0 = \min(\xi^2/4\alpha)$, where ξ is a characteristic size of the mesh elements and α is equal to $\alpha = \lambda/\rho C$. Subsequent time steps are updated according to $\Delta t_{n+1} = k\Delta t_n$ where k is a scaling factor that is chosen to be 0.25, 0.5, 1.0, 2.0, or 4.0 depending on the spatial and temporal behavior of the potential, sources, and boundary conditions.

Material Properties

Figure 11.2 shows the window for editing Block Label Properties including thermal conductivity components, volume power of heat source, specific heat, and density. The Anisotropic box is checked to specify different thermal conductivity components. The word 'none' will appear if zero is the thermal conductivity specified. Regions with zero thermal conductivity are excluded from the solution region.

Thermal conductivity may be defined as a function of temperature by checking Nonlinear in the Thermal Conductivity field group. Figure 11.3 shows the curve editor for specifying temperature-dependent thermal conductivity $\lambda(T)$. Heat sources may also be defined as a function of time in transient analysis

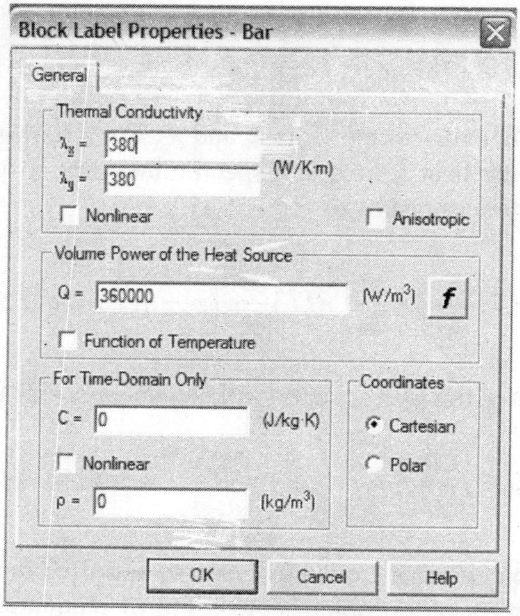

FIGURE 11.2 **Window for editing block label data for heat transfer problems: To open the block label window click on the corresponding block label in the tree diagram on the left of the screen.**

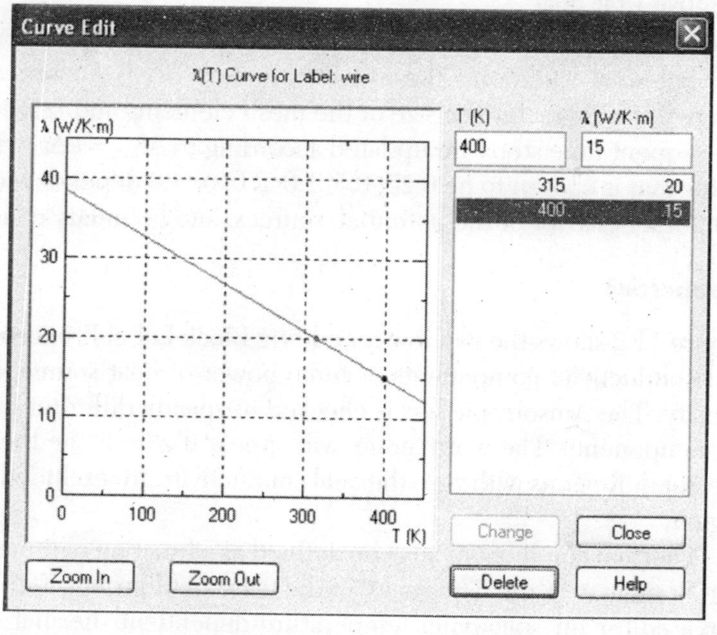

FIGURE 11.3 **Window for specifying temperature-dependent thermal conductivity. A cubic spline curve is fit to data entered in the λ-T table.**

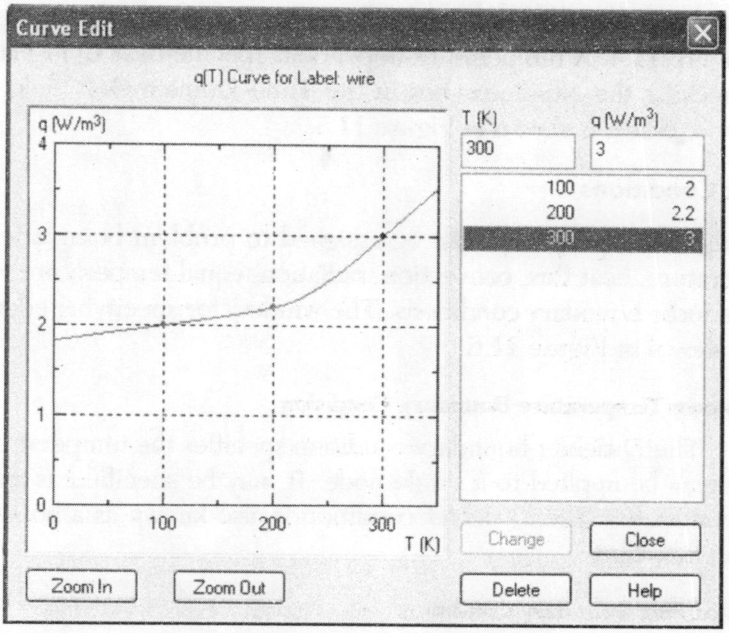

FIGURE 11.4 **Window for specifying a temperature-dependent volume heat source. A cubic spline curve is fit to data entered in the *q*-*T* table.**

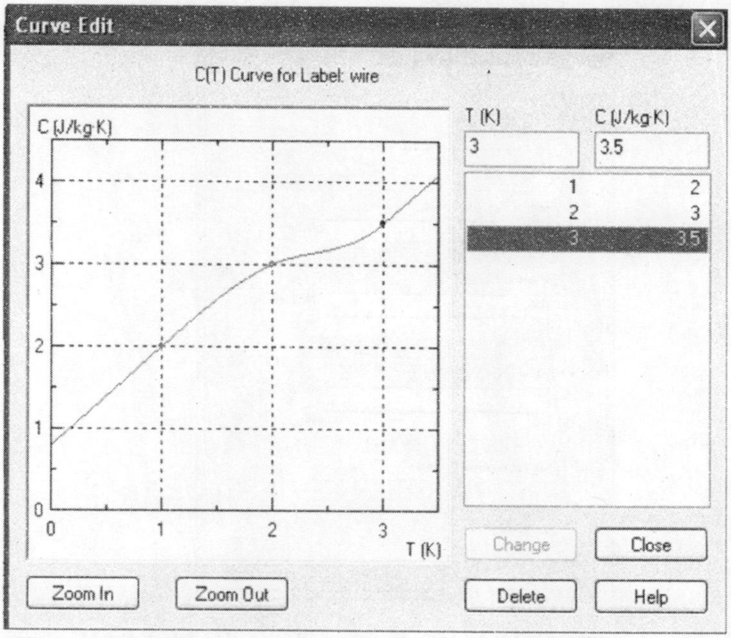

FIGURE 11.5 **Window for specifying temperature-dependent specific heat. A cubic spline curve is fit to data entered in the *C*-*T* table.**

using the formula editor or temperature using the $q(T)$ curve editor shown in Figure 11.4. A temperature-dependent specific heat $C(T)$ may be specified by checking the Nonlinear box in the Time-Domain Only field group. The $C(T)$ curve editor is shown in Figure 11.5.

Boundary Conditions

Edge label properties can be assigned to problem boundaries specifying temperature, heat flux, convection, radiation, equal temperature, and even and odd periodic boundary conditions. The window for specifying edge label properties is shown in Figure 11.6.

Known Temperature Boundary Condition

The Dirichlet boundary condition specifies the temperature on a boundary. It may be applied to a single node. It may be specified as a linear function of coordinates. The Dirichlet condition is also known as a boundary condition of the first kind.

Heat Flux Boundary Condition

The Neumann condition specifies the normal component of heat flux density F_n at outward boundaries of the solution region

$$F_n = -q_s \tag{11.29}$$

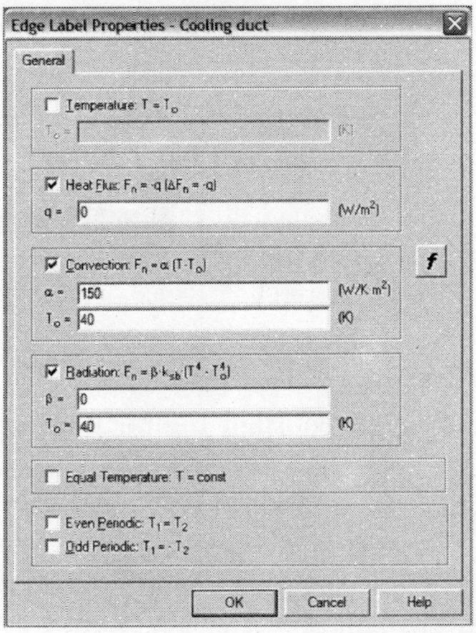

FIGURE 11.6 **Window for editing edge label data for heat transfer problems: To open the edge label window click on the corresponding edge label in the tree diagram on the left of the screen.**

and at interior boundaries

$$F_n^+ - F_n^- = -q_s. \quad (11.30)$$

The units of q_s are W/m^2. A value of $q_s = 0$ corresponds to the homogeneous boundary condition that is the default boundary condition on outward boundaries and is applied to outward boundaries used as symmetry planes of the model.

Convection Boundary Condition

Convective heat transfer is described by

$$F_n = \alpha(T - T_0), \quad (11.31)$$

the outward boundary of a region, where α is a film coefficient with units W/K · m^2 and T_0 is the temperature of the surrounding medium.

Radiation Boundary Condition

The radiated power per unit area is given by

$$F_n = k_{SB}\beta(T^4 - T_0^4), \quad (11.32)$$

where $k_{SB} = 5.67 \cdot 10^{-8}$ W/m^{2}K^4, β is the dimensionless emissivity that depends on the material type and T_0 is the ambient temperature. Heat flux, convection, and radiation boundary conditions may be applied to a single edge, where heat flows due to a combination of sources.

Constant Temperature Boundary Condition

The temperature on a boundary. Interiors of bodies with constant temperature may be excluded from the solution region since the temperature will have the same value at all points inside the boundary. The condition is analogous to the constant potential boundary condition applied to perfect electrical conductors in electrostatics.

Vertex Conditions

Vertex labels may be assigned a constant temperature or heat source that is coordinate dependent. Time-dependent temperature and heat sources may be associated with vertex labels in transient heat transfer problems. Figure 11.7 shows the window for editing vertex label data for heat transfer problems. Time- and coordinate-dependent values may be linked to all vertices sharing the same label entering formulas in the heat source or temperature box.

Calculated Values

Components of the heat flux density calculated from $\mathbf{F} = -\lambda\nabla T$ in x-y symmetry are

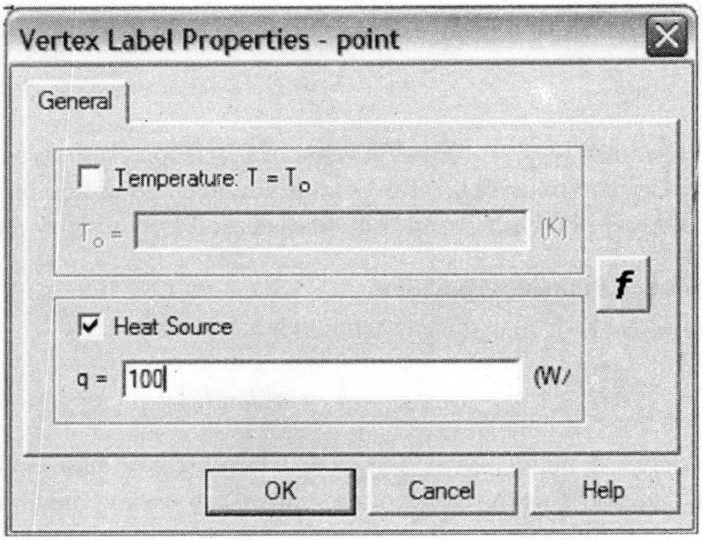

FIGURE 11.7 **Window for editing vertex label data for heat transfer problems: To open the vertex label window click on the corresponding vertex label in the tree diagram on the left of the screen.**

$$F_x = -\lambda_x \frac{\partial T}{\partial x} \quad F_y = -\lambda_y \frac{\partial T}{\partial y}. \tag{11.33}$$

In axial symmetry radial and z-component flux densities are calculated according to

$$F_z = -\lambda_z \frac{\partial T}{\partial z} \quad F_r = -\lambda_r \frac{\partial T}{\partial r}. \tag{11.34}$$

The total heat flux is calculated from the surface integral

$$\Phi = \iint_{\text{surf}} \mathbf{F} \cdot \mathbf{n} dS \tag{11.35}$$

over specified open or closed surfaces.

Steady State Heat Flow in a Sphere

As an example of steady state heat transfer we consider a metal sphere with a diameter of 100 cm half submerged in a liquid nitrogen bath at 77 K. The top of the sphere is exposed to air at 273 K. Both convection and radiation boundary conditions are applied to the top and bottom hemispherical surfaces. The metal has an isotropic thermal conductivity of $\lambda = 25$ W/k · m. Heat transfer

FIGURE 11.8 **Temperature intensity map and heat flux vectors for a metal sphere half submerged in liquid nitrogen. The heat flux vectors point from the high temperature surface in contact with air to the low temperature surface in contact with liquid nitrogen.**

convection coefficients are $\alpha = 100$ and 11.4 W/k · m^2 on the surfaces in contact with liquid nitrogen and air, respectively. The emissivity $\beta = 0.9$ is specified for radiation boundary conditions on each side of the sphere. Figure 11.8 shows the temperature intensity map and heat flux vectors inside the sphere. The heat flux vectors point from the high temperature surface in contact with air to the low temperature surface in contact with liquid nitrogen.

Transient Heat Flow in a Cylinder

We consider the transient heat flow after liquid nitrogen that is introduced into a copper cylinder. The initial temperature distribution of the cylinder is imported from a steady state heat transfer problem sharing the same model file. The copper cylinder has a 7.5 cm outer and a 5 cm inner radius. At $t = 0$, the cylinder is

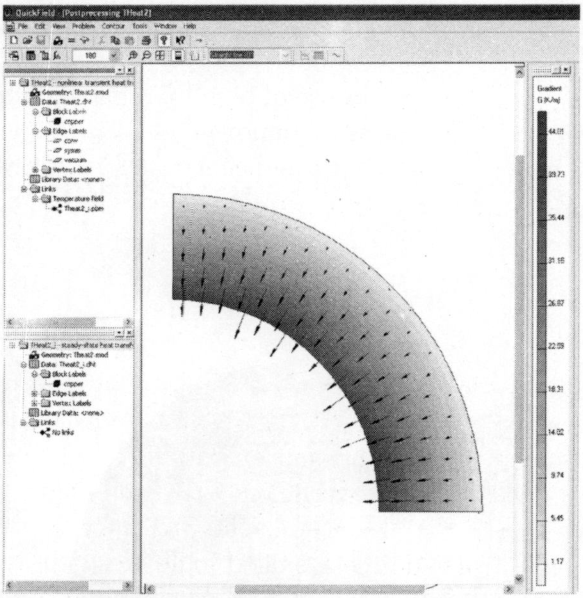

FIGURE 11.9 **Transient heat transfer problem of a copper cylinder initially at 273 K filled with liquid nitrogen at 77 K. The initial temperature distribution is imported from the steady state heat transfer problem. Flux density arrows and the intensity map of temperature gradient are shown 180 seconds after the nitrogen is introduced into the cylinder.**

filled with liquid nitrogen at 77 K. The outer boundary of the cylinder is at 273 K. Convection and radiation boundary conditions are applied to the inner and outer surfaces of the cylinder. Heat transfer convection coefficients are $\alpha = 100$ and 11.4 W/k · m^2 on the surfaces in contact with liquid nitrogen and air, respectively. The emissivity $\beta = 0.9$ is specified on inner and outer surfaces of the cylinder. Because of symmetry, only one quarter of the cylinder is modeled. The zero flux density boundary condition is applied to the flat edges. Flux density arrows and the intensity map of temperature gradient 180 seconds after nitrogen is introduced inside the cylinder as shown in Figure 11.9.

11.5 COUPLED TIME VARYING MAGNETIC AND HEAT TRANSFER PROBLEMS

Temperature distributions may be calculated in QuickField based on Joule heat sources in AC or Transient Magnetics problems. The coupled heat transfer and time varying magnetic problems must share the same model file. A link is created in the Heat Transfer Problem Properties to the AC or Transient Magnetics problem.

Modeling Hyperthermia in Oncology

Hyperthermia treatment makes use of high-frequency electromagnetic fields to inhibit cancer by locally elevating tumor temperature. The technique is often used in conjunction with radiation and chemotherapy. The Specific Absorption Rate gives a measure of energy absorption in tissue

$$\text{SAR} = \frac{1}{\rho_m}\sigma E^2, \tag{11.36}$$

Where ρ_m is mass density in kg/m^3, σ conductivity in S/m, and E is the root mean square electric field in V/m. The SAR may be calculated in AC Magnetics for axisymmetric tissue, tumor, and excitation coil geometries neglecting the displacement contribution. It turns out that conduction currents provide the main contribution to the absorption. The SAR may be calculated by dividing the QuickField calculated time averaged Joule heat density by the mass density Q/ρ_m. Joule heating calculated in the AC Magnetics can then be imported into Thermal Analysis to calculate the temperature and heat flux distribution with convection and radiation boundary conditions applied to the body surface.

11.6 COUPLED CURRENT FLOW AND HEAT TRANSFER PROBLEMS

Electric power losses calculated in AC and DC Conduction Analysis problems may be used as sources for Thermal Analysis. Temperature distributions are calculated from Joule heating. The coupled heat transfer and current flow problems must share the same model file. A link is created in the Heat Transfer Problem Properties to the AC or DC current flow problem.

Temperature Distribution in a Fuel Cell Stack

Temperature distribution in a PEM fuel cell may be obtained by importing Joule heat sources calculated in the DC Conduction module into the thermal analysis. Modeling the temperature distribution near the membrane may help identify regions susceptible to burn-through that could result in hydrogen and oxygen fuel gas cross over and catastrophic failure of the fuel cell stack.

The copper fuel cell components are assigned a resistivity of $\rho = 1.7 \cdot 10^{-8}\ \Omega \cdot$ m with thermal conductivity $\lambda = 398$ W/K. The graphite resistivity is taken as $\rho = 7.8 \cdot 10^{-6}\ \Omega \cdot$ m with a thermal conductivity of $\lambda = 50$ W/K. The gas-flow channels are assigned zero resistivity and the fuel cell membrane is taken as a nearly perfect electrical insulator with a resistive of $\rho = 10^3\ \Omega \cdot$ m. Note that the membrane adjacent to the gas-flow channels is conducting to protons and not electrons.

Radiation and convection boundary conditions are applied to the outer surface of the fuel cell with an ambient temperature of 273 K emissivity $\beta = 0.9$ and convection coefficient $\alpha = 5.678\, W/k \cdot m^2$ (equal to 1 Btu/ft^2hr^0F).

Fuel Cell in Open-circuit Mode

Figure 11.10 shows isothermal contours of the PEM fuel cell in open-circuit mode with zero net current flowing through the cell. Open-circuit mode is simulated with application of the zero normal current density boundary condition on the end caps of the fuel cell stack. Shaded parts correspond to regions with different thermal conductivity including copper, graphite, and the proton exchange membrane material. Isothermal contours, heat flux vectors, and an intensity map of heat flux in a section of the cylindrical PEM fuel cell stack is shown in Figure 11.11. Heat flux vectors are maximal in the copper region in this figure. Figure 11.12 shows the equipotentials and current flow vectors with intensity map of electric field strength in the same section of the cylindrical PEM fuel cell stack in open-circuit mode.

Fuel Cell with Zero Load

Figure 11.13 shows the current flow vectors with the intensity map of current density in a section of the cylindrical PEM fuel cell stack with a net current flowing through the stack. In this simulation a zero potential boundary condition is applied to the end caps of the fuel cell so that the current flow is perpendicular

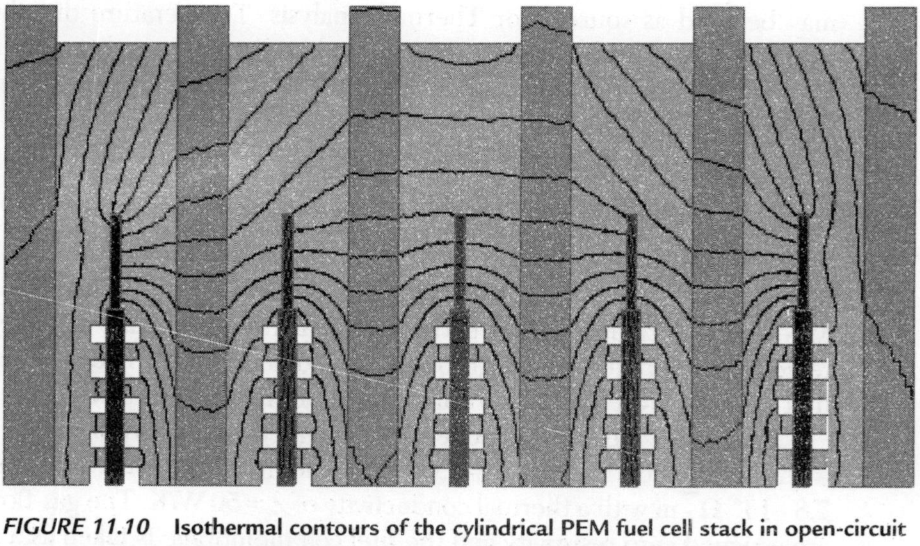

FIGURE 11.10 **Isothermal contours of the cylindrical PEM fuel cell stack in open-circuit mode. Shaded regions of different thermal conductivity include copper, graphite, and proton exchange membrane material.**

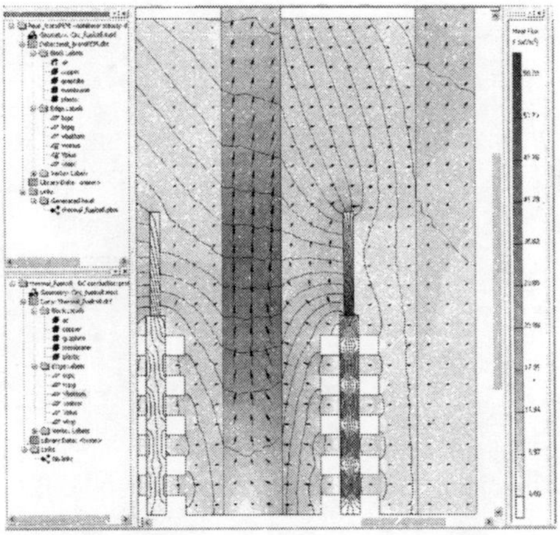

FIGURE 11.11 **Isothermal contours, heat flux vectors, and intensity map of heat flux in a section of the cylindrical PEM fuel cell stack in open-circuit mode with zero net current flowing through the stack. Regions shown include copper, graphite, gas-flow channels, and the PEM membrane. Heat flux vectors are maximal in the copper region.**

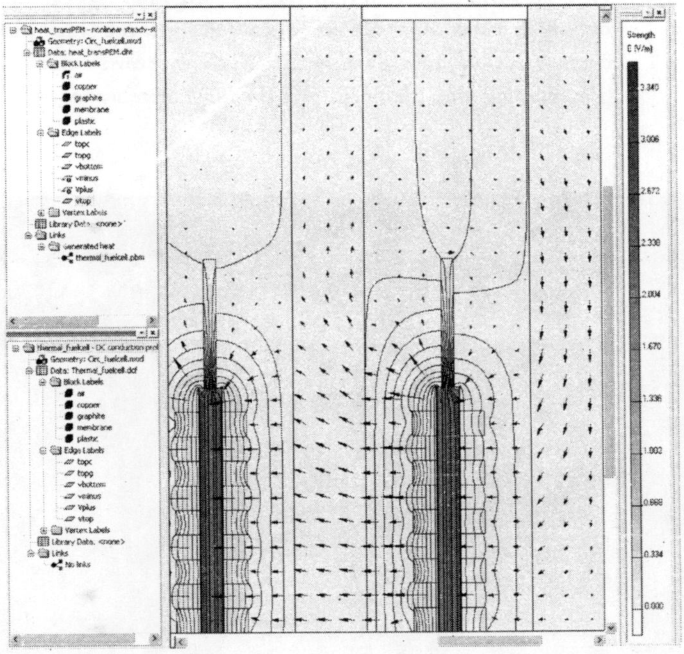

FIGURE 11.12 **Equipotentials and current flow vectors with intensity map of electric field strength in a section of the cylindrical PEM fuel cell stack in open-circuit mode with zero net current flowing through the stack. Regions shown include copper, graphite, gas-flow channels, and the PEM membrane.**

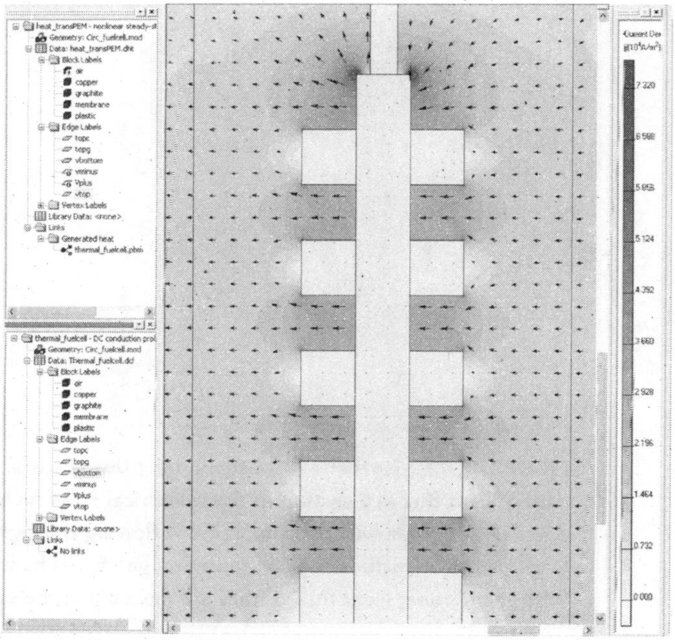

FIGURE 11.13 Current flow vectors with intensity map of current density in a section of the cylindrical PEM fuel cell stack with a net current flowing through the stack. Regions shown include copper, graphite, gas-flow channels, and the PEM membrane.

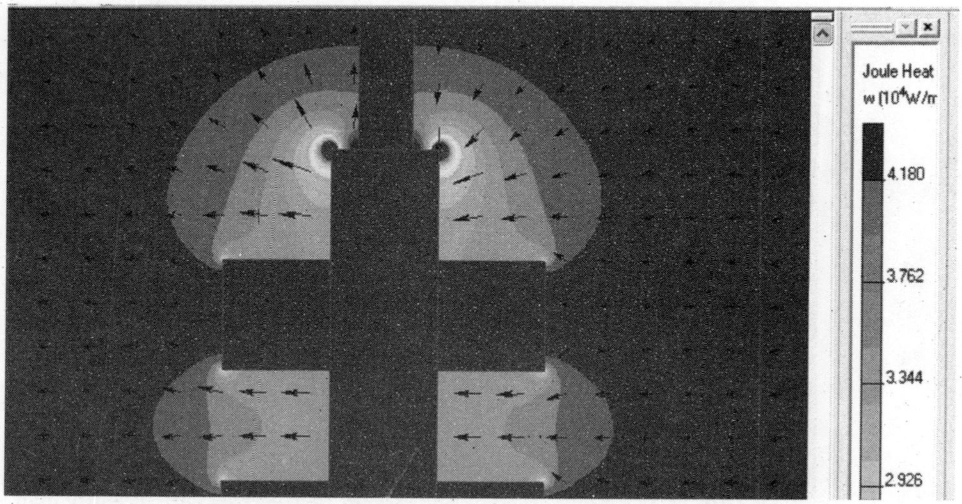

FIGURE 11.14 Current flow vectors with intensity map of Joule heat in a section of the cylindrical PEM fuel cell stack with a net current flowing through the stack. Regions shown include copper, graphite, gas-flow channels, and the PEM membrane.

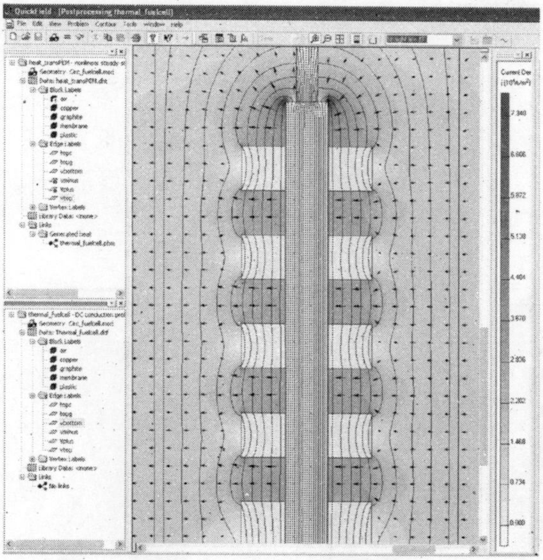

FIGURE 11.15 **Equipotentials and current flow vectors with intensity map of electric field strength in a section of a PEM fuel cell with a net current flowing through the stack. Regions shown include copper, graphite, gas-flow channels, and the PEM membrane.**

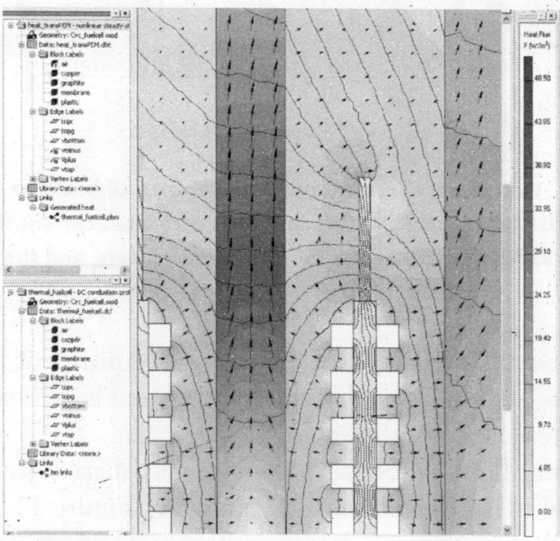

FIGURE 11.16 **Isothermal contours, heat flux vectors, and intensity map of heat flux in a section of the PEM fuel cell in open-circuit mode with a net current flowing through the cell. Regions shown include copper, graphite, gas-flow channels, and the PEM membrane. Heat flux vectors are maximal in the copper region.**

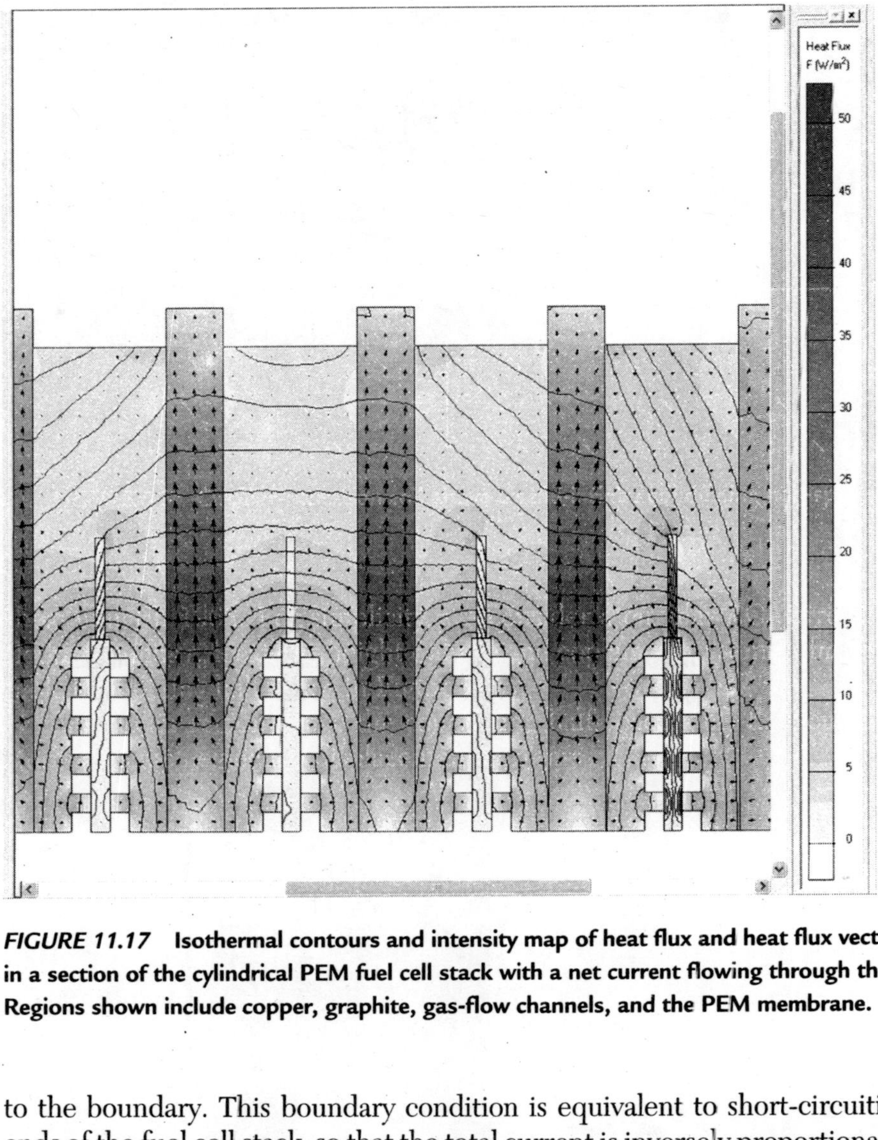

FIGURE 11.17 **Isothermal contours and intensity map of heat flux and heat flux vectors in a section of the cylindrical PEM fuel cell stack with a net current flowing through the cell. Regions shown include copper, graphite, gas-flow channels, and the PEM membrane.**

to the boundary. This boundary condition is equivalent to short-circuiting the ends of the fuel cell stack, so that the total current is inversely proportional to the internal cell resistance of the stack.

The current flow vectors with the intensity map of Joule heat in a section of the cylindrical PEM fuel cell stack is shown in Figure 11.14 with equipotentials and current flow vectors as shown in Figure 11.15. Isothermal contours, heat flux vectors, and the intensity map of heat flux are illustrated in Figure 11.16. The heat flux vectors are maximal in the copper region with higher thermal conductivity. Isothermal contours, the intensity map of heat flux, and heat flux vectors in a section of the cylindrical PEM fuel cell stack are shown in Figure 11.17 with a net current flowing through the cell.

REFERENCES

H. S. Carslaw and J. C. Jaeger. (1959). *Conduction of Heat in Solids, 2nd edition*, Oxford University Press, London.

Stanley J. Farlow. (1993). *Partial Differential Equations for Scientists and Engineers*, Dover Publications, Inc., New York.

Morteza Shafii-Mousavi. (1991). *The ESSENTIALS of Laplace Transforms*, Research and Education Association.

QuickField User's Guide: Tera Analysis Ltd. (2007) *Chapter 2: Introductory Guide; Chapter 9: Theoretical Description.*

EXERCISES

11.1 Solve the heat equations with boundary conditions in Section 11.2 by separation of variables substituting a trial solution $T(x,t) = X(x)\Theta(t)$ into the differential equation.

11.2 Newton's law of cooling $\frac{dT}{dt} = \beta(T_{env} - T)$ gives an approximation for the temperature of a body in contact with an environmental reservoir with temperature T_{env}. Write a MATLAB code to solve the first-order differential equation for $T(t)$ given the initial condition $T(t = 0) = T_0$. Plot $T(t)$ for $T_0 > T_{env}$ and $T_0 < T_{env}$.

11.3 Use Poisson's integral formulation to calculate the temperature distribution in the half-space given the $y = 0$ boundary temperature

$$G(x') = \begin{cases} T_0 & x' \le 0 \\ 0 & x' > 0 \end{cases}.$$

Use MATLAB to plot the isothermal contours and heat flux vectors in the region $-1 \le x \le 1$ and $0 \le y \le 1$.

11.4 Use Poisson's integral formulation to calculate the temperature distribution in the half-space given the $y = 0$ boundary temperature

$$G(x') = T_0 \sin(2\pi x'/L) \quad -\infty < x' < \infty.$$

Use MATLAB to plot the isothermal contours and heat flux vectors in the region $-L \le x \le L$ and $0 \le y \le L$.

11.5 Use Poisson's integral formulation to calculate the temperature distribution in a circle given the $r = R$ boundary temperature

$$F(\theta') = T_0 \cos(\theta') \quad 0 < \theta' < 2\pi.$$

Use QuickField to plot the isothermal contours and heat flux vectors in a circle of radius R maintained at this boundary temperature.

11.6 Two semicircular boundaries of a circular plate with thermal conductivity λ are maintained at temperatures T_1 and T_2. Calculate the steady state temperature distribution over the plate and compare the QuickField solution with the analytical solution obtained by applying Poisson's integral theorem.

11.7 A semi-infinite strip with thermal conductivity λ is maintained at temperatures T_1, T_2, and T_3 on each edge. Calculate the temperature distribution using QuickField and compare to the conformal mapping solution obtained by transformation of the strip to a half-space followed

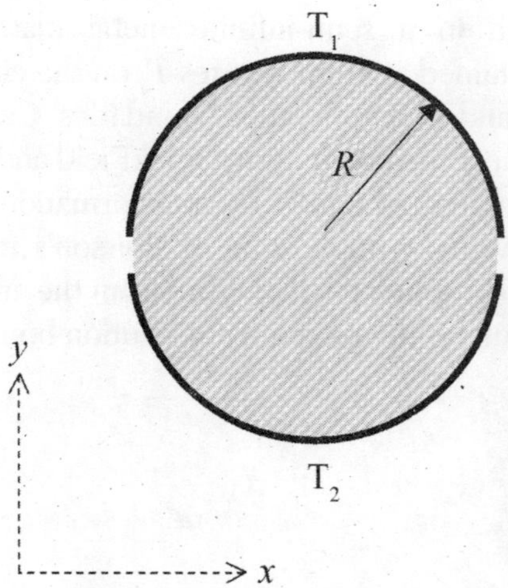

FIGURE E-11.6 Circular plate with thermal conductivity λ. Geometry: x-y symmetry.

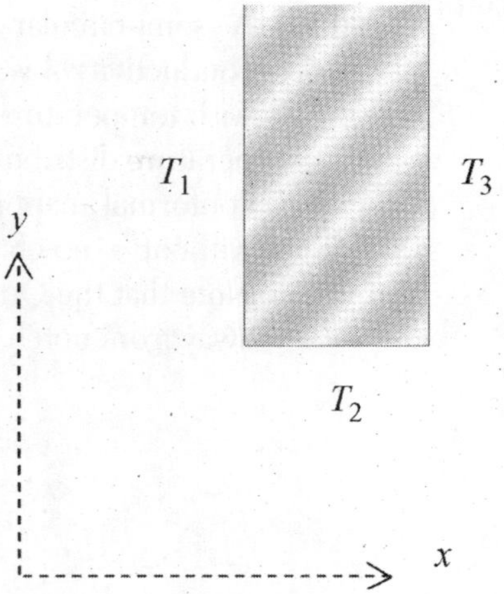

FIGURE E-11.7 Semi-infinite strip with thermal conductivity λ. Geometry: x-y symmetry.

by application of Poisson's integral formula. Note that the QuickField solution will deviate from the analytical solution away from the lower edge because of the proximity of the upper edge in the FEM model.

11.8 A triangular metal plate with thermal conductivity λ is maintained at temperatures T_1, T_2, and T_3 along the edges L_1, L_2, and L_3 at angles θ_1, θ_2, and θ_3. Calculate the temperature distribution using QuickField and compare to the conformal mapping solution obtained by transformation of the triangular region to a semi-infinite plate with temperatures specified on the new boundary $y = 0$ and then making use of Poisson's integral formula. Repeat for temperatures T_1, T_2, and T_3 along the sides of an equilateral triangular lamina.

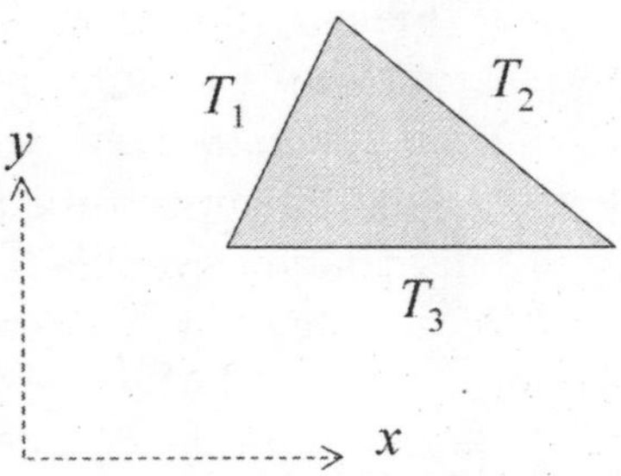

FIGURE E-11.8 Triangular metal plate with thermal conductivity λ. Geometry: x-y symmetry.

11.9 A semi-circular notch in a semi-infinite metal plate with thermal conductivity λ is maintained at temperatures T_1 on the circular boundary with temperatures T_2 and T_3 on the other boundaries. Calculate the temperature distribution near the notch using QuickField and compare to the conformal mapping solution obtained by transformation to a half-space without a notch followed by application of Poisson's integral formula. Note that the QuickField solution will deviate from the analytical solution away from notch because of the proximity of solution boundary edges.

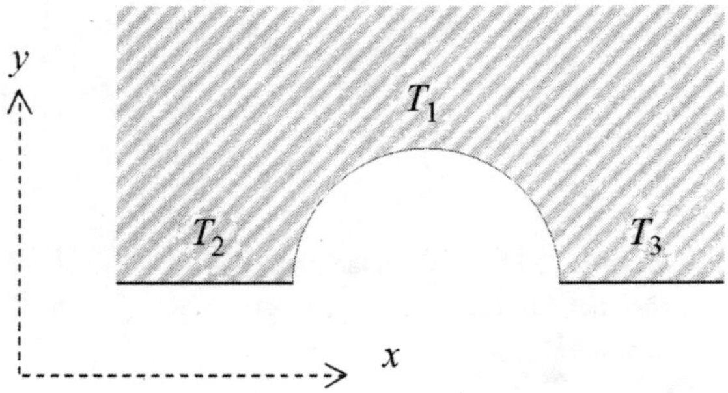

FIGURE E-11.9 **Semi-circular notch in a semi-infinite metal plate with thermal conductivity λ. Geometry: *x-y* symmetry.**

11.10 A circular hole in a metal plate with thermal conductivity λ is maintained at temperatures T_1 on the circular boundary of the first quadrant and at T_2 on the remaining boundary spanned by a 270-degree arc. Calculate the temperature distribution near the hole using QuickField and compare to the conformal mapping solution obtained by transformation of the region to a circle followed by application of Poisson's integral formula. Note that the QuickField solution will deviate from the analytical solution away from the hole because of the proximity of solution boundary edges.

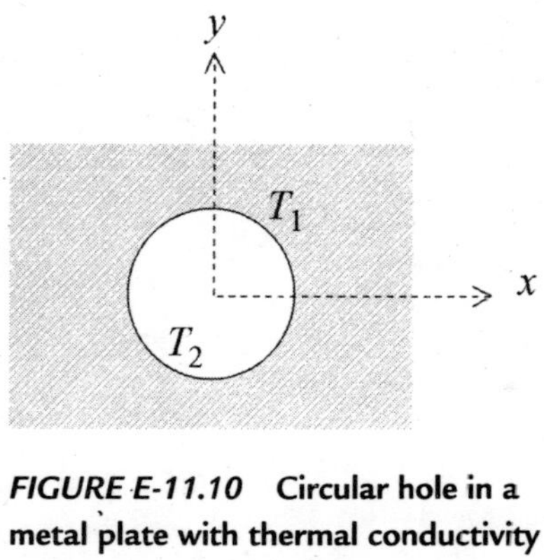

FIGURE E-11.10 **Circular hole in a metal plate with thermal conductivity λ. Geometry: *x-y* symmetry.**

11.11 An L-shaped metal plate with thermal conductivity λ is maintained at temperatures T_1 and T_2 on each edge. Calculate the temperature distribution near the bend using QuickField and compare to the conformal mapping solution obtained by transformation of the region to a half-plane followed by application of Poisson's integral formula. Note that the QuickField solution will deviate from the analytical solution away from the bend because of the proximity of solution boundary edges.

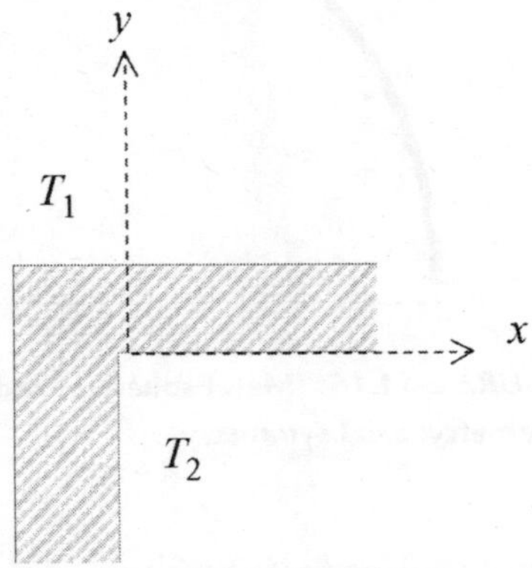

FIGURE E-11.11 **L-shaped metal plate with thermal conductivity λ. Geometry: *x-y* symmetry.**

11.12 Consider the one-dimensional heat transfer problem, where the $x = 0$ plane of a semi-infinite heat conductor is held at a temperature T_0 during the time $a \le t \le b$ described by the difference of Heaviside functions $T(0,t) = T_0\left[\Theta(t-a) - \Theta(t-b)\right]$. Calculate $T\ (x,t)$ using the Laplace transform method. Use MATLAB to plot the spatial temperature distribution at several times after the initial pulse.

11.13 Consider the one-dimensional problem, where the $x = 0$ plane of a semi-infinite heat conductor is subject to an instantaneous temperature pulse described by $T(0,t) = T_0\delta(t - a)$. Calculate $T\ (x,t)$ using the Laplace transform method. Use MATLAB to plot the spatial temperature distribution at several times after the delta function pulse.

11.14 Consider the one-dimensional heat transfer problem, where the $x = 0$ plane of a semi-infinite heat conductor is varied as $T(0,t) = T_0\sin(\omega t)$. Calculate $T(x,t)$ using the Laplace transform method. Use MATLAB to plot the spatial temperature distribution at several times.

11.15 A metal sphere of radius R initially at 300 K is immersed in liquid nitrogen at 77 K. The sphere has thermal conductivity λ. Calculate the temperature distribution inside the sphere as a function of time. Make a plot of the temperature at the center of the sphere as a function of time.

11.16 The upper hemisphere of a metal sphere of radius R and thermal conductivity λ is held at temperature T_1 with the lower half at temperature T_2. Calculate the steady state temperature distribution inside the sphere.

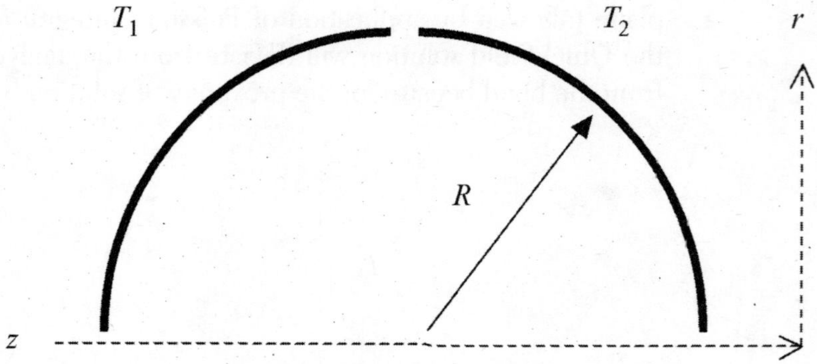

FIGURE E-11.16 **Metal sphere of radius *R* and thermal conductivity λ. Geometry: axial symmetry.**

11.17 A Grande coffee from Starbucks is initially at 350 K. Model the cooling of the coffee in axial symmetry as a function of time. Apply the radiation boundary condition to the sides and bottom of the cup with convection and radiation boundary conditions to the top surface of the liquid.

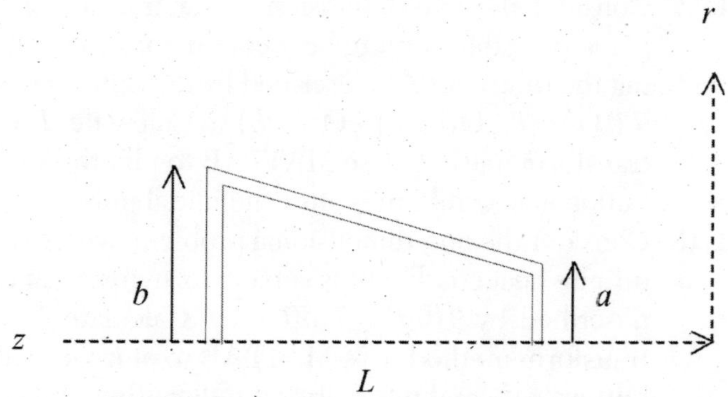

FIGURE E-11.17 **Grande coffee from Starbucks. Geometry: axial symmetry.**

11.18 A silver coin is dropped in a liquid nitrogen bath at time $t = 0$. Calculate the temperature distribution inside the coin modeled in axial symmetry as a function of time. Make a plot of the temperature at the center of the coin as a function of time.

11.19 A rectangular metal plate is initially at a temperature T_1. One corner of the plate is held at temperature $T_2\Theta(t)$. Calculate the temperature distribution in the plate as a function of time.

11.20 One side of rectangular metal plate is held at a temperature T_1. One corner of the plate on the opposite side is held at temperature T_2. Calculate the steady state temperature distribution in the plate. Plot the isothermal contours and heat flux vectors.

11.21 Einstein's molar specific heat formula in (cal/mole K) is given by

$$C(T) = 3R\left(\frac{T_E}{T}\right)\frac{e^{\frac{T_E}{T}}}{\left(e^{\frac{T_E}{T}} - 1\right)^2},$$

where T_E is the Einstein temperature (1300 K for a diamond) $R = 5.97/3$ cal/mol · K. Calculate the temperature distribution in a cylindrical diamond rod with ends at 77 and 4 K with the radiation boundary condition applied to the sides of the diamond. Use the above formula to enter several values in the $C(T)$ curve editor.

11.22 Show that the heat flux vector **F** satisfies

$$\nabla \cdot \mathbf{F} + \rho c\frac{\partial T}{\partial t} = q.$$

11.23 Show that the function $T(x,t) = xt^{-3/2}\exp\left(\frac{x^2}{4at}\right)$ satisfies the one-dimensional heat equation

$$\frac{\partial T}{\partial t} = \mathrm{a}\frac{\partial^2 T}{\partial x^2}.$$

Use MATLAB to create a mesh plot $T(x,t)$.

11.24 Write a cylindrical coordinate Monte Carlo program in MATLAB to calculate the steady state temperature distribution in a cylinder with thermal conductivity λ maintained at temperature T_1 on the sides and T_2 on the end caps.

11.25 Write a Monte Carlo program in MATLAB using 3-D Cartesian coordinates to calculate the steady state temperature distribution in a cube with thermal conductivity λ maintained at temperature T_1 on one side and at T_2 on the other five sides.

11.26 Coupled AC Magnetic and Thermal Analysis: Calculate the temperature distribution in the spherical tumor embedded in a cylindrical thorax. Calculate the SAR in the tumor as a function of frequency of excitation frequency.

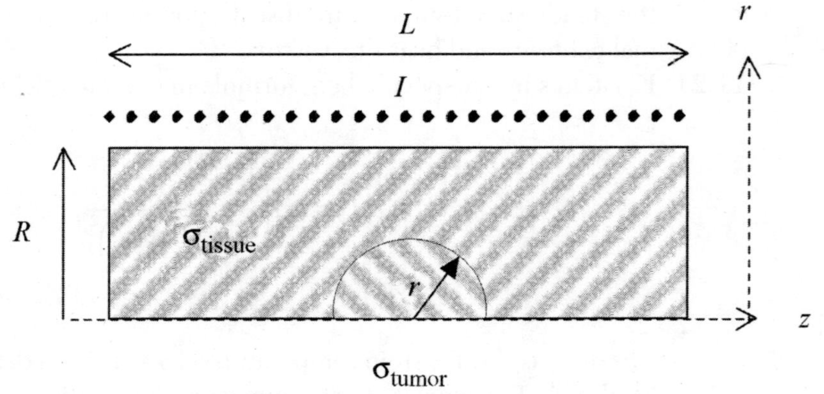

FIGURE E-11.26 **Spherical tumor imbedded in a cylindrical tissue region surrounded by an AC induction coil. Geometry: axial symmetry.**

11.27 Coupled AC Current and Thermal Analysis: Calculate the temperature distribution in the spherical tumor embedded in a cylindrical thorax under electrical stimulation. Calculate the SAR in the tumor as a function of frequency of excitation frequency.

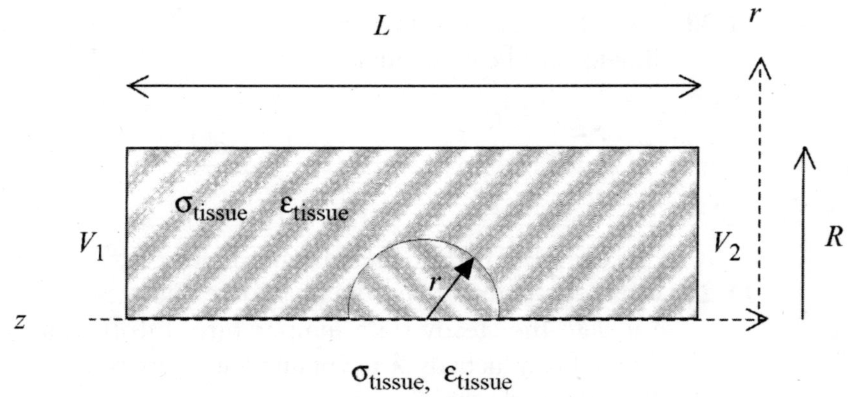

FIGURE E-11.27 **Spherical tumor imbedded in a cylindrical tissue region with AC voltage excitation. Geometry: axisymmetry.**

11.28 Coupled AC magnetic and Thermal Analysis: Calculate the temperature distribution in the human brain containing a tumor resulting from an

AC excitation supplied by a single coil. The cranium and tissue regions surround the brain. Calculate the SAR in the brain tumor as a function of excitation frequency.

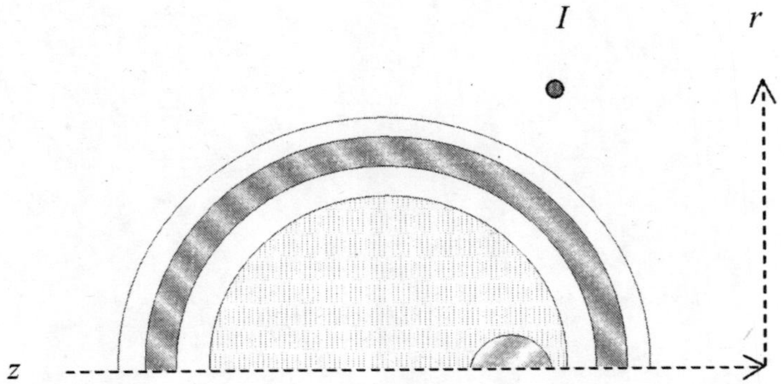

FIGURE E-11.28 **Current loop surrounding the brain: Geometry axisymmetry.**

Chapter 12 STRESS ANALYSIS

In This Chapter

- Stress and Strain
- Beams
- Stress Analysis in QuickField
- Coupled Thermal and Stress Analysis
- Stress Analysis with Electric Forces
- Stress Analysis with Magnetic Forces
- Coupled Electric Current, Thermal and Stress Analysis

12.1 STRESS AND STRAIN

Young's Modulus

In 1807 Thomas Young developed a theory for characterizing material bodies deformed by external forces. Two decades later, C. Navier established a mathematical form of Young's theory in 1826. In one dimension, the stress σ is the force applied to a body divided by its area $\sigma = F/A$. Strain ε, a dimensionless factor, is the change in length ΔL divided by the original length L_0 or $\varepsilon = \Delta L/L_0$. Young's modulus, often referred to as the elastic modulus, is defined as the ratio of the stress divided by the strain

$$E = \frac{\sigma}{\varepsilon} \tag{12.1}$$

for a body in tension or compression. Young's modulus is an intrinsic property of a given material that is independent of its shape. The units of E are force divided by area N/m^2 or Pa. The relation between Young's modulus and Hooke's law $F = k\Delta L$ relating the linear stretch of a body with force constant k under a given load for sufficiently small ΔL may be obtained considering

$$E = \frac{F/A}{\Delta L/L_0}. \tag{12.2}$$

Solving for F gives

$$F = \left(\frac{EA}{L_0}\right)\Delta L \tag{12.3}$$

so that the familiar spring constant is Young's modulus times the cross-sectional area divided by the initial length.

Shear Modulus

The shear modulus S is defined by the shear stress $\sigma_\perp = F_\perp/A$ divided by the shear strain $\varepsilon_\perp = \gamma$ in radians

$$S = \frac{\sigma_\perp}{\varepsilon_\perp} = \frac{F_\perp/A}{\gamma}. \tag{12.4}$$

The sheer modulus is roughly two to three times smaller than Young's modulus in most materials.

Young's modulus is given for various engineering materials in Appendix 4 and for biological tissues in Appendix 5.

Poisson's Ratio

Material bodies pulled in one direction generally tend to contract in the orthogonal direction as shown in Figure 12.1. The negative transverse contraction strain perpendicular to the load divided by the axial, or extension strain in the direction of the load gives Poisson's ratio

$$\upsilon = -\frac{\varepsilon_{\text{trans}}}{\varepsilon_{\text{axial}}}. \tag{12.5}$$

Most materials have υ values between $0 - 0.5$ with many metals $\upsilon = 0.27 - 0.35$.

Color images for many of the figures may be found on the CD-ROM.

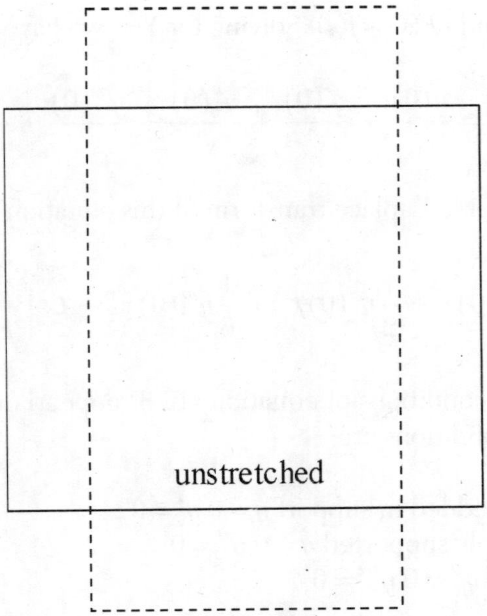

FIGURE 12.1 **Transverse and axial strains of a bar stretched in the vertical direction illustrating Poisson's ratio.**

12.2 BEAMS

One of the simplest analytical structural problems is that of a one-dimensional beam under a distance dependent loading force $F(x)$. The beam deflection is described by the fourth order differential equation

$$EI\frac{d^4y}{d^4x} = F(x), \tag{12.6}$$

where $y(x)$ is the vertical displacement and product EI is the flexural rigidity. The flexural rigidity is defined as Young's modulus E times the area moment of inertia I. The area (or second) moment of inertia is a measure of an object's resistance to bending (not to be confused with the moment of inertia inversely proportional to angular acceleration). The Laplace transform of the beam equation gives

$$s^4Y(s) - s^3y(0) - s^2y'(0) - sy''(0) - y'''(0) = \frac{1}{EI}f(s) \tag{12.7}$$

with $Ly(x) = Y(s)$ and $LF(x) = f(s)$. Solving for $Y(s)$ we have

$$Y(s) = \frac{y(0)}{s} + \frac{y'(0)}{s^2} + \frac{y''(0)}{s^3} + \frac{y'''(0)}{s^4} + \frac{1}{EI}\frac{f(s)}{s^4}. \quad (12.8)$$

Taking the inverse Laplace transform of this equation gives

$$y(x) = y(0) + y'(0)x + \frac{1}{2}y''(0)x^2 + \frac{1}{6}y'''(0)x^3 + L^{-1}\left(\frac{1}{EI}\frac{f(s)}{s^4}\right). \quad (12.9)$$

The boundary conditions of equation (12.8) depend on the type of support. These boundary conditions are:

I. Beam imbedded in support $y = 0$ $y' = 0$
II. Beam simply supported $y = 0$ $y'' = 0$
III. Beam free $y'' = 0$ $y''' = 0$

Figure 12.2 illustrates several different support configurations on each end of the beam. Simple load configurations include the delta function and Heaviside step function loading at various beam locations.

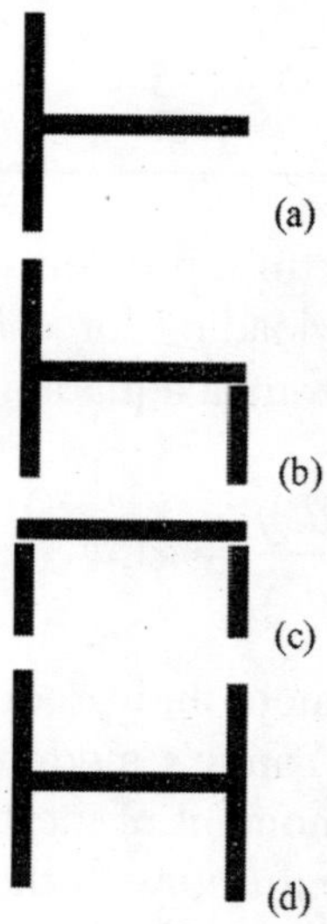

FIGURE 12.2 **(a) Beam with one end embedded and one end free; (b) Beam with both ends embedded; (c) Beam embedded at one end with one end simply supported; (d) Beam with both ends simply supported.**

12.3 STRESS ANALYSIS IN QUICKFIELD

Linear stress analysis in QuickField is based on Navier's equations of elasticity. Stress and strain calculations may be performed in x-y and axial symmetry.

X-Y Symmetry

In two-dimensional x-y symmetry we may define a displacement field as

$$\mathbf{d} = \delta_x \mathbf{i} + \delta_y \mathbf{j}. \tag{12.10}$$

The three independent components of the strain are related to the displacement field

$$[\varepsilon] = \begin{Bmatrix} \varepsilon_x \\ \varepsilon_y \\ \gamma_{xy} \end{Bmatrix} = \begin{Bmatrix} \dfrac{\partial \delta_x}{\partial x} \\ \dfrac{\partial \delta_y}{\partial y} \\ \dfrac{\partial \delta_x}{\partial y} + \dfrac{\partial \delta_y}{\partial x} \end{Bmatrix} \tag{12.11}$$

with corresponding stress components

$$[\sigma] = \begin{Bmatrix} \sigma_x \\ \sigma_y \\ \tau_{xy} \end{Bmatrix}. \tag{12.12}$$

The equilibrium equations relate the stress components and the volume force vector $\mathbf{f} = f_x \mathbf{i} + f_y \mathbf{j}$

$$\frac{\partial \sigma_x}{\partial x} + \frac{\partial \tau_{xy}}{\partial y} = -f_x \tag{12.13}$$

$$\frac{\partial \tau_{xy}}{\partial x} + \frac{\partial \sigma_y}{\partial y} = -f_y. \tag{12.14}$$

The stress-strain relation for linear elasticity is given by

$$[\sigma] = [D] \cdot [\varepsilon - \varepsilon_0], \tag{12.15}$$

where ε_0 is the initial thermal strain. The matrix of elastic coefficients $[D]$ will involve components of Young's modulus and Poisson's ratio for the various situations given below:

(1) Plane stress and isotropic material

$$[D] = \frac{E}{1 - v^2}\begin{bmatrix} 1 & v & 0 \\ v & 1 & 0 \\ 0 & 0 & \dfrac{1 - v}{2} \end{bmatrix} \tag{12.16}$$

(2) Plane stress and orthotropic material

$$[D] = \begin{bmatrix} \dfrac{1}{E_x} & -\dfrac{v_{yx}}{E_y} & 0 \\ -\dfrac{v_{yx}}{E_y} & \dfrac{1}{E_y} & 0 \\ 0 & 0 & \dfrac{1}{S_{xy}} \end{bmatrix}^{-1} \tag{12.17}$$

(3) Plane strain and isotropic material

$$[D] = \frac{E(1 - v)}{(1 + v)(1 - 2v)}\begin{bmatrix} 1 & \dfrac{v}{1 - v} & 0 \\ \dfrac{v}{1 - v} & 1 & 0 \\ 0 & 0 & \dfrac{1 - 2v}{2(1 - v)} \end{bmatrix} \tag{12.18}$$

(4) Plain strain and orthotropic material

$$[D] = \begin{bmatrix} \dfrac{1}{E_x} - \dfrac{v_{zx}^2}{E_z} & -\dfrac{v_{yx}}{E_y} - \dfrac{v_{zx}v_{zy}}{E_z} & 0 \\ -\dfrac{1}{E_y} - \dfrac{v_{zx}v_{zy}}{E_z} & \dfrac{1}{E_y} - \dfrac{v_{zy}^2}{E_z} & 0 \\ 0 & 0 & \dfrac{1}{S_{xy}} \end{bmatrix}^{-1} \tag{12.19}$$

E_x E_y E_z are Young's modulus components of orthotropic materials along the x-, y-, and z-axis, respectively. v_{yx} v_{zx} v_{zy} are Poisson's ratios for orthotropic materials; v is the Poisson ratio for isotropic materials and S_{xy} is the shear modulus.

Axisymmetry

The displacement vector field is given by

$$\mathbf{d} = \delta_r \hat{\mathbf{r}} + \delta_z \hat{\mathbf{z}}. \tag{12.20}$$

The strain components are given by their corresponding derivatives of the displacement

$$[\varepsilon] = \begin{Bmatrix} \varepsilon_z \\ \varepsilon_r \\ \varepsilon_\theta \\ \gamma_{rz} \end{Bmatrix} = \begin{Bmatrix} \dfrac{\partial \delta_z}{\partial z} \\ \dfrac{\partial \delta_r}{\partial r} \\ \dfrac{\delta_r}{r} \\ \dfrac{\partial \delta_z}{\partial r} + \dfrac{\partial \delta_r}{\partial z} \end{Bmatrix}, \tag{12.21}$$

where ε_θ is the out-of-plane strain caused by radial deformation. The stress components are given by

$$[\sigma] = \begin{Bmatrix} \sigma_z \\ \sigma_r \\ \sigma_\theta \\ \tau_{rz} \end{Bmatrix}, \tag{12.22}$$

which obey the axisymmetric equilibrium equation

$$\frac{1}{r}\frac{\partial (r\sigma_r)}{\partial r} + \frac{\partial \tau_{rz}}{\partial z} = -f_r \tag{12.23}$$

$$\frac{1}{r}\frac{\partial (r\tau_{rz})}{\partial r} + \frac{\partial \sigma_z}{\partial z} = -f_z, \tag{12.24}$$

where $\mathbf{f} = f_r\hat{\mathbf{r}} + f_z\hat{\mathbf{z}}$. The stress-strain components obey $[\sigma] = [D]\cdot[\varepsilon - \varepsilon_0]$ with elastic coefficient matrices for the cases:

(1) Axisymmetric and isotropic material

$$[D] = \frac{E(1-v)}{(1+v)(1-2v)}\begin{bmatrix} 1 & \frac{v}{1-v} & \frac{v}{1-v} & 0 \\ \frac{v}{1-v} & 1 & \frac{v}{1-v} & 0 \\ \frac{v}{1-v} & \frac{v}{1-v} & 1 & 0 \\ 0 & 0 & 0 & \frac{1-2v}{2(1-v)} \end{bmatrix} \tag{12.25}$$

(2) Axisymmetric and orthotropic material

$$[D] = \begin{bmatrix} \frac{1}{E_z} & -\frac{v_{rz}}{E_r} & -\frac{v_{\theta z}}{E_\theta} & 0 \\ -\frac{v_{rz}}{E_r} & \frac{1}{E_r} & -\frac{v_{\theta r}}{E_\theta} & 0 \\ -\frac{v_{\theta z}}{E_\theta} & -\frac{v_{\theta r}}{E_\theta} & \frac{1}{E_\theta} & 0 \\ 0 & 0 & 0 & \frac{1}{G_{zr}} \end{bmatrix}^{-1}, \tag{12.26}$$

where E_r E_θ E_z are Young's moduli of orthotropic materials along the r-, θ-, and z-axis, respectively. v_{rz} $v_{\theta z}$ and $v_{\theta r}$ are Poisson's ratios for orthotropic materials; v is the Poisson ratio for isotropic materials and G_{zr} is the shear modulus.

Thermal Strain

Thermal strain for plane stress and isotropic materials

$$[\varepsilon_0] = \begin{Bmatrix} \alpha \\ \alpha \\ 0 \end{Bmatrix}\Delta T \tag{12.27}$$

Thermal strain for plane stress and orthotropic materials

$$[\varepsilon_0] = \begin{Bmatrix} \alpha_x \\ \alpha_y \\ 0 \end{Bmatrix}\Delta T \tag{12.28}$$

Plane strain for isotropic materials

$$[\varepsilon_0] = (1 + v)\begin{Bmatrix} \alpha \\ \alpha \\ 0 \end{Bmatrix}\Delta T \tag{12.29}$$

Plane strain for orthotropic materials

$$[\varepsilon_0] = \begin{Bmatrix} \alpha_x + v_{zx}\alpha_z \\ \alpha_y + v_{zy}\alpha_z \\ 0 \end{Bmatrix}\Delta T \tag{12.30}$$

For axisymmetric problems and isotropic materials

$$[\varepsilon_0] = \begin{Bmatrix} \alpha \\ \alpha \\ \alpha \\ 0 \end{Bmatrix}\Delta T \tag{12.31}$$

For axisymmetric problems and orthotropic materials

$$[\varepsilon_0] = \begin{Bmatrix} \alpha_z \\ \alpha_r \\ \alpha_\theta \\ 0 \end{Bmatrix}\Delta T \tag{12.32}$$

Material Properties and Loading Sources

Windows for editing block label data for stress analysis problems are shown in Figures 12.3 through 12.5: To open the block label window click on the corresponding block label in the tree diagram on the left of the screen. Select Elasticity, Loads, or Allowable Stresses Tabs.

Elasticity data includes Young's Modulus components, Poisson's Ratios, and Shear Modulus. Only two of these quantities are independent for isotropic materials. The Anisotropic box is checked to specify orthotropic materials, seven independent quantities. If "None" is selected in Young's modulus then blocks corresponding to the label will not be included in the calculation. Enter a value of zero and "None" will appear.

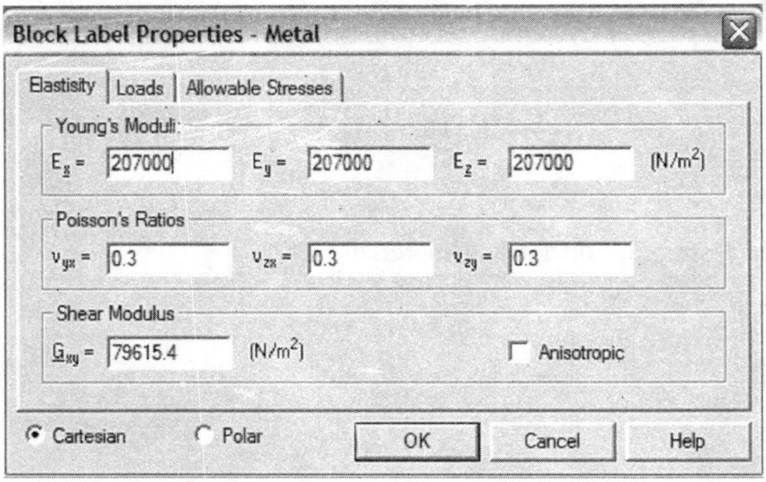

FIGURE 12.3 **Windows for editing elasticity block label data: To open the block label window click on the corresponding block label in the tree diagram on the left of the screen. Select the Elasticity tab and enter Young's Modulus components, Poisson's Ratios, and the Shear Modulus (if Anisotropic).**

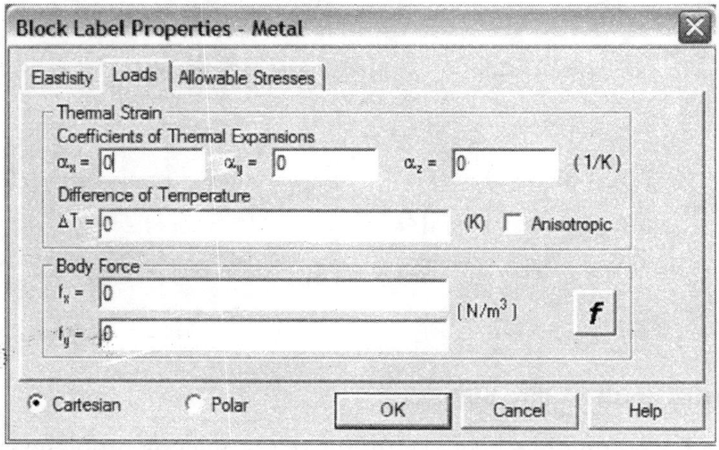

FIGURE 12.4 **Windows for editing block label loads data: To open the block label window click on the corresponding block label in the tree diagram on the left of the screen. Select Loads and enter Coefficients of Thermal Expansion, Difference of Temperature, and Body Force with any formula dependence.**

Loads include Thermal Strain and Body Forces. Thermal loading is treated differently for coupled and uncoupled problems. For uncoupled thermal loading, the difference in temperature is specified between strained and unstrained states.

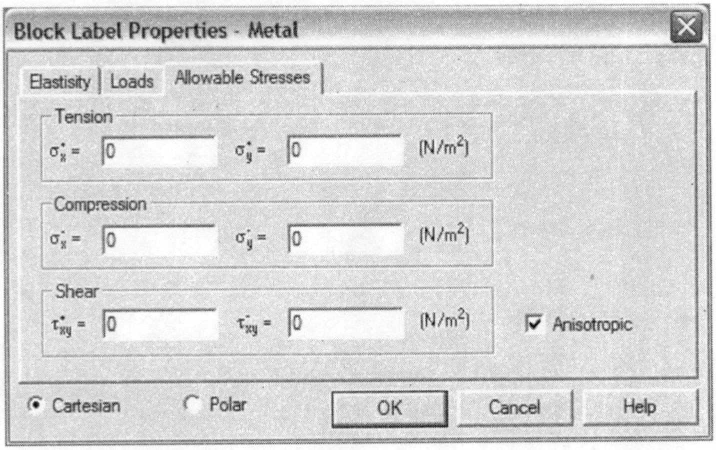

FIGURE 12.5 **Windows for editing allowable stresses block label data: This data does not influence the solution of the stress analysis problem but is useful for calculating various stress criteria. Enter Allowable Tension, Compression, and Sheared components.**

Allowable Stresses data including allowable tension and compression may be specified. This data does not affect the solution of the problem but is used in post processing to calculate Mohr-Coulomb, Drucker-Prager, and Hill criteria. Entering allowable stresses is optional if these criteria are not required.

Boundary Conditions

The Edge Label Properties window for editing edge label data for stress analysis problems is shown in Figure 12.6: To open the edge label window click on the corresponding edge label in the tree diagram on the left of the screen. Select values for Prescribed Displacements, Normal Pressure, and Surface Force components. To apply displacements of a boundary along a given axis check the boxes corresponding to the desired displacement.

Vertex Conditions

The window for editing vertex label data is shown in Figure 12.7. To open the vertex label window click on the corresponding vertex label in the tree diagram on the left of the screen. Vertex properties include Rigid Constrains along a given axis, External Force components applied to a point, and Elastic Support components.

Calculated Values

Post-processing results include the displacement field with magnitudes.

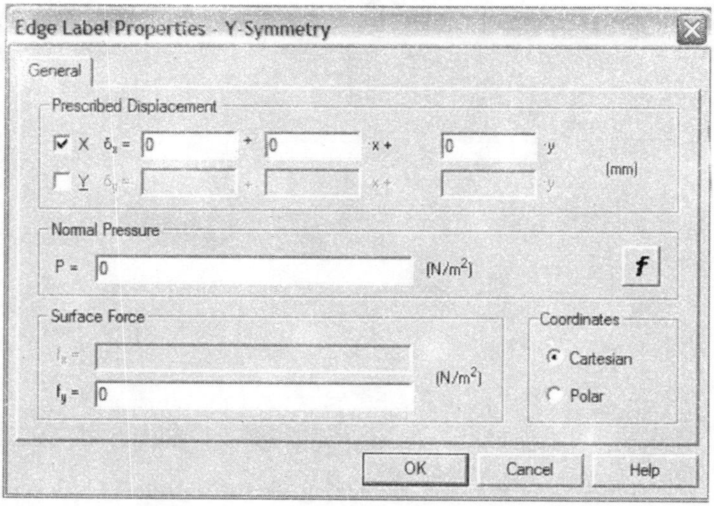

FIGURE 12.6 **Window for editing edge label data: To open the edge label window click on the corresponding edge label in the tree diagram on the left of the screen. Enter Prescribed Displacements, Normal Pressure with any functional coordinate dependence, and Surface Force components.**

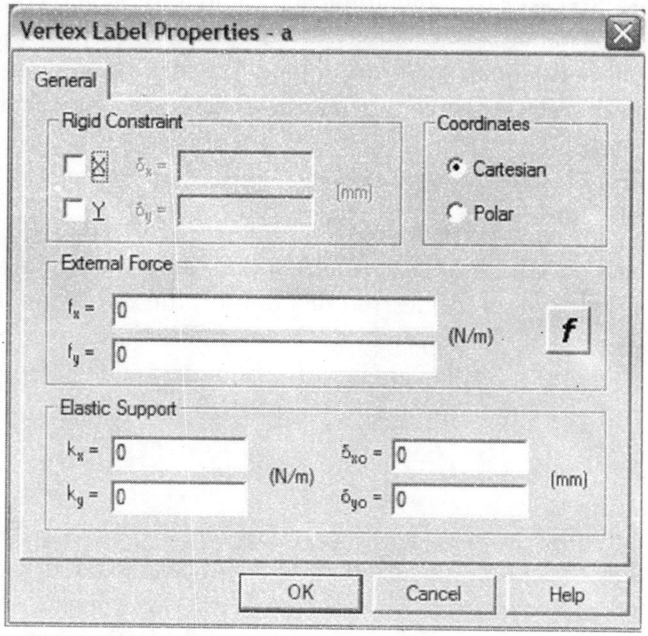

FIGURE 12.7 **Window for editing vertex label data: To open the vertex label window click on the corresponding vertex label in the tree diagram on the left of the screen. Enter Rigid Constraints, External Force components (with any coordinate functional dependence), and Elastic Support components.**

$$d = \sqrt{\delta_x^2 + \delta_y^2} \tag{12.33}$$

for *x-y* symmetry and

$$d = \sqrt{\delta_r^2 + \delta_z^2} \tag{12.34}$$

for axial symmetry. The von Mises stress

$$\sigma_e = \sqrt{\frac{(\sigma_1 - \sigma_2)^2 + (\sigma_2 - \sigma_3)^2 + (\sigma_3 - \sigma_1)^2}{2}} \tag{12.35}$$

is related to the energy of deformation. The maximum shear is given by the Tresca criterion

$$\sigma_e = \sigma_1 - \sigma_3. \tag{12.36}$$

The Mohr-Coulomb criterion is calculated as

$$\sigma_e = \sigma_1 - \chi\sigma_3, \tag{12.37}$$

where

$$\chi = \frac{[\sigma_+]}{[\sigma_-]} \tag{12.38}$$

and $[\sigma_+]$ and $[\sigma_-]$ denote tensile and compressive allowable stress. The Drucker-Prager criterion is calculated as

$$\sigma_i = (1 + \chi)\sigma_i - \frac{\sqrt{\chi} - \chi}{1 + \sqrt{\chi}}\bar{\sigma} + \frac{1}{[\sigma_-]}\left(\frac{1 - \sqrt{\chi}}{1 + \sqrt{\chi}}\bar{\sigma}\right)^2, \tag{12.39}$$

where

$$\sigma_i = \sqrt{\frac{(\sigma_1 - \sigma_2)^2 + (\sigma_2 - \sigma_3)^2 + (\sigma_3 - \sigma_1)^2}{2}} \tag{12.40}$$

and

$$\bar{\sigma} = \frac{\sigma_1 + \sigma_2 + \sigma_3}{3}. \tag{12.41}$$

The Hill failure index for orthotropic materials is

$$\text{F. I.} = \frac{\sigma_1^2}{X_1^2} - \frac{\sigma_1\sigma_2}{X_1^2} + \frac{\sigma_2^2}{X_2^2} + \frac{\tau_{12}^2}{S_{12}^2}, \tag{12.42}$$

where σ_1, σ_2, and τ_{12} are the computed stresses in the material directions and

$$X_1 = \begin{cases} X_1^T & \text{if } \sigma_1 > 0 \\ X_1^C & \text{if } \sigma_1 < 0 \end{cases} \tag{12.43}$$

$$X_2 = \begin{cases} X_2^T & \text{if } \sigma_2 > 0 \\ X_2^C & \text{if } \sigma_2 < 0 \end{cases}, \tag{12.44}$$

where X_1^T, X_2^T, and X_1^C, X_2^C are tensile and compressive allowable stresses

$$S_{12} = \begin{cases} S_{12}^+ & \text{if } \tau_{12} > 0 \\ S_{12}^- & \text{if } \tau_{12} < 0 \end{cases}, \tag{12.45}$$

where S_{12}^+ and S_{12}^- are allowable sheer stresses.

Integral Values

Calculated integral quantities include the vector force acting on a volume, which is given by the integral

$$\mathbf{F} = \oiint_{\text{surf}} (s \cdot \mathbf{n})\, dS \tag{12.46}$$

over the surface of the volume, where σ is the stress tensor and $\mathbf{n}$ is the unit normal to the surface. The torque acting on a volume is given by the surface integral

$$\mathbf{T} = \oiint_{\text{surf}} \mathbf{r} \times (s \cdot \mathbf{n})\, dS \tag{12.47}$$

and $\mathbf{r}$ is the radius vector to the point of integration. Note that there is no torque in axial symmetric problems.

Stress Distribution Near a Crack

Figure 12.8 (a) shows an intensity map of the von Mises stress distribution in a 90 × 182 cm metal plate with a crack under vertical loading. Figure 12.8 (b) shows the finite element mesh near the crack showing an increased mesh density in this region. The metal plate has a Young's modulus $E = 2.07 \cdot 10^5\,\text{N/m}^2$,

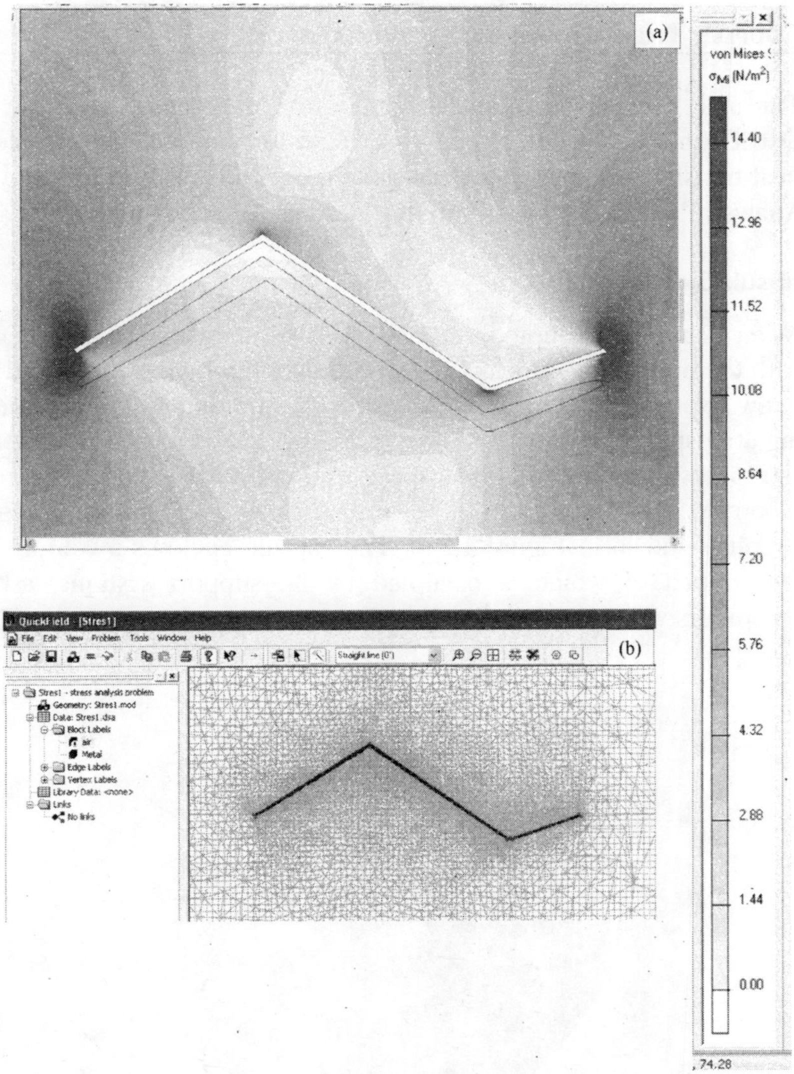

FIGURE 12.8 **(a) Crack in a metal plate under loading in the vertical direction showing intensity map of von Mises stress and deformed boundary (×100). Crack dimensions are roughly 32 mm × 0.4 mm. (b) Finite element mesh with increased mesh density near the crack.**

a Poisson ratio of $v = 0.3$ in each orthogonal direction, and a shear modulus $G_{zr} = 7.96 \cdot 10^4\,N/m^2$. The crack has 32×0.4 mm dimensions with zero Young's modulus. The top of the plate is held fixed while a downward vertical load of $-1.0\,N/m^2$ is applied to the bottom of the plate. The figure shows that the stress is greatest near the crack edges. The deformed crack boundary is shown ×100.

12.4 COUPLED THERMAL AND STRESS ANALYSIS

Thermal stresses may be calculated in QuickField based on temperature distributions calculated in the Heat Transfer module. The coupled stress and heat transfer problems share the same model file. A link is created in the Stress Analysis Problem Properties to the Thermal Analysis problem file.

Space Capsule Simulations

As an example of coupled thermal and stress analysis, we consider a spacecraft with the outer dimensions of the Constellation programs proposed by the Orion Crew Exploration Vehicle (CEV). An artist's rendition of the Orion capsule during atmospheric reentry is shown in Figure 12.9. Orion is designed to carry six astronauts to the international space station by 2014 and four astronauts to the moon by 2020. A mockup of the capsule to be carried into space by the Ares rocket is shown in Figure 12.10. The capsule encloses a volume of 700 cubic feet, much of which is occupied by life-support systems, astronaut gear, computers, avionics, and thermal protection systems as shown in Figure 12.11.

FIGURE 12.9 **Artist's rendering of a Crew Exploration Vehicle's (CEV) during atmospheric re-entry (obsolete configuration) Image credit: NASA.**

FIGURE 12.10 Mockup of Orion crew module built by the Fabrication Branch at NASA's Dryden Flight Research Center. Image credit: NASA.

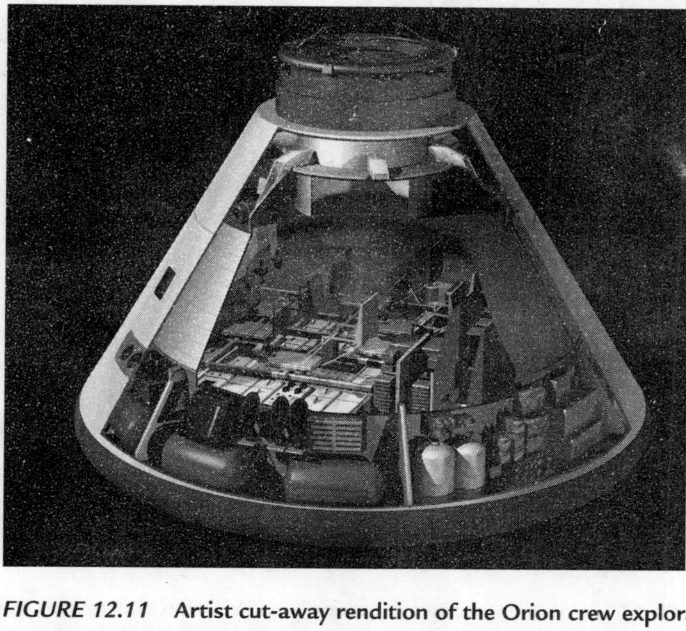

FIGURE 12.11 Artist cut-away rendition of the Orion crew exploration vehicle's crew module showing the outer shell, inner pressure vessel, and equipment bays. Image Credit: NASA.

The Orion CEV is similar in design to the Apollo capsule with some components such as the outer shell, heat shield, and inner pressure shell that may be simulated in axial symmetry. Current capsule designs are roughly 3.62 meters in height with a maximum radius of 2.25 meters and frustrated cone angle of 32.5 degrees.

Launch Abort System

An artist's rendition of the Orion crew module to be used for launch abort and landing systems testing is shown in Figure 12.12. The integrated launch abort system, crew module, service module, and spacecraft adapter is depicted in Figure 12.13 along with the pad abort flight test depiction in Figure 12.14.

Atmospheric Reentry

We consider an idealized model of atmospheric reentry of the space vehicle with axisymmetric temperature and stress distribution. The gas dynamics surrounding the capsule is complicated requiring detailed modeling of the shock

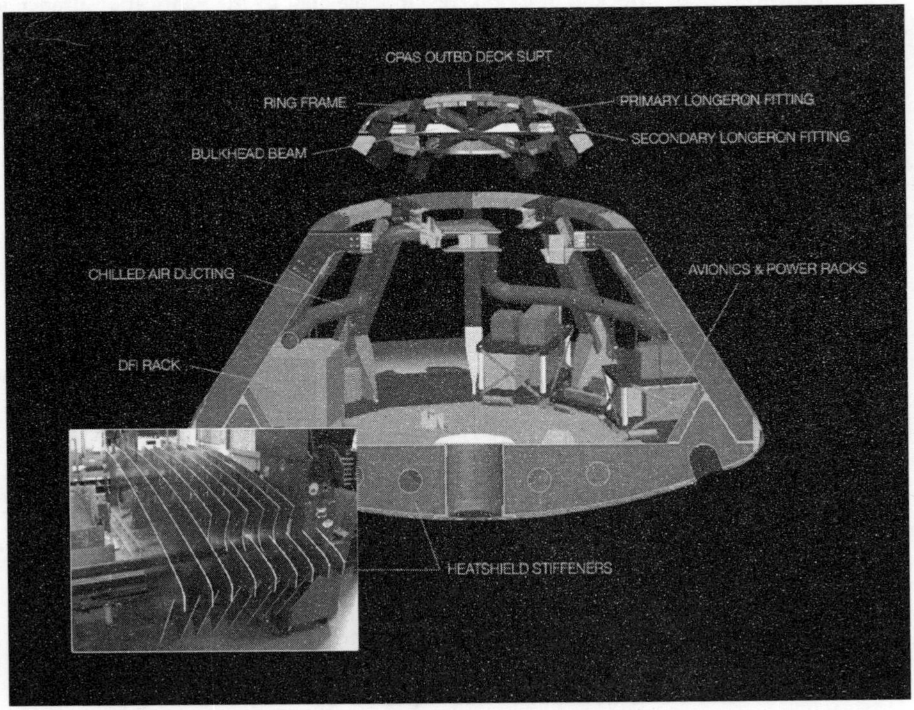

FIGURE 12.12 **Artists cut-away rendering of the Orion boilerplate crew module that will be used in Launch Abort System and landing systems test at White Sands Missile Range, N.M. Image credit: NASA.**

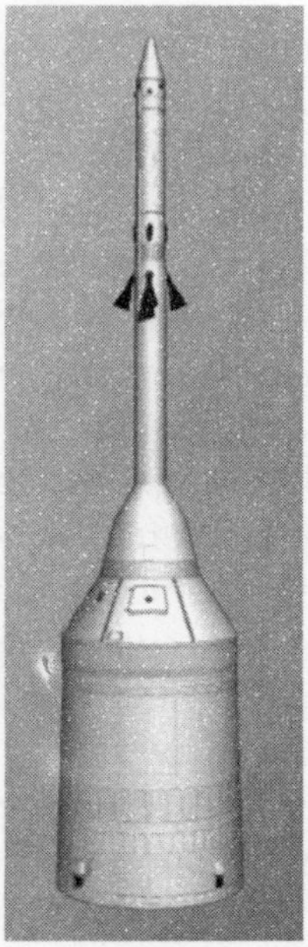

FIGURE 12.13 **Artist rendering of integrated Orion crew exploration vehicle including the launch abort system, crew module, service module, and spacecraft adapter. Image credit: NASA.**

FIGURE 12.14 **Artist rendering of a pad abort flight test of the Orion Launch Abort System. Image credit: NASA.**

layer that forms on the front surface of the thermal protection systems heat shield. Computational fluid dynamics including real-gas modeling are required to account for the breakdown and recombination of air molecules in this region. Evidently friction is not the primary heat source in the sock layer but is second to radiative heating from recombining molecules. In addition, ablation of the thermal protection system can influence the gas dynamics. It turns out that under certain conditions the peak shock layer temperature in Kelvins is roughly equal to the reentry speed in meters per second.

Figure 12.15 shows a simplified axisymmetric finite element model of the space capsule with boundaries between different structural elements. So far only the heat shield support is meshed and there is no separation between the outer shell and inner pressure shell. A close-up of the capsule edge is shown in Figure 12.16 illustrating (a) the ablative heat shield, (b) thermal protection systems carbon-fiber composite matrix, (c) air bag compartment, (d) air bag deployment system compartment, (e) titanium cabin shell, and (f) crew compartment. Figure 12.17 shows the temperature distribution in the boilerplate capsule model with 3000 K specified on the heat shield boundary. This temperature distribution is then imported into the Stress Analysis module, where von Mises stress distribution and deformed boundary of the capsule are plotted in Figure 12.18.

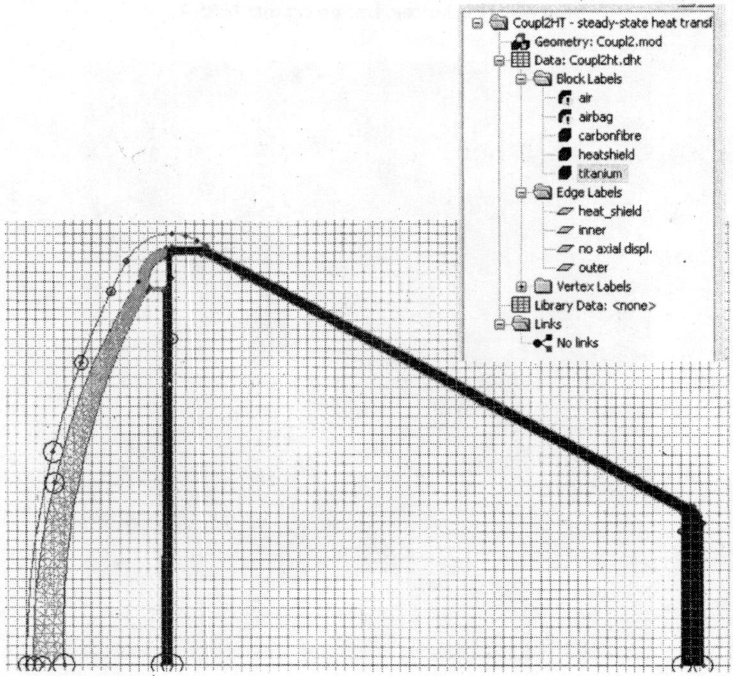

FIGURE 12.15 Space capsule finite element model (under construction) showing different structural elements. So far only the heat shield support is meshed.

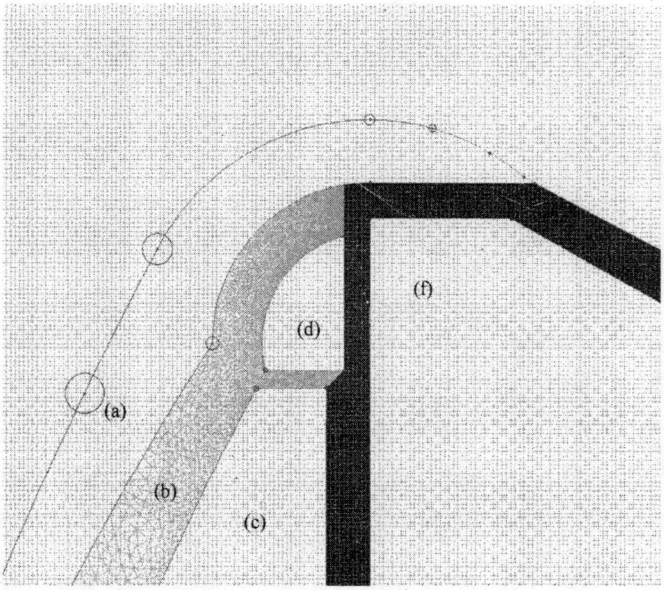

FIGURE 12.16 **(a) Ablative heat shield; (b) thermal protection systems carbon-fiber composite matrix; (c) air bag compartment; (d) air bag deployment system compartment; (e) titanium cabin shell; (f) crew compartment.**

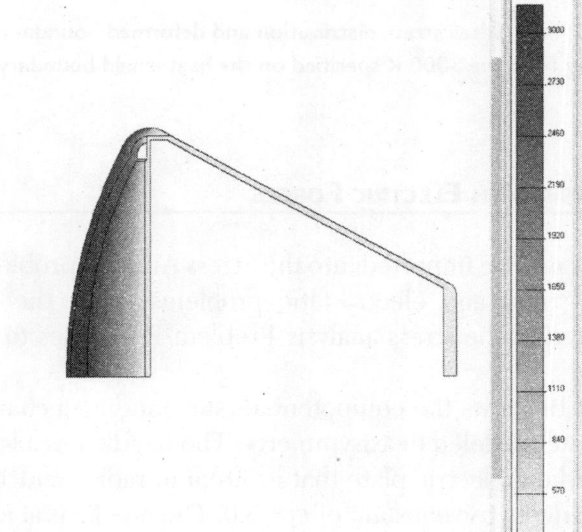

FIGURE 12.17 **Temperature distribution in the Orion Crew Exploration Vehicle model with 3000 K specified on the heat shield boundary.**

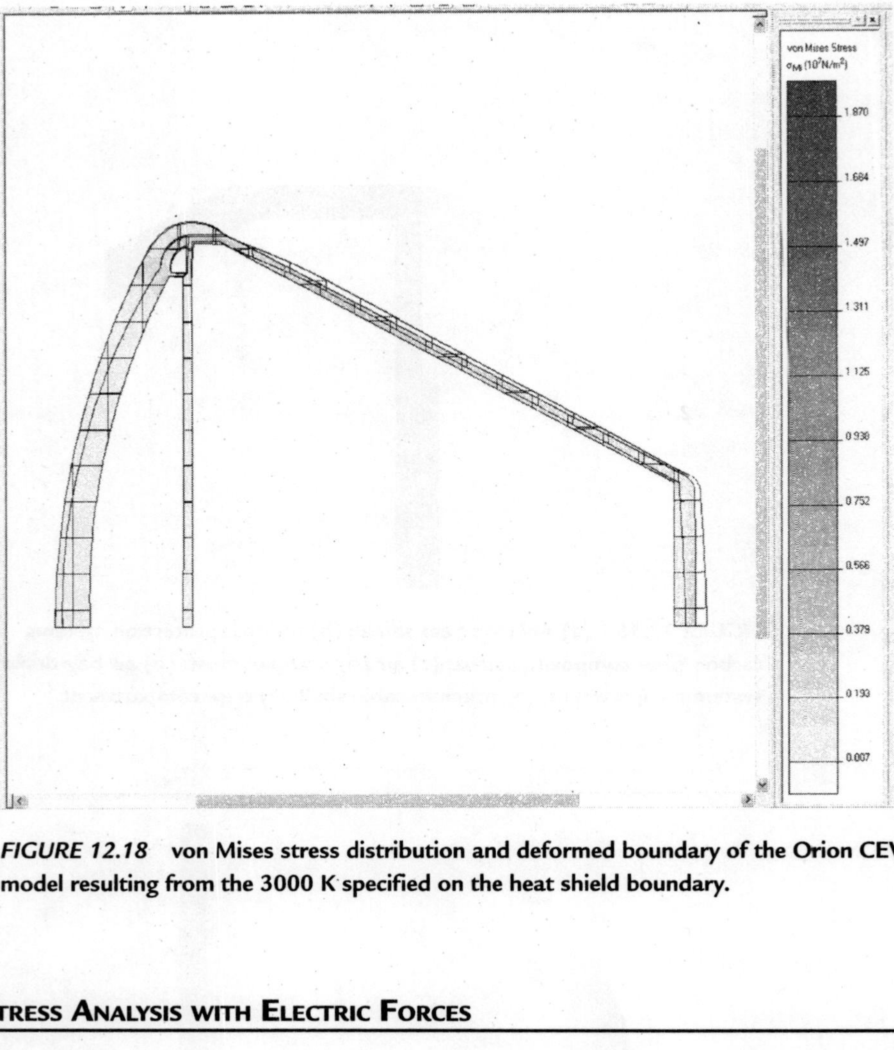

FIGURE 12.18 **von Mises stress distribution and deformed boundary of the Orion CEV model resulting from the 3000 K specified on the heat shield boundary.**

12.5 STRESS ANALYSIS WITH ELECTRIC FORCES

Electric forces may be imported into the Stress Analysis problems in QuickField. The coupled stress and electrostatic problems share the same model file. A link is created in the stress analysis Problem Properties to the Electrostatics problem file.

Figure 12.19 shows the equipotentials surrounding a charged needle above a dielectric plate modeled in axisymmetry. The needle has a length of 20 cm and is 1 cm above the dielectric plate that is 30 cm in radius and 1.0 cm in thickness with a relative dielectric constant of $\varepsilon_r = 5.0$. The needle is at a potential of 100 V with respect to the edges of the solution region held at zero potential.

The dielectric plate has a Young's modulus $E = 10^6$ N/m^2 and a Poisson ratio of $\upsilon = 0.335$ in each orthogonal direction and a shear modulus $G_{zr} = 3.7 \cdot 10^5$ N/m^2. Figure 12.20 shows a color map of von Mises stress and displacement $\times 2 \cdot 10^8$.

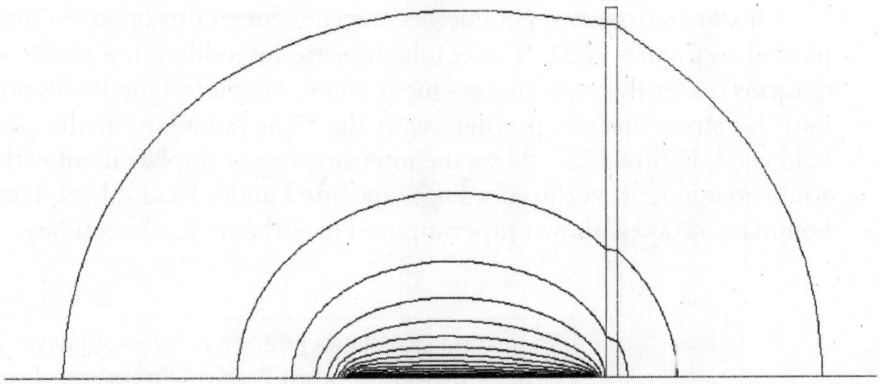

FIGURE 12.19 **Equipotentials surrounding a charged needle above a dielectric plate modeled in axisymmetry.**

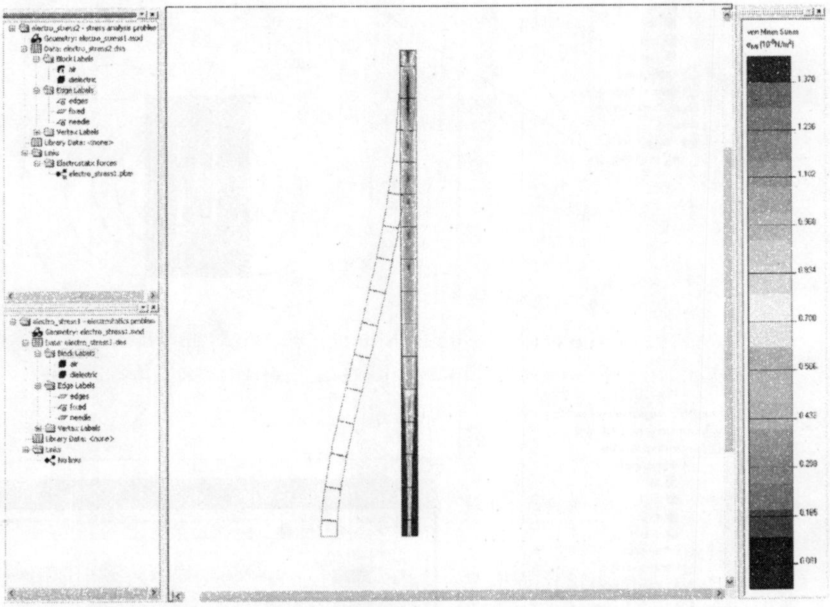

FIGURE 12.20 **Intensity map of von Mises stress and deformed boundary (×2.0 · 10^8) of a dielectric plate under the charged needle shown in Figure 12.19.**

12.6 STRESS ANALYSIS WITH MAGNETIC FORCES

Magnetic forces calculated in Magnetostatics, AC Magnetics, or Transient Magnetic modules may be used as sources in the Stress Analysis module. The coupled stress and magnetic problems share the same model file. A link is created in the stress analysis Problem Properties to the corresponding magnetics problem file.

Flux lines from two conductors carrying current in opposite directions are plotted in Figure 12.21. The conductors are imbedded in a plastic matrix that deforms under the repulsive magnetic force. Magnetostatic forces are imported into the stress analysis problem with the right boundary of the plastic casing held fixed. Figure 12.22 shows the intensity map of displacement with stress and strain components at the coordinate specified under local values. The deformed boundary ×2000 is shown superimposed over the original boundary.

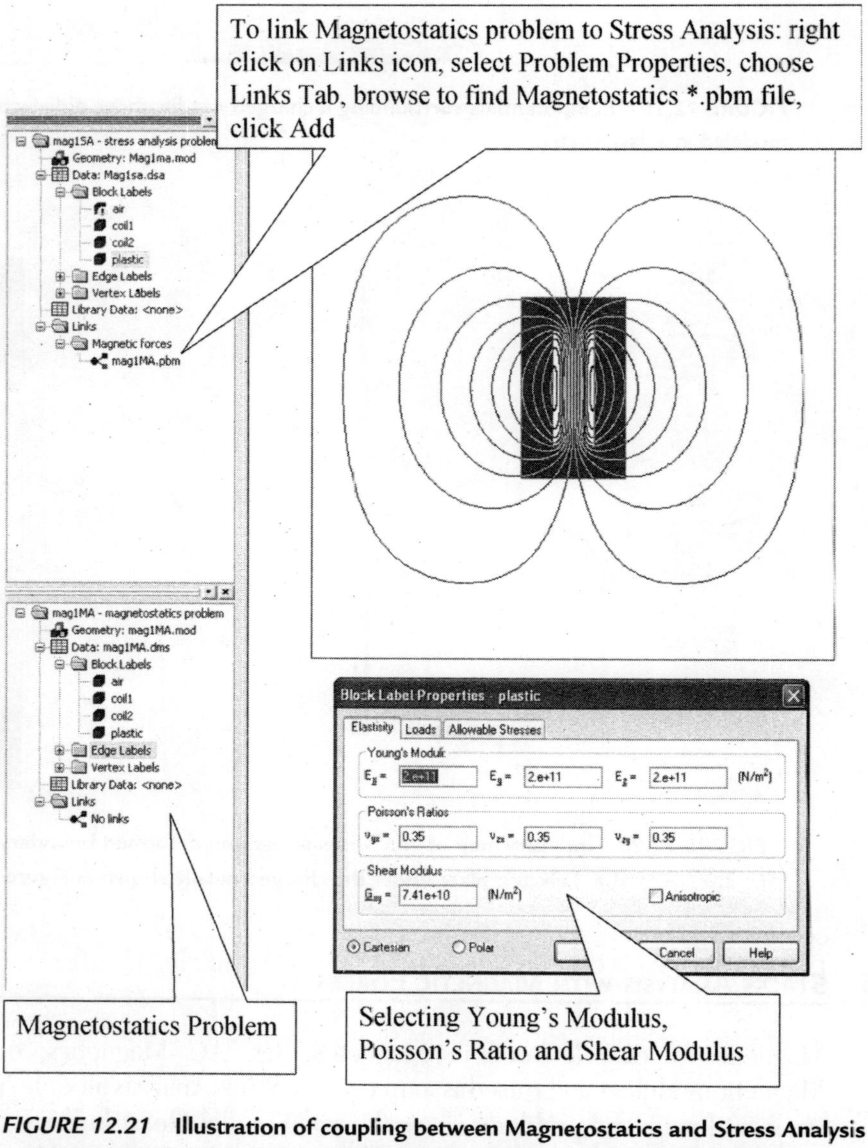

FIGURE 12.21 **Illustration of coupling between Magnetostatics and Stress Analysis problems. Flux lines result from slabs carrying current in opposite directions.**

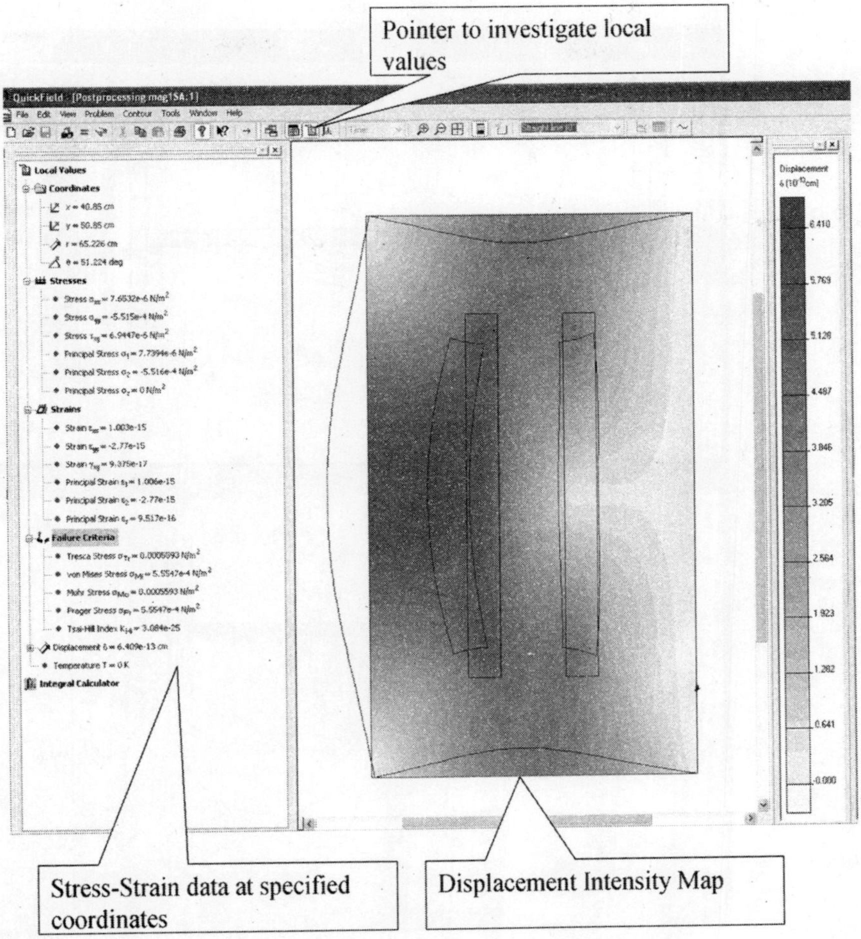

FIGURE 12.22 **Displacement field and deformed boundary (×2000) resulting from the repulsive magnetic force between current carrying conductors. Stress and strain values are at the coordinates specified above.**

We consider a magnet below a highly I-beam structure in Figure 12.23 as a second example of coupled magnetostatics and stress analysis. The flux density is roughly 7.0 mT in the gap between the magnet and the I-beam. The magnetic coercive force equals to 1000 A/m oriented 90 degrees from the x-axis. The I-beam has a width of 68 cm and a height of 32 cm. Thickness of central section is 2 cm and a 2-cm gap separates the magnet and beam. The relative permeability of the I-beam is $\mu_r = 10^4$. Young's modulus of the I-beam is $E = 7 \cdot 10^{10}$ N/m^2 with a shear modulus $G_{xy} = 2.62 \cdot 10^{10}$ N/m^2 and a Poisson ratio $v = 0.335$ in each orthogonal direction. The intensity map of von Mises stress is shown with the deformed boundary $\times 7 \cdot 10^7$.

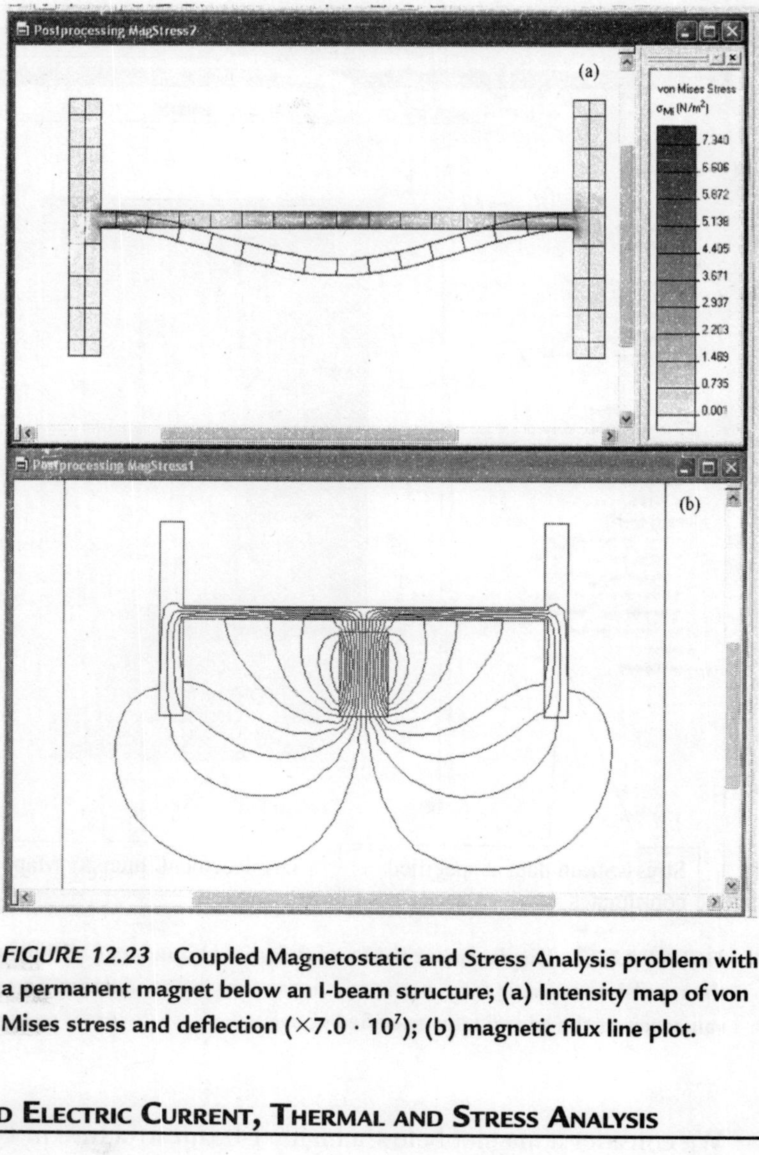

FIGURE 12.23 **Coupled Magnetostatic and Stress Analysis problem with a permanent magnet below an I-beam structure; (a) Intensity map of von Mises stress and deflection ($\times 7.0 \cdot 10^7$); (b) magnetic flux line plot.**

12.7 COUPLED ELECTRIC CURRENT, THERMAL AND STRESS ANALYSIS

The thermal field generated by electric currents (AC, DC, or Induction currents generated in Transient or Time-harmonic Magnetics) may also be used as sources in the Stress Analysis module. The stress, thermal, and electric (AC or DC) or transient magnetic (transient or time-harmonic) problems share the same model file. A link is created in the stress analysis Problem Properties to the corresponding thermal problem file, which in turn is linked to the electric (or transient magnetic) problem file.

As an example of a coupled current flow, heat transfer and stress analysis problem we consider the current flow through a resistive wire with a disk-shaped

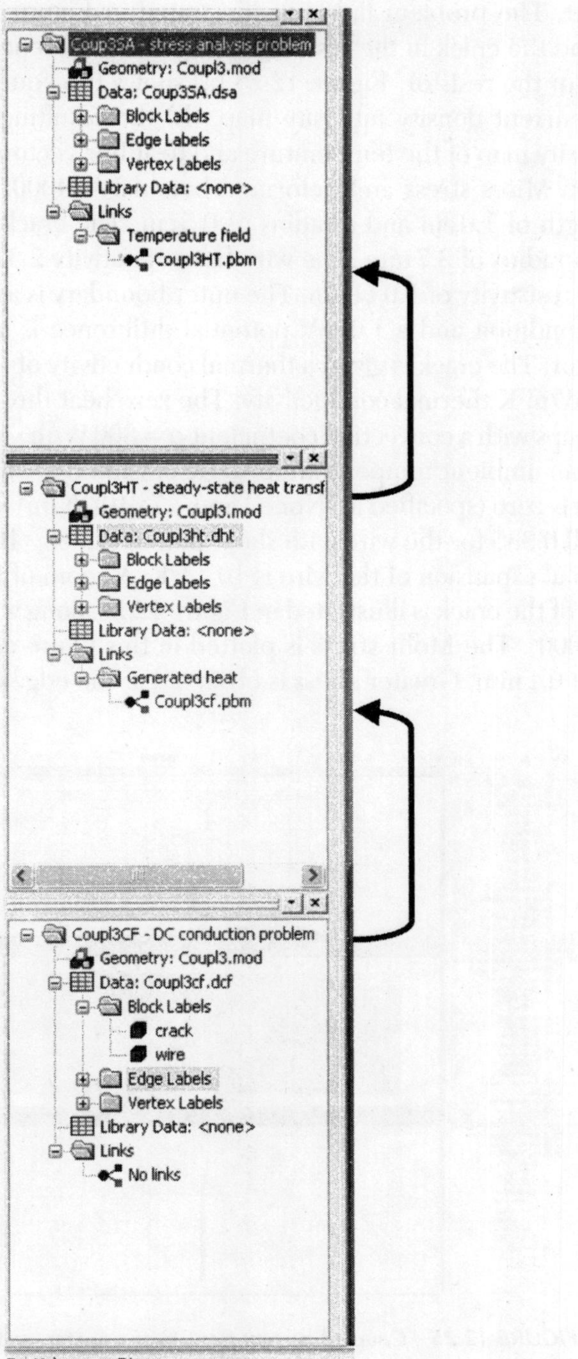

FIGURE 12.24 **Links between DC Conduction and steady state Heat Transfer problems and link between Heat Transfer and Stress Analysis.**

defect. The problem links are illustrated in Figure 12.24. Current is deviated around the crack in the resistor. A potential difference of 10^{-3}V is applied to the ends of the resistor. Figure 12.25 (a) shows the equipotentials, current vectors, and current density intensity map, (b) the resulting thermal field showing an intensity map of the temperature and heat flux vectors, and (c) the intensity map of von Mises stress and deformed boundary ×4000. The resistive element has a length of 1.0 cm and a radius of 0.5 cm. The crack has a thickness of 0.1 mm and a radius of 3.7 mm. The wire has a resistivity $2.37 \cdot 10^{-8}\Omega \cdot$ m and the crack has a resistivity of $1.0\ \Omega \cdot$ m. The outer boundary is assigned a zero current density condition and a 1.0 mV potential difference is applied to each side of the resistor. The crack is given a thermal conductivity of 1 W/m K while the wire has a 20 W/m K thermal conductivity. The zero heat flux condition is applied on the end caps with a convection coefficient $\alpha = 800\,\text{W/m} \cdot \text{K}^2$ on a cylindrical boundary with an ambient temperature of 20 K outside the wire. Young's modulus of the crack is zero (specified as 'None') and $7.7 \cdot 10^{10}\,\text{N/m}^2$ for the wire. Poisson's ratios are all 0.335 for the wire with shear modulus $2.88 \cdot 10^{10}\,\text{N/m}^2$. The coefficient of thermal expansion of the wire is 10^{-6} 1/K. A zoom of the region surrounding the edge of the crack is illustrated in Figure 12.26 along with the deformed boundary (×20000). The Mohr stress is plotted in this figure corresponding to a width of about 0.1 mm. Greater stress is observed at the edge of the crack as before.

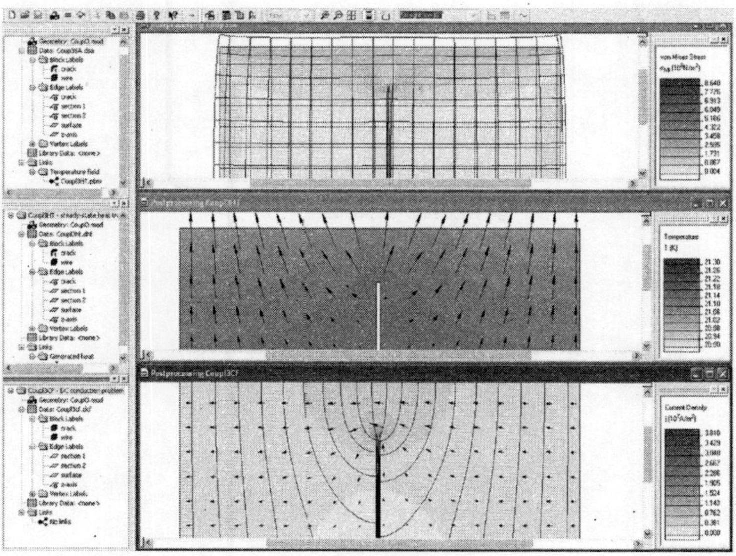

FIGURE 12.25 **Coupled current flow, heat transfer, and stress analysis problem; (a) current flow through a resistive wire with a disk-shaped defect showing equipotentials, current vectors, and current density intensity map; (b) resulting thermal field showing an intensity map of the temperature and heat flux vectors; (c) intensity map of von Mises stress and deformed boundary (×4000).**

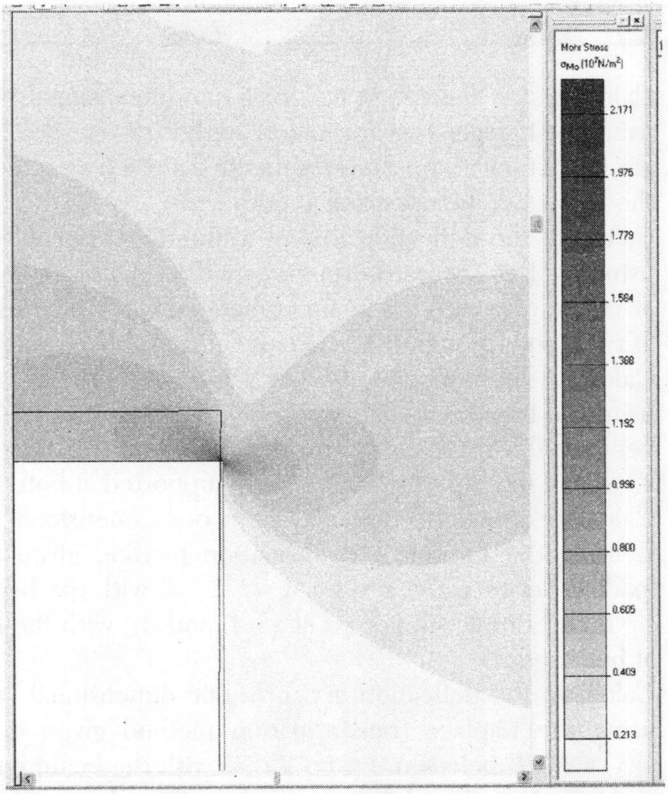

FIGURE 12.26 **Zoom of corner of crack showing deformed boundary (×20000) and Mohr stress. Width of region show is about 0.1 mm.**

REFERENCES

S. P. Timoshenko and J. N. Goodier. (1970). *Theory of Elasticity 3rd edition*, McGraw-Hill.

Walter D. Pilkey. (1997). *Peterson's Stress Concentration Factors 2nd edition*, John Wiley & Sons.

QuickField User's Guide: Tera Analysis Ltd. (2007). *Chapter 2: Introductory Guide, Chapter 10: Theoretical Description, Chapter 11: Examples.*

William T. Thomson. (1962). *Laplace Transformation, 2nd edition*, Prentice-Hall. *Chapter 5: Structural Applications*.

Raymond J. Roark and Warren C. Young. (1975). *Formulas for Stress and Strain, 5th edition*, McGraw-Hill.

Roland Glaser. (2001). *Biophysics*, Springer-Verlag. *Chapter 2.5.3: Mechanical Properties of Biological Membranes.*

EXERCISES

12.1 Calculate the deflection $y(x)$ of a one-dimensional beam of length L using the Laplace transformation method given a delta function loading at $x = L/2$ (a) with the beam embedded at $x = 0$ and free at $x = L$ (b) with the beam simply supported at both ends.

12.2 Calculate the deflection $y(x)$ of a one-dimensional beam of length L using the Laplace transformation method given a delta function loading at $x = L/2$; (a) with the beam embedded at $x = 0$ and simply supported at $x = L$ and (b) with the beam embedded at both ends.

12.3 Calculate the deflection $y(x)$ of a one-dimensional beam of length L using the Laplace transformation method given a theta function loading between $0 \leq x \leq L/2$; (a) with the beam embedded at $x = 0$ and free at $x = L$ and (b) with the beam simply supported at both ends.

12.4 Calculate the deflection $y(x)$ of a one-dimensional beam of length L using the Laplace transformation method given a theta function loading between $0 \leq x \leq L/2$ $x = L/2$; (a) with the beam embedded at $x = 0$ and simply supported at $x = L$ and (b) with the beam embedded at both ends.

12.5 Calculate the deflection $y(x)$ of a one-dimensional beam of length L using the Laplace transformation method given a ramped loading $w(x) = F_0x/L$ between $0 \leq x \leq L$; (a) with the beam embedded at $x = 0$ and free at $x = L$ and (b) with the beam simply supported at both ends.

12.6 Calculate the deflection $y(x)$ of a one-dimensional beam of length L using the Laplace transformation method given a ramped loading $w(x) = F_0x/L$ between $0 \leq x \leq L$; (a) with the beam embedded at $x = 0$ and simply supported at $x = L$ and (b) with the beam embedded at both ends.

12.7 Model the stress distribution in a notched bar subject to loading in the tensile testing machine shown in Figure E-12.7. Compare the stress distribution for both stretching and compression. Use Label Mover to plot the maximum von Mises stress as a function of loading in the bar.

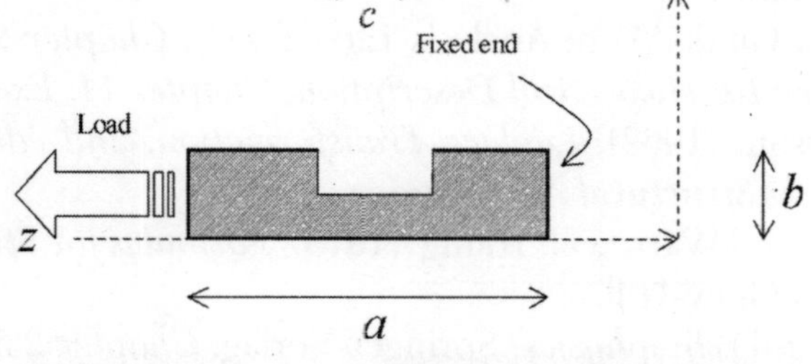

FIGURE E-12.7 **Notched bar in a tensile testing machine. Geometry: axisymmetry.**

12.8 Coupled thermal and stress analysis: A steel bar shown in Figure E-12.8 is half submerged in a liquid nitrogen bath at $t = 0$. Model the temperature and stress distribution in the bar as a function of time.

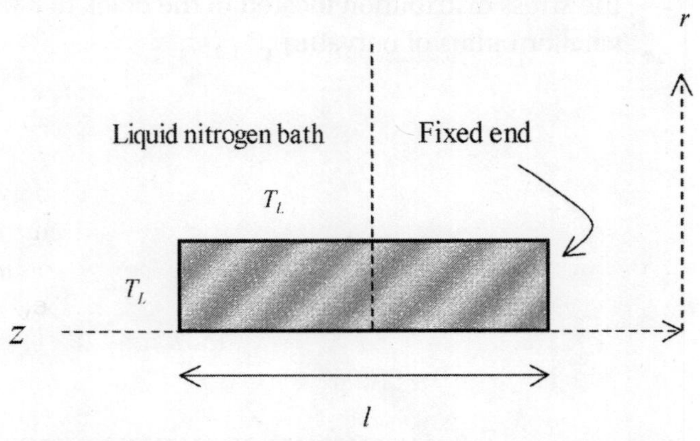

FIGURE E-12.8 **Metal bar half submerged in a liquid nitrogen bath. Geometry: axisymmetry.**

12.9 A submarine with an elliptical cross-section has a stress crack running lengthwise along the ship shown in Figure E-12.9. Calculate the stress distribution in the hull and near the crack if the submarine is at a depth corresponding to an exterior pressure of 25 atmospheres with an internal pressure of 1 atmosphere. Compare the stress distribution locating the crack in a portion of the hull with a smaller radius of curvature.

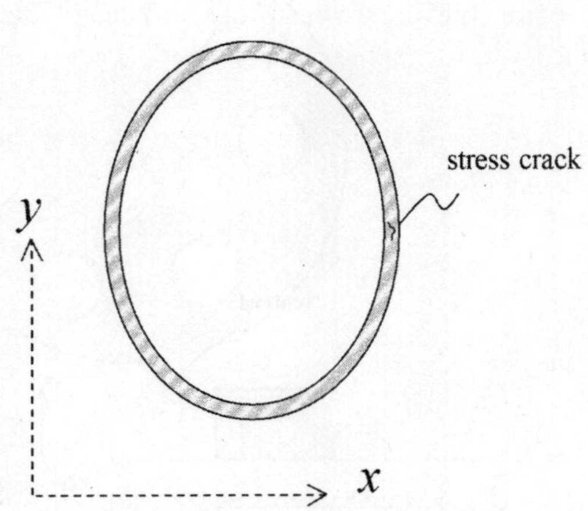

FIGURE E-12.9 **Cross-section of a submarine with a stress crack running parallel to the hull.Geometry: *x-y* symmetry.**

12.10 A submarine with a stress crack running along a circumference of the ship is shown in Figure E-12.10. Calculate the stress distribution in the hull and near the crack if the submarine is at a depth corresponding to 25 atmospheres with an internal pressure of 1 atmosphere. Compare the stress distribution located in the crack in a portion of the hull with a smaller radius of curvature.

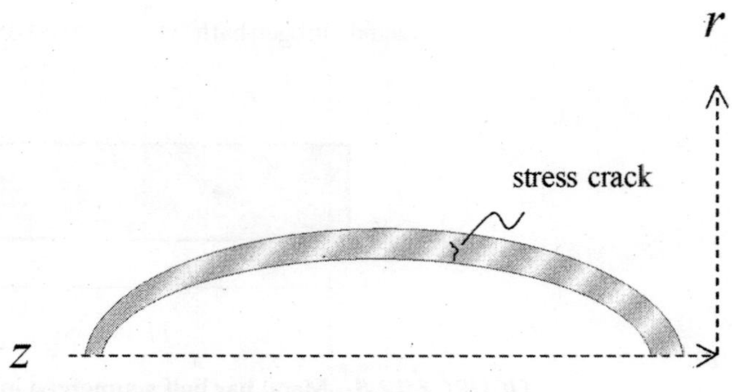

FIGURE E-12.10 **Submarine with a stress crack spanning circumference of the hull. Geometry: axisymmetry.**

12.11 Coupled thermal and stress analysis. A plastic cylinder with machined holes shown in Figure E-12.11 is removed from a liquid nitrogen bath. Calculate the thermal and stress distribution as a function of time.

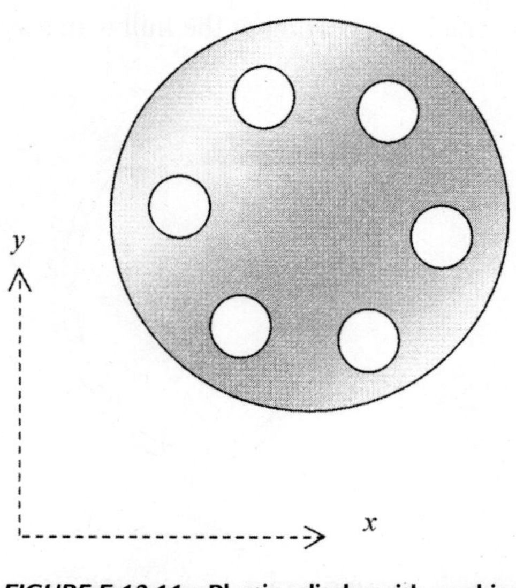

FIGURE E-12.11 **Plastic cylinder with machined holes. Geometry: *x-y* symmetry.**

12.12 Coupled magnetostatic and stress analysis. A plastic cylinder with machined holes shown in Figure E-12.12 contains a cable carrying a current I. Calculate the stress distribution in the cable for various current configurations.

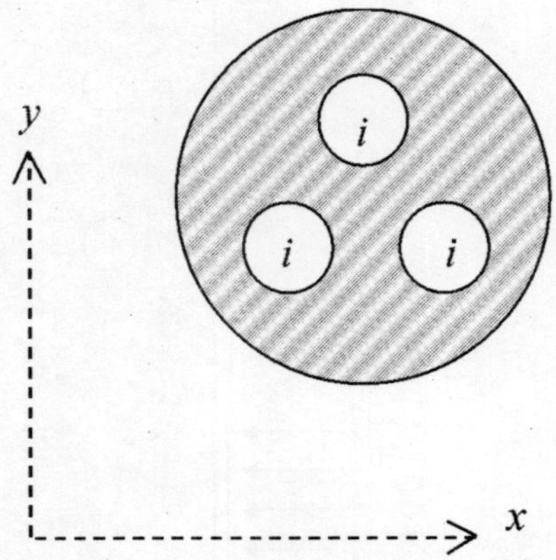

FIGURE E-12.12 **Insulated bundle of current carrying cables.Geometry: *x-y* symmetry.**

12.13 Coupled current, thermal and stress analysis. A cylindrical wire in Figure E-12.13 is immersed in liquid nitrogen and carries a current of 100 amperes. Calculate the thermal and stress distribution in the wire.

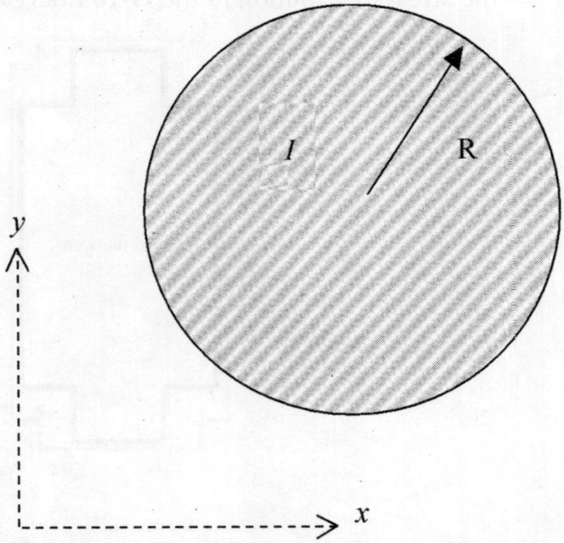

FIGURE E-12.13 **Current carrying wire of radius *R*. Geometry: *x-y* symmetry.**

12.14 Calculate the stress distribution near the rivets in the plate pictured in Figure E-12.14 for various current load configurations on the plate. Specify the rivet boundaries to be contours with zero x- and y-displacement.

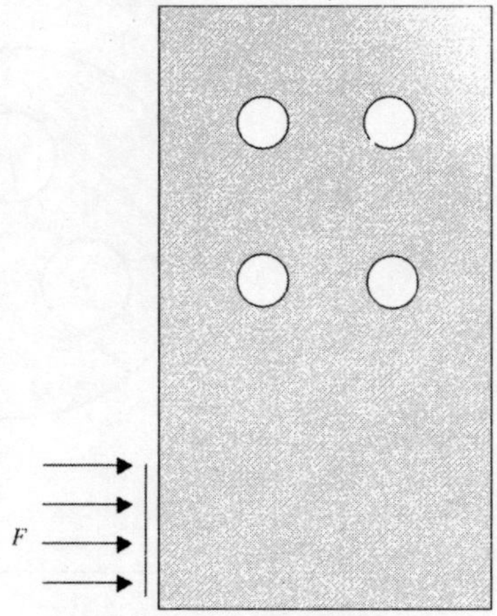

FIGURE E-12.14 **Plate with four rivets under shear loading.Geometry: *x-y* symmetry.**

12.15 Design a dual-walled G-10 fiberglass liquid nitrogen Dewar such as that shown in Figure E-12.15 with a vacuum jacket in x-y symmetry. Apply radiation boundary conditions to the inner surfaces of the Dewar. Apply radiation and a convection boundary condition to the outer surface of the Dewar. Perform a coupled thermal and stress analysis to investigate the stress distribution in the G-10 fiberglass components.

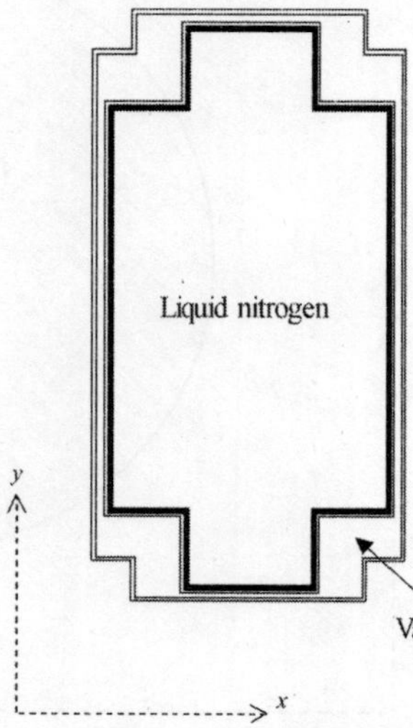

FIGURE E-12.15 **Liquid nitrogen Dewar (thermos bottle) modeled in *x-y* symmetry.**

12.16 Perform 12.15 modeling a cylindrical Dewar in axisymmetry.

12.17 Model a permanent magnet near a superconducting structure as shown in Figure E-12.17. Calculate the von Mises stress distribution and displacement field in the superconductor (superconductor modeled with relative permeability of 1.0 e-6).

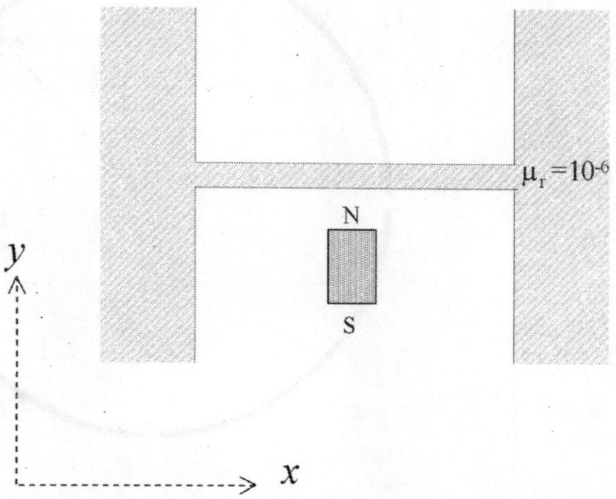

FIGURE E-12.17 **Permanent magnet near a superconducting structure.**

12.18 Coupled Magnetostatic and Stress Analysis: Model an Atomic Force Microscope (AFM) probe as shown in Figure E-12.18. Calculate the von Mises stress distribution and displacement field in the cantilever.

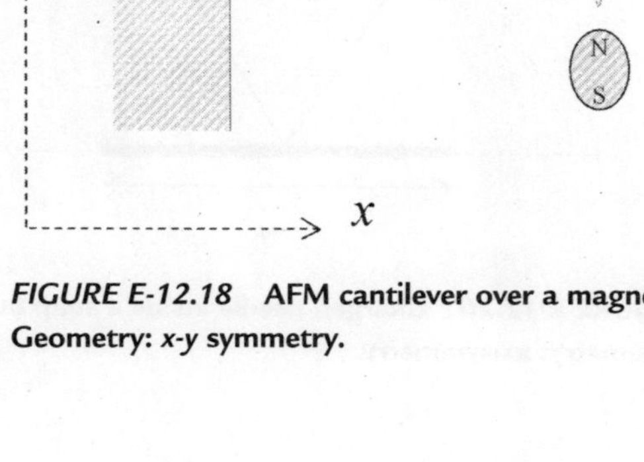

FIGURE E-12.18 **AFM cantilever over a magnetic dipole. Geometry: *x-y* symmetry.**

12.19 Coupled Electrostatic and Stress Analysis: Plot the von Mises stress distribution and displacement field in a circular dielectric plate maintained at four boundary potentials as shown in Figure E-12.19.

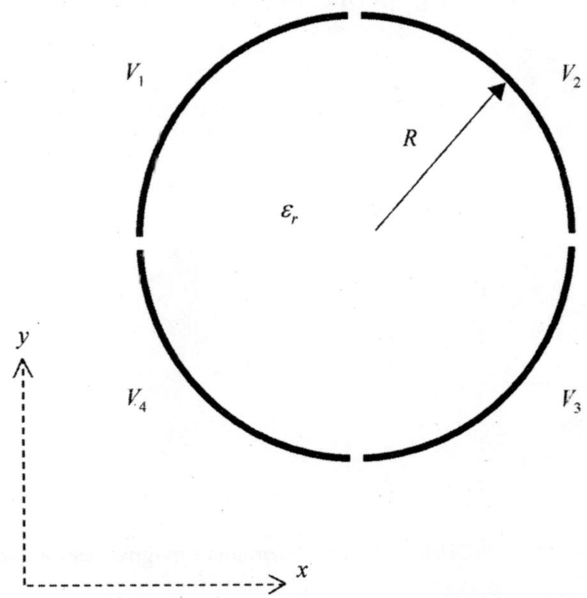

FIGURE E-12.19 **Dielectric circular plate maintained at four potentials. Geometry: *x-y* symmetry.**

12.20 Coupled Electrostatic and Stress analysis: Model a charged needle inside of a soap bubble (or dielectric membrane) as shown in Figure E-12.20 in axial symmetry. Plot the von Mises stress distribution and the displacement field.

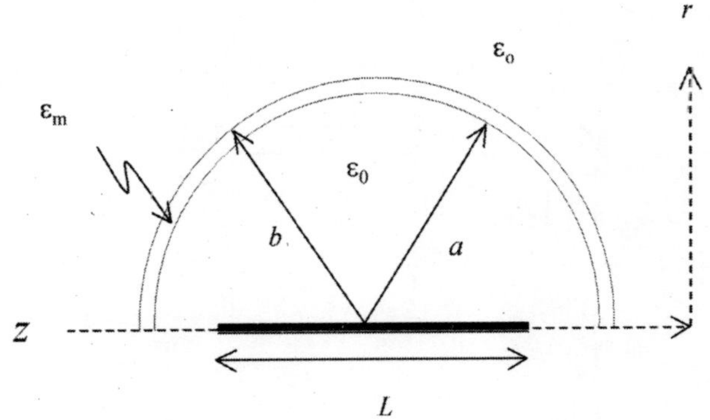

FIGURE E-12.20 **Charged needle inside a soap bubble or dielectric membrane. Geometry: axisymmetry.**

12.21 Coupled Magnetostatic and Stress Analysis: Calculate the stress distribution in a permeable half-space due to a permanent magnet in contact with the metal as shown in Figure E-12.21. Plot the flux lines, von Mises stress distribution, and the displacement field.

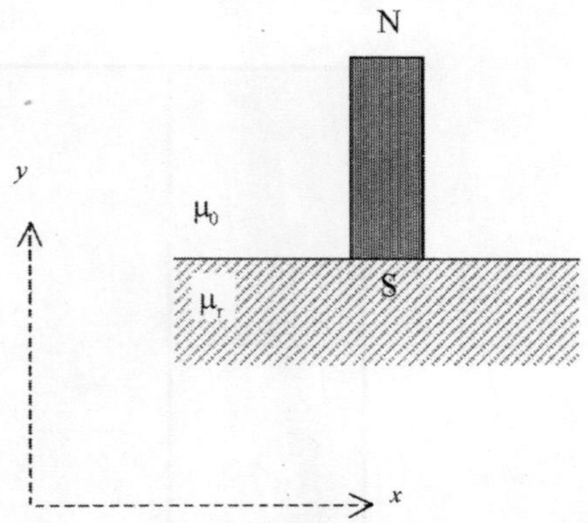

FIGURE E-12.21 **Permanent magnet in contact with a highly permeable half-space. Geometry: *x-y* symmetry.**

12.22 Model a metal part with a conical hole leading to a cylindrical shaft that opens into a disk-shaped cavity as shown in Figure E-12.22. The conical hole, shaft, and cavity are filled with liquid nitrogen at 77 K. The surface of the part is 300 K. Calculate the resulting von Mises stress distribution in the material. Apply radiation and convection boundary conditions to the outer edge of the part.

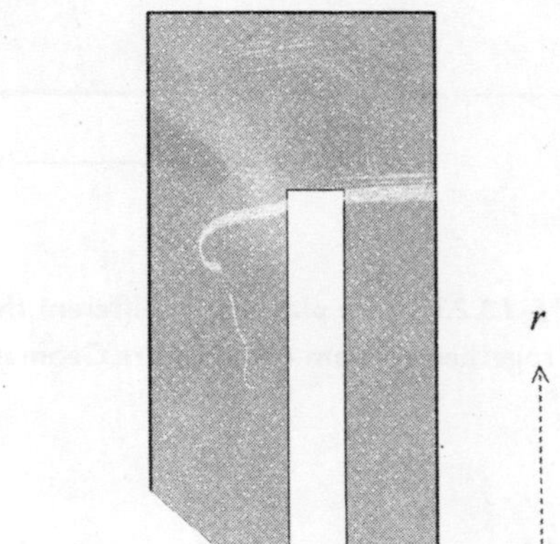

FIGURE E-12.22 **Metal part with cavity filled with liquid nitrogen. Geometry: axisymmetry.**

12.23 Two plates with different thermal expansion coefficients are bonded together at room temperature as shown in Figure E-12.23. Calculate the stress distribution in the plates if they are suddenly immersed in liquid nitrogen at 77 K.

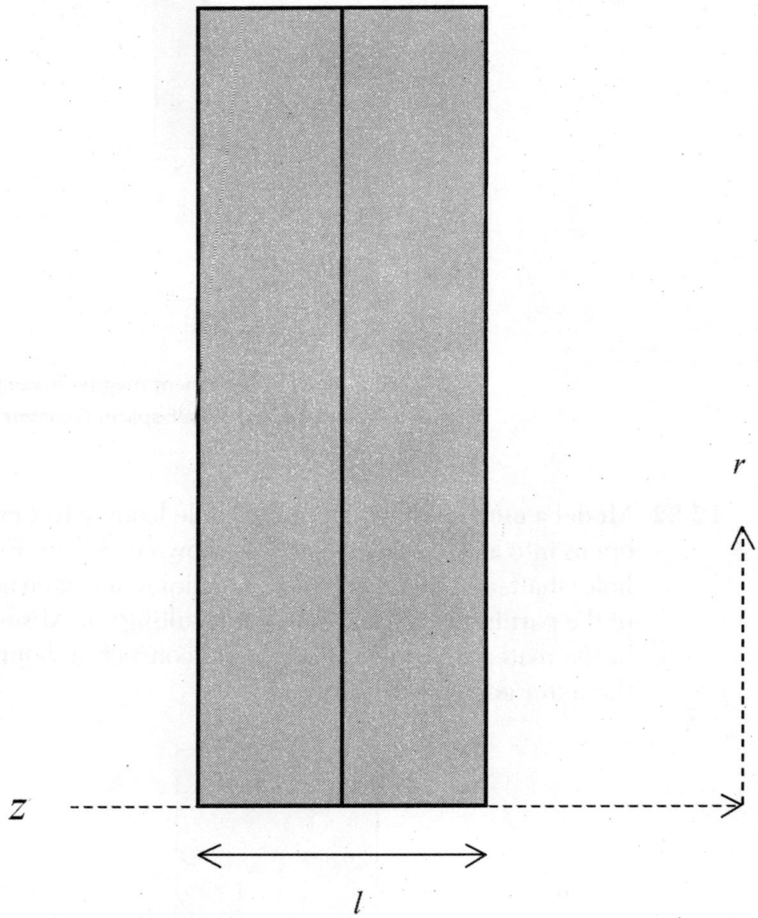

FIGURE E-12.23 **Two plates with different thermal expansion coefficients bonded together at room temperature.Geometry: axisymmetry.**

12.24 Coupled Magnetostatic and Stress Analysis: A permanent magnet is positioned in a highly permeable shell as shown in Figure E-12.24. Calculate the von Mises stress distribution in the shell as well as the displacement field. Repeat the calculation with a superconducting shell by assigning a very small value of relative permeability to the shell.

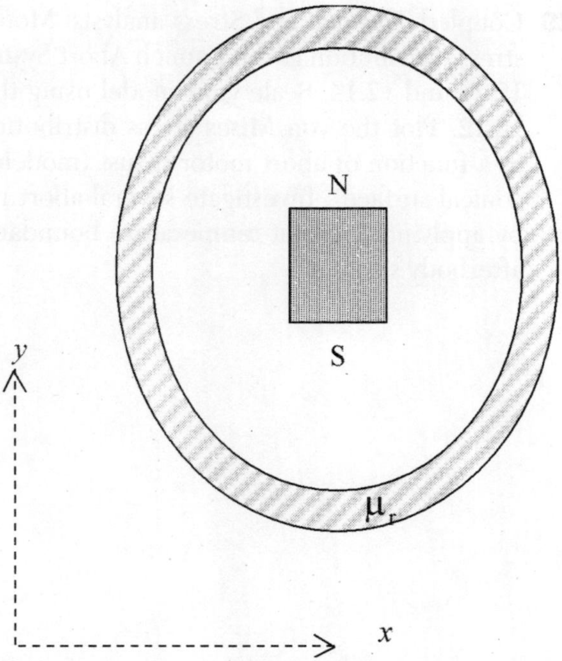

FIGURE E-12.24 **Magnet inside a highly permeable shell. Geometry: *x-y* symmetry.**

12.25 Coupled Magnetostatic and Stress analysis: Model a magnet inside a highly permeable cylindrical tube with a central partition in axial symmetry. Calculate the von Mises stress distribution inside the mumetal as well as the displacement field.

12.26 Coupled Magnetostatic and Stress analysis: Model a magnet inside a superconducting tube with a central partition in axial symmetry. Calculate the stress distribution inside the superconductor. Plot the maximum von Mises stress translating the magnet along the *z*-axis using Label Mover.

12.27 Coupled Electrostatic and Stress analysis: Model a charged needle puncturing a dielectric membrane in axial symmetry. Plot the von Mises stress distribution and the displacement field.

12.28 Coupled Thermal and Stress analysis: Model the coupled thermal and stress distribution of the Orion crew exploration vehicle during atmospheric re-entry. Plot the peak von Mises stress in the capsule as a function of re-entry speed taking the heat shield outer temperature equal to the re-entry speed in meter per second between 1.0–7.0 km per second using Label Mover. Include the separation between the outer shell and inner pressure vessel according to Figure 12.11 in your simulation.

12.29 Coupled Thermal and Stress analysis: Model the coupled thermal and stress distribution in the Launch Abort Systems test pictured in Figures 12.13 and 12.14. Scale your model using the artist rendition in Figure 12.12. Plot the von Mises stress distribution and deformed boundary as a function of abort motor thrust (modeled as load on the frustrated conical surface). Investigate several abort motor exhaust temperatures by applying constant temperature boundary conditions on the conical afterbody surface.

Chapter 13 Electrical Circuits

In This Chapter

- Electrical Current and Ohm's Law
- Kirchhoff's Laws
- Resistances in Parallel and Series
- Capacitances in Parallel and Series
- Impedances in AC Circuits
- Transient Currents in Electrical Circuits
- Electrical Circuit Modeling in QuickField
- Filter Applications
- Electrical Circuit Models of Biological Systems
- Chua's Chaotic Circuit

13.1 Electrical Current and Ohm's Law

Electrical current I is defined as the total charge that flows through a conductor per unit time or

$$I = \frac{dQ}{dt}. \tag{13.1}$$

The SI unit of current is the Ampere (A) defined as a Coulomb (C) per second. Benjamin Franklin first adopted the convention that current flows in the direction of positive charge. The motion of negative charge to the left is equivalent to a positive current flow to the right. The charge carriers in a circuit are usually negatively charged electrons, hence the direction of current is taken as opposite to the direction of electron flow. Current flows in proportion to the potential drop V across a conductor and in inverse proportion to the materials resistance R according to Ohm's law

$$V = IR. \tag{13.2}$$

Electrical resistance results from the scattering of electrons by the crystal lattice, phonons, and impurities in a conductor. The unit of resistance is the Ohm (Ω) defined as a volt per Ampere. Experimentally Ohm found that resistance is proportional to the length L and inversely proportional to the cross-sectional area S of a conductor

$$R = \rho\frac{L}{S}. \tag{13.3}$$

The resistivity ρ is a material property that is geometry independent with S.I. units of $\Omega \cdot$ m. Most metals are good conductors of electricity. Resistivity decreases with lower temperature and becomes identically zero below a transition temperature T_c in superconductors. In poorly conducting nonmetals, or insulators, electrons are strongly bound to positive nuclear charges and are not free to move.

13.2 KIRCHHOFF'S LAWS

Kirchhoff developed the laws governing electrical circuits in 1859.

Loop Rule

Kirchhoff's loop rule states that the sum of voltage drops around any closed loop is zero. This rule may be applied to each loop in complex circuits with multiple loops.

Junction Rule

Kirchhoff's junction, or node rule, states that the sum of currents entering a junction must equal the sum of currents leaving the junction. The Junction Rule results from conservation of electric charge. The total charge entering a circuit junction must equal the charge leaving the junction per unit time.

We may use Kirchhoff's rules to calculate the currents flowing in each segment of the three-loop circuit shown in Figure 13.1. Application of the loop and junction rules gives three equations and three unknowns.

$$\left.\begin{aligned} i_1 + i_2 &= i_3 \\ i_4 + i_6 &= i_5 \\ i_2 + i_6 &= i_5 \end{aligned}\right\} \text{node rule}$$

$$\left.\begin{aligned} 1V - i_3 \cdot 1\Omega - i_1 \cdot 1\Omega &= 0 \\ 1V + 1V - i_3 \cdot 1\Omega - i_4 \cdot 1\Omega &= 0 \\ 1V + 1V - i_6 \cdot 1\Omega &= 0 \end{aligned}\right\} \text{loop rule} \tag{13.4}$$

We may solve this system of equations for each of the six currents symbolically in MATLAB

```
E1='i1+i2=i3'
E2='i4+i6=i5'
E3='i2+i6=i5'
E4='1-i3-i1=0'
E5='2-i3-i4=0'
E6='2-i6=0'
[i1,i2,i3,i4,i5,i6]=solve(E1,E2,E3,E4,E5,E6)
```

Alternatively, the system (13.4) may be written in matrix form as

$$\begin{bmatrix} 1 & 1 & -1 & 0 & 0 & 0 \\ 0 & 0 & 0 & 1 & -1 & 1 \\ 0 & 1 & 0 & 0 & -1 & 1 \\ 1 & 0 & 1 & 0 & 0 & 0 \\ 0 & 0 & 1 & 1 & 0 & 0 \\ 0 & 0 & 0 & 0 & 0 & 1 \end{bmatrix} \begin{bmatrix} i_1 \\ i_2 \\ i_3 \\ i_4 \\ i_5 \\ i_6 \end{bmatrix} = \begin{bmatrix} 0 \\ 0 \\ 0 \\ 1 \\ 2 \\ 2 \end{bmatrix} \tag{13.5}$$

ON THE CD Color images for many of the figures may be found on the CD-ROM.

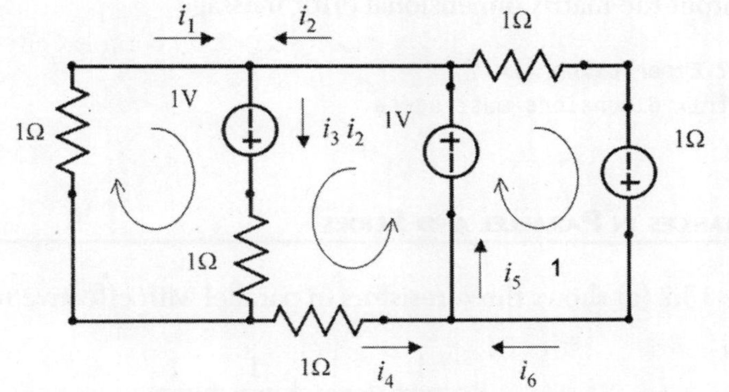

FIGURE 13.1 **Three-loop circuit with resistances and DC voltage sources.**

or $\mathbf{AI} = \mathbf{V}$, where $\mathbf{A}$ is the coefficient matrix and $\mathbf{V}$ is a column vector of voltage values. We may solve for the column vector of current values calculating $\mathbf{I} = \mathbf{A}^{-1}\mathbf{V}$ in MATLAB

```
A=[1 1 -1 0 0 0;
   0 0 0 1 -1 1;
   0 1 0 0 -1 1;
   1 0 1 0 0 0;
   0 0 1 1 0 0;
   0 0 0 0 0 1]
V=[0 0 0 1 2 2]';
I=inv(A)*V
```

This gives the numerical output

```
I =
     0
     1
     1
     1
     3
     2
```

corresponding to $i_1 = 0$, $i_2 = 1$, $i_3 = 1$, $i_4 = 1$, $i_5 = 3$, $i_6 = 2$ A. Note that using the forward division in MATLAB

```
I = A/V
```

will give the same result and uses the more efficient Gaussian elimination method requiring fewer floating point operations. However, the MATLAB command

```
I = V/A
```

will output the matrix dimensional error message

```
??? Error using ==> /
Matrix dimensions must agree.
```

13.3 RESISTANCES IN PARALLEL AND SERIES

Figure 13.2 (a) shows three resistors in parallel with effective resistance

$$\frac{1}{R_{\text{eff}}} = \frac{1}{R_1} + \frac{1}{R_2} + \frac{1}{R_3}. \tag{13.6}$$

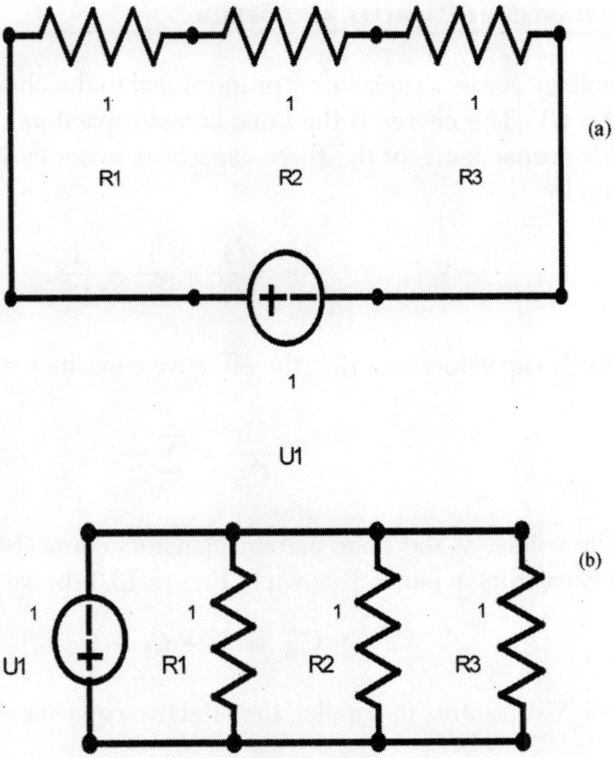

FIGURE 13.2 **(a) Resistors in series and (b) parallel.**

The voltage is the same across each resistor connected in parallel. If we have N resistors in parallel, the effective resistance is given by the sum

$$\frac{1}{R_{\text{eff}}} = \sum_{i=1}^{N} \frac{1}{R_i}. \tag{13.7}$$

If a voltage V is applied across three resistors in series as shown in Figure 13.2 (b) the effective resistance is

$$R_{\text{eff}} = R_1 + R_2 + R_3. \tag{13.8}$$

For N resistors in series, the current through each resistor is the same and we have

$$R_{\text{eff}} = \sum_{i=1}^{N} R_i. \tag{13.9}$$

13.4 CAPACITANCES IN PARALLEL AND SERIES

The voltage across a capacitor is proportional to the charge stored on the capacitor $Q = CV$. The charge is the same across capacitors connected in series. The effective capacitance of the three capacitors in series shown in Figure 13.3 (a) is given by

$$\frac{1}{C_{\text{eff}}} = \frac{1}{C_1} + \frac{1}{C_2} + \frac{1}{C_3}. \tag{13.10}$$

For N capacitors in series, the effective capacitance is given by the sum

$$\frac{1}{C_{\text{eff}}} = \sum_{i=1}^{N} \frac{1}{C_i}. \tag{13.11}$$

The voltage is the same across capacitors connected in parallel. Given the three capacitors in parallel shown in Figure 13.3 (b) we have

$$C_{\text{eff}} = C_1 + C_2 + C_3. \tag{13.12}$$

For N capacitors in parallel, the effective capacitance is given by the sum

$$C_{\text{eff}} = \sum_{i=1}^{N} C_i. \tag{13.13}$$

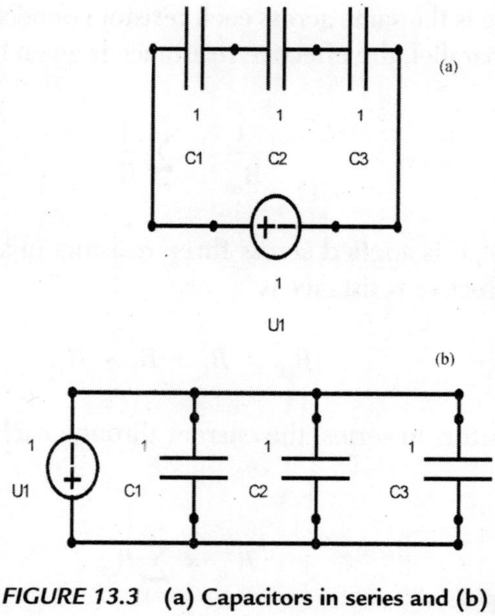

FIGURE 13.3 **(a) Capacitors in series and (b) parallel.**

13.5 IMPEDANCES IN AC CIRCUITS

Inductive Impedance

Faraday's law predicts that the voltage, or electromotive force, induced in a coil is proportional to the time rate of change of magnetic flux Φ threading the coil. Since $\Phi = LI$, where L is the coil inductance in Henrys (1 H = 1 V · s/A), we have $V = LdI / dt$. For AC circuits, this gives $V = i\omega LI$ so that $V = IZ_L$, where the inductive impedance $Z_L = i\omega L$ has units of Ohms. Since I is inversely proportional to Z_L, the effect of an inductor is to attenuate, or "choke" high-frequency current, hence inductors are often referred to as chokes.

Capacitive Impedance

Since the charge on a capacitor $Q = CV$ is proportional to its voltage, the current $I = i\omega CV$ so that $V = IZ_C$, where the capacitive impedance $Z_C = 1/i\omega C$ has units of Ohms. Capacitors, therefore, attenuate low-frequency currents.

Resistive Impedance

The impedance of a resistor is $Z_R = R$, so that resistors impede currents equally at all frequencies.

Impedances in series and parallel may be calculated in the same way as resistors in series and parallel. If we have N impedances in parallel, the effective impedance is

$$\frac{1}{Z_{\text{eff}}} = \sum_{i=1}^{N} \frac{1}{Z_i}. \tag{13.14}$$

For N impedances in series, the effective impedance is

$$Z_{\text{eff}} = \sum_{i=1}^{N} Z_i. \tag{13.15}$$

The effective impedances of a few common circuits are listed below.

RC series circuit:

$$Z_{\text{eff}} = R + \frac{1}{i\omega C} \tag{13.16}$$

RL series circuit:

$$Z_{\text{eff}} = R + i\omega L \tag{13.17}$$

LC series circuit:

$$Z_{\text{eff}} = i\omega L + \frac{1}{i\omega C} \tag{13.18}$$

RLC series circuit:

$$Z_{\text{eff}} = R + i\omega L + \frac{1}{i\omega C} \tag{13.19}$$

The effective impedance of the above circuits may be written as $Z_{\text{eff}} = R + iX$, where X is the reactance. The complex admittance is $Y^* = 1/Z_{\text{eff}}$ so that

$$\text{Re}\,Y^* = \frac{R}{R^2 + X^2} \quad \text{and} \quad \text{Im}\,Y^* = \frac{-X}{R^2 + X^2}. \tag{13.20}$$

The frequency response of an AC circuit may be plotted in the complex impedance or admittance plane.

13.6 Transient Currents in Electrical Circuits

The transient response of electrical circuits can be modeled using ordinary differential equations. The Laplace transform method of solution may then be applied to calculate the transient currents resulting from both periodic and pulsed voltage sources in some cases.

RL Circuit

Kirchhoff's loop rule applied to the RL series combination driven by a voltage $V(t)$ gives the differential equation

$$L\frac{dI}{dt} + IR = V(t). \tag{13.21}$$

with voltage drops across the inductor LdI/dt and resistor IR. Taking the Laplace transform of both sides of (13.21) gives

$$LsI(s) + RI(s) = V_0 \frac{\omega}{\omega^2 + s^2}, \tag{13.22}$$

where the voltage $V(t) = V_0 \sin \omega t$ and with initial condition $I(0) = 0$ that corresponds to the voltage source switched on at $t = 0$. Solving for $I(s)$

$$I(s) = V_0 \frac{\omega}{\omega^2 + s^2}(Ls + R)^{-1}. \tag{13.23}$$

Taking the inverse Laplace transform of this expression gives the current as a function of time

$$I(t) = V \frac{L\omega\left(e^{-\frac{Rt}{L}} - \cos\omega t\right) + R\sin\omega t}{R^2 + L^2\omega^2}, \tag{13.24}$$

which is zero at $t = 0$. The total current is a sum of transient and steady state currents given by

$$I_{\text{trans}}(t) = V \frac{L\omega}{R^2 + L^2\omega^2} e^{-\frac{Rt}{L}} \tag{13.25}$$

$$I_{\text{steady}}(t) = V \frac{R\sin\omega t - L\omega\cos\omega t}{R^2 + L^2\omega^2}. \tag{13.26}$$

The current will oscillate according to (13.26) after the initial transients described by (13.25) attenuate.

LC Circuit

For an LC series combination driven by a time-dependent voltage source, Kirchhoff's law gives the differential equation

$$L\frac{dI}{dt} + \frac{q}{C} = V(t). \tag{13.27}$$

Using the relation $I = dQ/dt$ this may be written as a second-order differential equation for the charge with $V(t) = V_0 \sin\omega t$

$$L\frac{d^2q}{dt^2} + \frac{q}{C} = V_0 \sin\omega t. \tag{13.28}$$

Taking the Laplace transform of both sides of (13.19) gives

$$Ls^2Q(s) + \frac{1}{C}Q(s) = V_0 \frac{\omega}{\omega^2 + s^2}, \tag{13.29}$$

where the voltage source is again switched on at $t = 0$ and

$$Q(s) = V_0 \frac{\omega}{\omega^2 + s^2}\left(Ls^2 + \frac{1}{C}\right)^{-1}. \tag{13.30}$$

Taking the inverse Laplace transform gives the charge as a function of time

$$q(t) = \frac{CV_0}{LC\omega^2 - 1}\left[\sqrt{LC}\,\omega \sin\left(\frac{t}{\sqrt{LC}}\right) - \sin\omega t\right]. \tag{13.31}$$

The current $I(t)$ is obtained by differentiation of (13.31), where we can see that both the current and charge are zero at $t = 0$.

RLC Circuit

Kirchhoff's law gives the differential equation for voltage drops around the RLC series circuit

$$L\frac{dI}{dt} + IR + \frac{q}{C} = V(t). \tag{13.32}$$

This may be written as a second-order differential equation for the charge again with $V(t) = V_0 \sin \omega t$

$$L\frac{d^2q}{dt^2} + R\frac{dq}{dt} + \frac{q}{C} = V_0 \sin\omega t. \tag{13.33}$$

Applying the Laplace transform to both sides of this equation gives

$$Ls^2Q(s) + RsQ(s) + \frac{1}{C}Q(s) = V_0\frac{\omega}{\omega^2 + s^2}, \tag{13.34}$$

where $Q(s) = L\, q(t)$ and $q(0) = \dot{q}(0) = 0$

$$Q(s) = V_0\frac{\omega}{\omega^2 + s^2}\left(Ls^2 + Rs + \frac{1}{C}\right)^{-1} \tag{13.35}$$

The charge $q(t)$ is then obtained from $L^{-1}Q(s)$ and the current from dq/dt. Alternatively, equation (13.23) may be expressed as an integro-differential equation

$$L\frac{dI}{dt} + IR + \frac{1}{C}\int_0^t I(t')dt' = V(t) \tag{13.36}$$

with a subsidiary equation

$$LsI(s) + RI(s) + \frac{1}{C}\frac{I(s)}{s} = V_0 \frac{\omega}{\omega^2 + s^2}. \tag{13.37}$$

The Laplace transformed current is then

$$I(s) = V_0 \frac{\omega}{\omega^2 + s^2} \frac{Cs}{1 + RCs + LCs^2} \tag{13.38}$$

so that $I(t)$ is determined from the inverse Laplace transform as before.

Multiple Loop Circuits

The Laplace Transform method may be applied to multiple loop circuits such as the resistance R in series with the LC parallel combination shown in Figure 13.4. Applying Kirchhoff's law to both loops gives the system of integro-differential equations

$$\left.\begin{aligned} RI_1(t) + \frac{1}{C}\int_0^t I_2(t')dt' &= V(t) \\ \frac{1}{C}\int_0^t I_2(t')dt' - L\frac{dI_3(t)}{dt} &= 0 \end{aligned}\right\} \text{loop rule.} \tag{13.39}$$

$$I_1(t) = I_2(t) + I_3(t) \qquad \text{node rule}$$

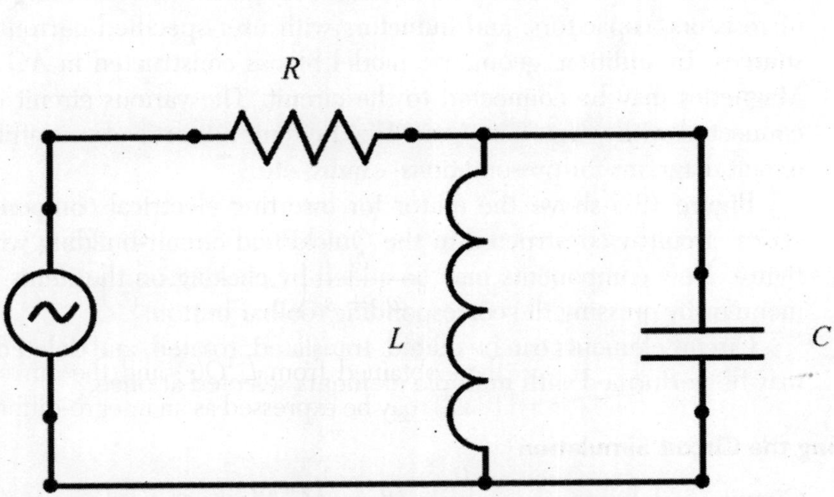

FIGURE 13.4 **Resistance in series with LC parallel combination.**

Taking the Laplace transform of each of these equations yields

$$\begin{aligned} &RI_1(s) + \frac{1}{C}\frac{I_2(s)}{s} = V(s) \\ &\frac{1}{C}\frac{I_2(s)}{s} - LsI_3(s) = 0 \\ &I_1(s) = I_2(s) + I_3(s) \end{aligned} \tag{13.40}$$

for an arbitrary voltage $V(t)$ switched on at $t = 0$. Solving for the Laplace transformed currents gives

$$\begin{aligned} I_1(s) &= V(s)\frac{1 + LCs^2}{R + Ls + RLCs^2} \\ I_2(s) &= V(s)\frac{LCs^2}{R + Ls + RLCs^2}. \\ I_3(s) &= V(s)\frac{1}{R + Ls + RLCs^2} \end{aligned} \tag{13.41}$$

The currents in each branch are then obtained by the inverse Laplace transform of each of these equations.

13.7 ELECTRICAL CIRCUIT MODELING IN QUICKFIELD

Building the Circuit

Passive electric circuits may be simulated in QuickField including combinations of resistors, capacitors, and inductors with user-specified current and voltage sources. In addition, geometric model blocks constructed in AC or Transient Magnetics may be connected to the circuit. The various circuit elements are connected with wires. The QuickField circuit editor is also useful for creating circuit diagrams for presentations, exams, etc.

Figure 13.5 shows the editor for inserting electrical components. An *LC* series circuit is constructed in the QuickField circuit-building window in this figure. New components may be added by clicking on the items in the *Insert* menu or by pressing the corresponding toolbar button.

Circuit elements can be edited, translated, rotated, and deleted. Operations may be performed with multiple elements selected at once.

Running the Circuit Simulation

Circuit models may be calculated in AC Magnetics or Transient Magnetics modules. In Transient Magnetics, if no functional time dependence is explicitly

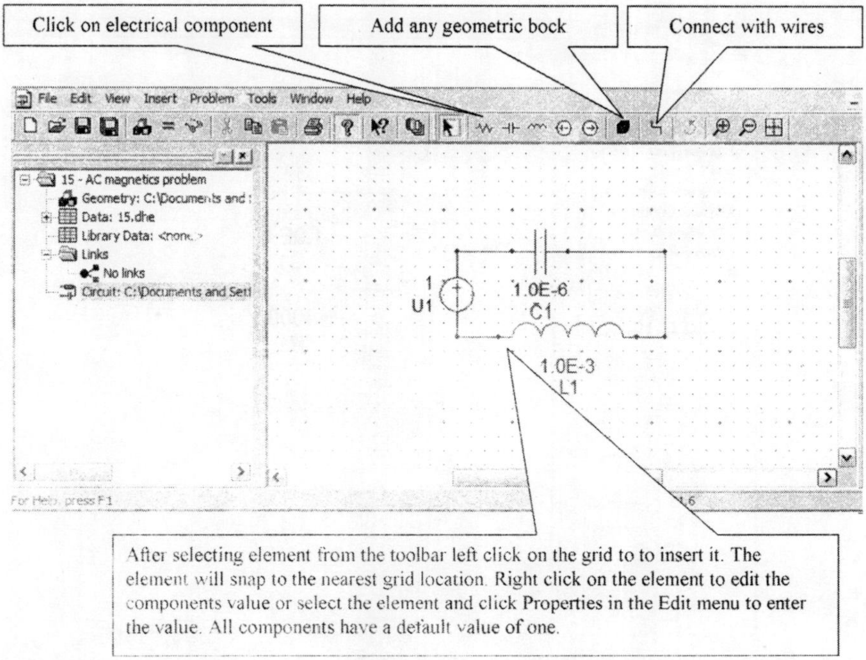

FIGURE 13.5 **Constructing a circuit in the QuickField circuit-building window.**

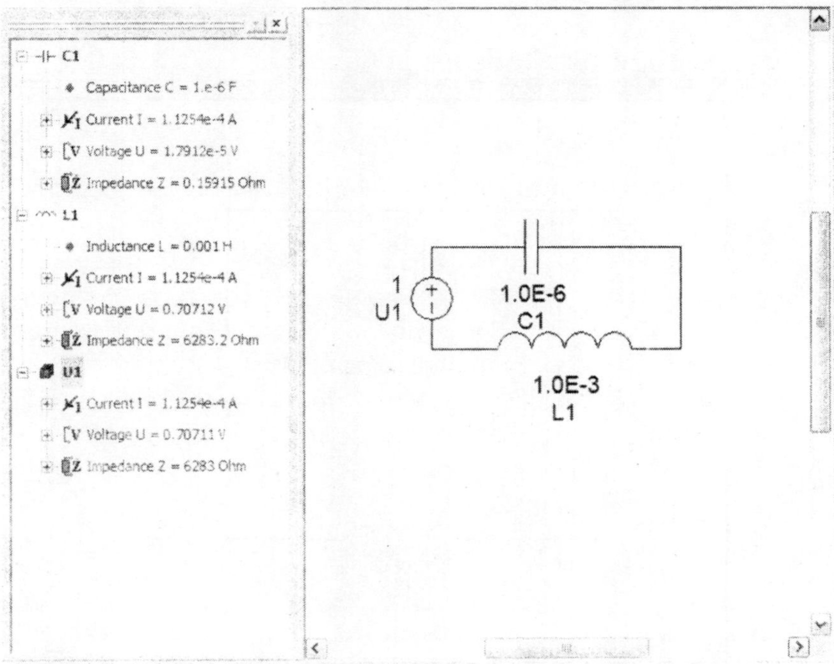

FIGURE 13.6 **Calculated current, voltages, and impedances through each LC circuit element at 10^6 Hz.**

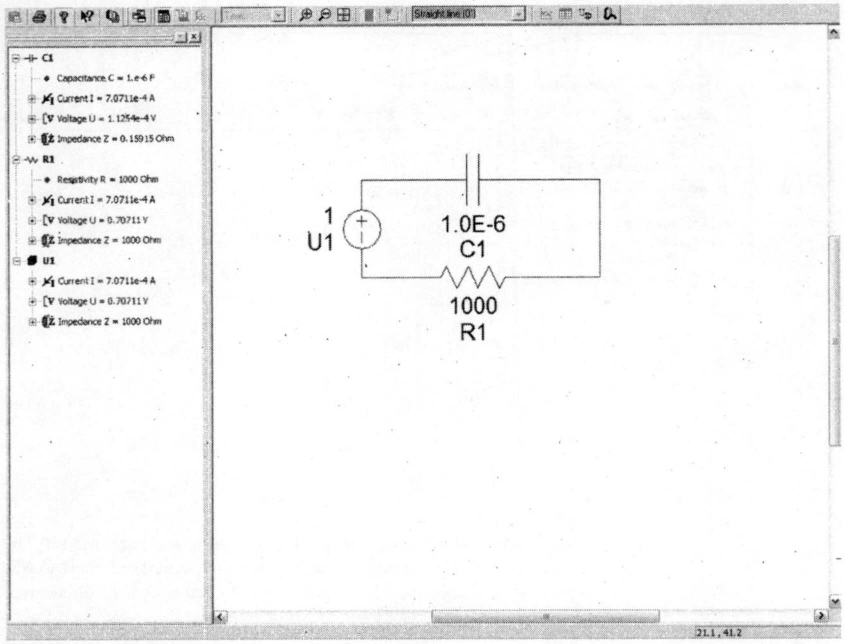

FIGURE 13.7 **Calculated current, voltages, and impedances through each RC circuit element at 10^6 Hz.**

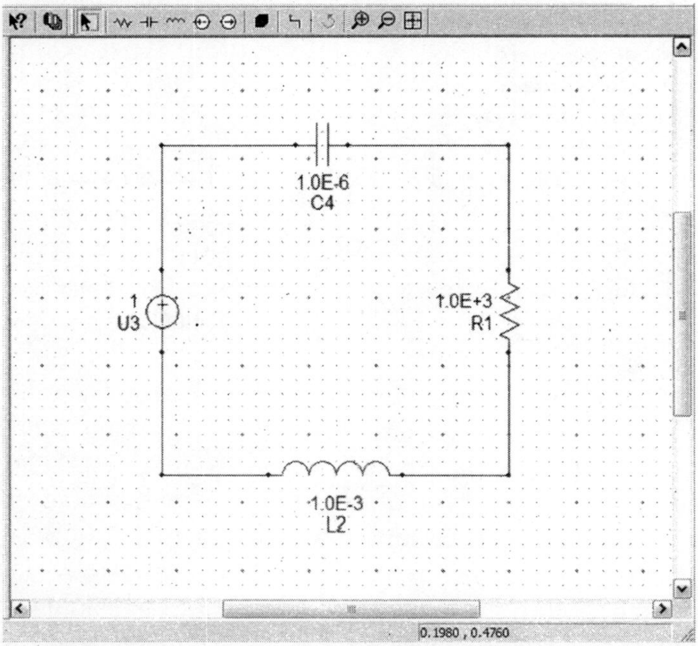

FIGURE 13.8 **RLC circuit constructed in the QuickField circuit-building window.**

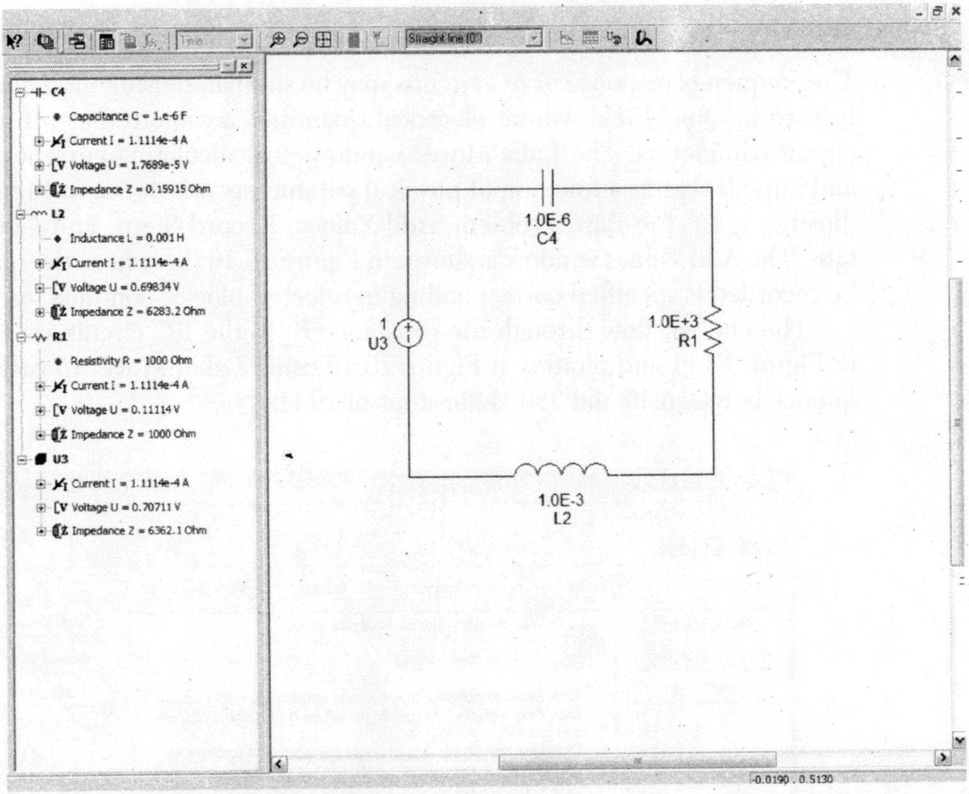

FIGURE 13.9 **Calculated current, voltages, and impedances through each RLC circuit element at 10^6 Hz.**

specified any voltage or current source will have a theta function time dependence as a given voltage or current switched on at $t = 0$.

In order to run the circuit simulation in QuickField, it is necessary to include a geometric model file*. mod in addition to the circuit with a*. qcr extension. It is not possible to run a circuit simulation without a model file. A model file must be included even if the model file and the circuit are not connected. To simulate a circuit by itself, a dummy model file must be created, where no connection is made to the circuit.

Clicking the equal sign on the toolbar or selecting Solve from the Problem menu on the toolbar solves the circuit. Results of the circuit simulation are viewed by right clicking the post-processing window and selecting Coupled Circuit or by selecting Coupled Circuit from the View menu on the toolbar. As an example, calculated currents, voltages, and impedances for LC and RC circuit elements at 10^6 Hz are shown in Figures 13.6 and 13.7, respectively. An LRC circuit is constructed in Figure 13.8 in the QuickField circuit-building window with calculated current, voltages, and impedances at 10^6 Hz as shown in Figure 13.9.

Using Label Mover

The frequency response of AC circuits may be simulated using the Label Mover feature in QuickField, where electrical quantities are plotted as a function of circuit parameters. The LabelMover windows for calculating current, voltages, and impedances as a function of physical parameters are shown in Figure 13.10 illustrating (a) the Base Problem, Add Values, Record Steps, and Get Results tabs. The Add Values window is shown in Figure 13.10 (b), where the quantity to be recorded is specified corresponding to selected blocks, contours, or vertices.

The current flow through the resistance R_1 in the RC circuit are tabulated in Figure 13.11 and plotted in Figure 13.12 using Label Mover to vary the frequency between 10 and 380 Hz in steps of 10 Hz.

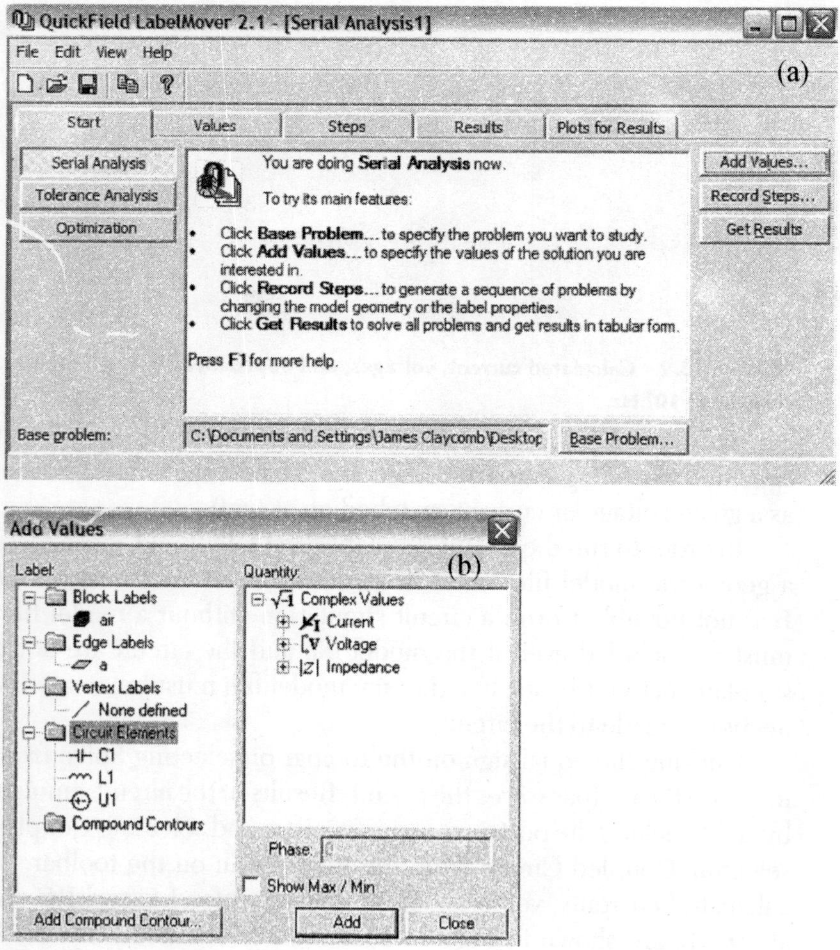

FIGURE 13.10 **Label Mover windows for plotting quantities as a function of varied parameters.**

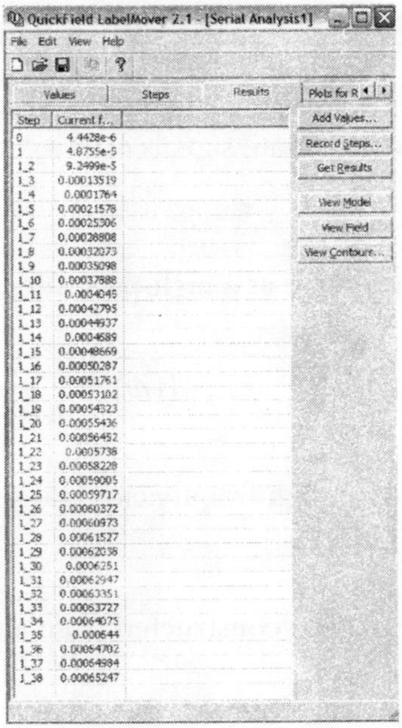

FIGURE 13.11 **RC circuit current values tabulated in Label Mover for frequencies between 10 and 380 Hz in steps of 10 Hz.**

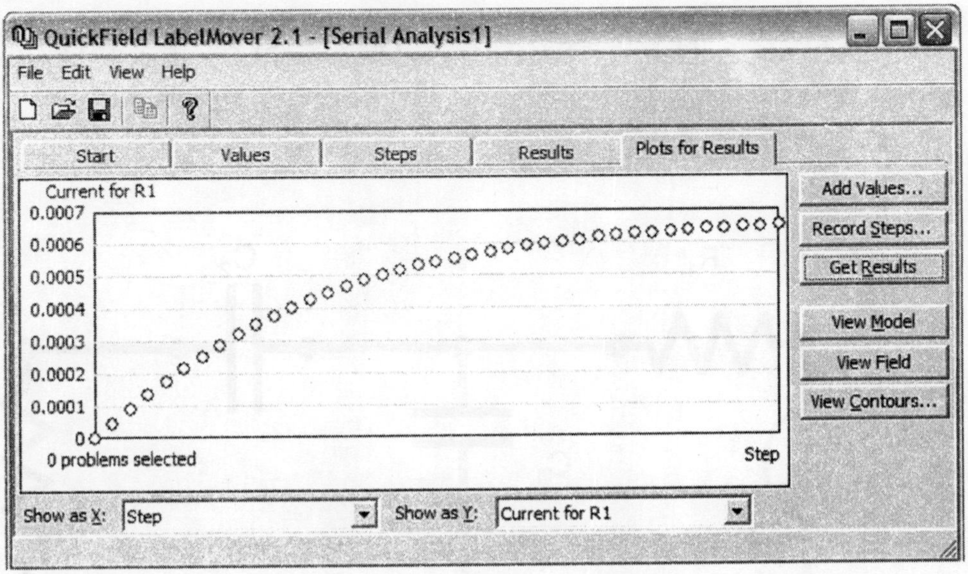

FIGURE 13.12 **Current in the RC circuit vs. frequency plotted in Label Mover for frequencies between 10 and 380 Hz in steps of 10 Hz.**

13.8 FILTER APPLICATIONS

Combinations of resistors, capacitors, and inductors may be used to construct filter circuits that attenuate signals over a certain range of frequencies.

High-pass Filter

An RC circuit may serve as a high-pass filter, where the current

$$I(\omega) = \frac{V}{R + \frac{1}{i\omega C}} \tag{13.42}$$

is equal to $I = V/R$ at high frequencies and zero at DC.

Low-pass Filter

A low-pass filter may be constructed from an RL series combination, where the current

$$I(\omega) = \frac{V}{R + i\omega L} \tag{13.43}$$

is equal to $I = V/R$ at DC and is attenuated at high frequencies.

Band-pass Filter

A band-pass filter may be constructed from an RLC series combination, where the current

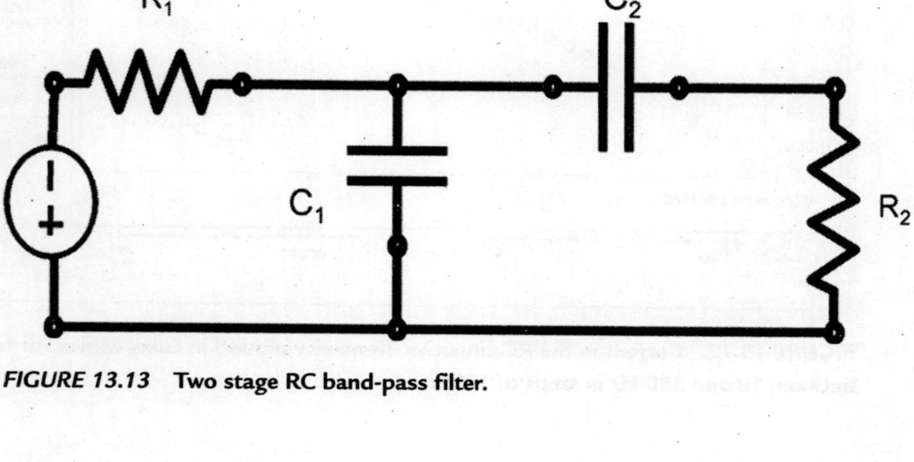

FIGURE 13.13 **Two stage RC band-pass filter.**

$$I(\omega) = \frac{V}{R + i\omega L + \frac{1}{i\omega C}} \tag{13.44}$$

is maximal at the resonant frequency $\omega = 1/\sqrt{LC}$ and is attenuated both at high and low frequencies. A band-pass filter may also be formed from concatenated RC circuits as shown in Figure 13.13.

13.9 Electrical Circuit Models of Biological Systems

Equivalent circuit models may be constructed to simulate the electrical response of cells and membranes to external fields and direct electrode stimulation. The AC and transient response of electrically excited cells and tissue can be modeled in QuickField in many situations.

Modeling the Passive Response of Neurons to Electrical Stimuli

A nerve cell consists of a long axon surrounded by a myelin sheath. The myelin sheath terminates on the cell body that surrounds the nucleus. Dendrites are shorter (mm-length) tentacle-like branches surrounding the nerve cell nucleus and are essential in signaling between neurons.

Active signal transport occurs along the axon by propagating action potentials. In 1952 Hodgkin and Huxley developed a model of action potentials mediated by voltage-gated ion channels across the cell membrane. Both active and passive transport is required for the propagation of nerve impulses. Dendrites do not have ion channels and therefore respond passively to signal voltages that travel roughly 50 times slower and just a few millimeters compared to action potentials traveling up to a meter.

A network of resistors and capacitors is used to model the passive response of nerve cells to electrical stimuli. Figure 13.14 shows the dendrite equivalent circuit, where R_i is the resistance per unit length and R_m and C_m are the membrane resistance and capacitance per unit length. R_0 is the resistance outside the dendrite. Circuit elements corresponding to the dendrite are contained in the boxed region in this figure.

Figure 13.15 shows the transient current and voltage across the second capacitor in the passive transport network due to a step voltage of one mV applied at $(x, t) = (0, 0)$. The simulation is conducted in Transient Magnetics with the defaulted step voltage. All resistances and capacitances in this simulation are set equal to 1 mΩ and 1 μF. Only three segments are included in this simplified model. Additional segments may be added in the QuickField circuit editor

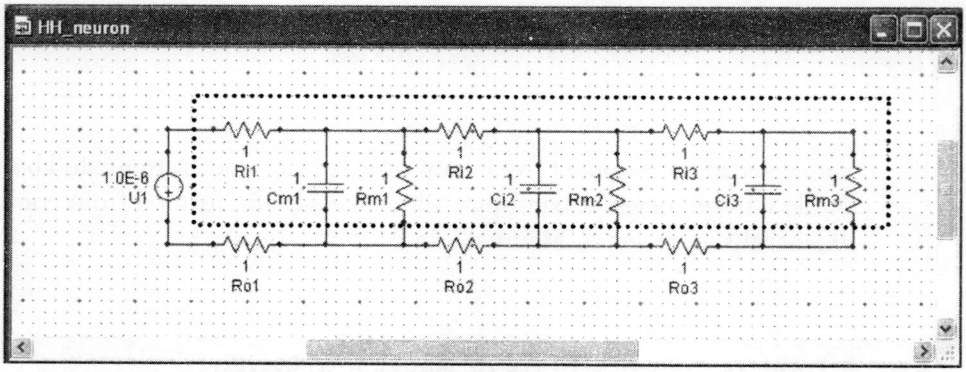

FIGURE 13.14 **Passive circuit model of a dendrite (contained in the boxed region) R_i resistance per unit length, R_m and C_m is the membrane resistance and capacitance per unit; respectively the length. R_0 is the resistance outside the dendrite.**

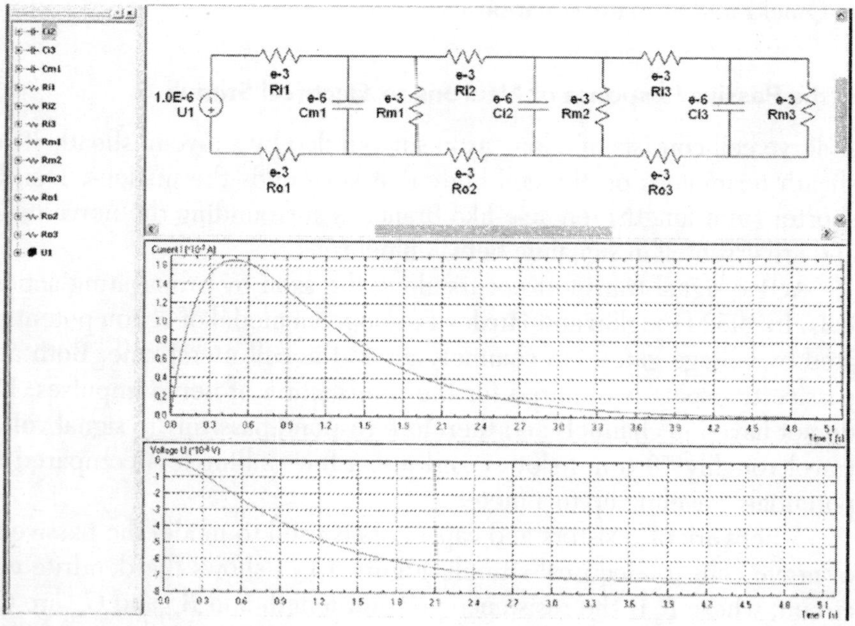

FIGURE 13.15 **Transient current and voltage across the second capacitor in the passive transport network due to a step voltage of one mV applied at $t = 0$.**

by cutting and pasting similar sections of the model. For an infinite chain, the voltage $v(x, t)$ is described by the cable equation

$$\frac{R_m}{R_i}\frac{\partial^2 v(x,t)}{\partial x^2} - R_m C_m \frac{\partial v(x,t)}{\partial t} = v(x,t). \tag{13.45}$$

This equation can be written simply as

$$\lambda^2 \frac{\partial^2 v(x,t)}{\partial x^2} - \tau \frac{\partial v(x,t)}{\partial t} = v(x,t). \tag{13.46}$$

The parameter $\lambda = (R_m/R_i)^{1/2} = (r\rho_m/2\rho_i)^{1/2}$, where ρ_i and ρ_m have units of resistivity and resistivity times length, respectively. The time constant is given by $\tau = R_m C_m$. The solution of (13.35) can be written as a product of exponential and complementary error functions

$$v(x,t) = \frac{1}{2}\begin{pmatrix} V_0 \\ \lambda R_i I_0 \end{pmatrix} \left\{ e^{-x/\lambda} \operatorname{erfc}\left(\frac{x/\lambda}{2\sqrt{t/\tau}} - \sqrt{t/\tau} \right) \pm e^{x/\lambda} \operatorname{erfc}\left(\frac{x/\lambda}{2\sqrt{t/\tau}} + \sqrt{t/\tau} \right) \right\}, \tag{13.47}$$

where the upper and lower prefactors and signs correspond to a unit step voltage V_0 or step current I_0 applied at $(x,t) = (0,0)$, respectively.

Debye Relaxation Model

The Debye relaxation model describes the AC response of systems such as living cell suspensions and polymers consisting of dipole populations to alternating fields. There is a time delay between the excitation and the dipole response characterized by a single relaxation time τ in the Debye model. The dielectric relaxation is described by a frequency-dependent complex permittivity

$$\varepsilon^*(\omega) = \varepsilon_\infty + \frac{\Delta\varepsilon}{1 + i\omega\tau}, \tag{13.48}$$

where $\Delta\varepsilon = \varepsilon_s - \varepsilon_\infty$ with respective static DC and high-frequency permittivity values ε_s and ε_∞. The complex permittivity is often represented on a Cole-Cole plot of ε' vs. ε'' where $\varepsilon^* = \varepsilon' - i\varepsilon''$. In terms of the real part of the permittivity and conductivity, we have the relation $\varepsilon^* = \varepsilon + i\sigma/\omega$ where $\varepsilon = \operatorname{Re}\varepsilon^*$ and $\operatorname{Im}\varepsilon^* = \sigma/\omega$. Figure 13.16 shows the Debye single dispersion equivalent circuit model drawn in QuickField consisting of an RC-C parallel combination.

Maxwell-Wagner Model

The Maxwell-Wagner effect models processes that occur at the interface between different dielectric regions giving rise to the β-dispersion in living cell suspensions. Spatial variations of dielectric and conducting matter can give rise to this effect in external electric fields. In general, the Maxwell-Wagner effect

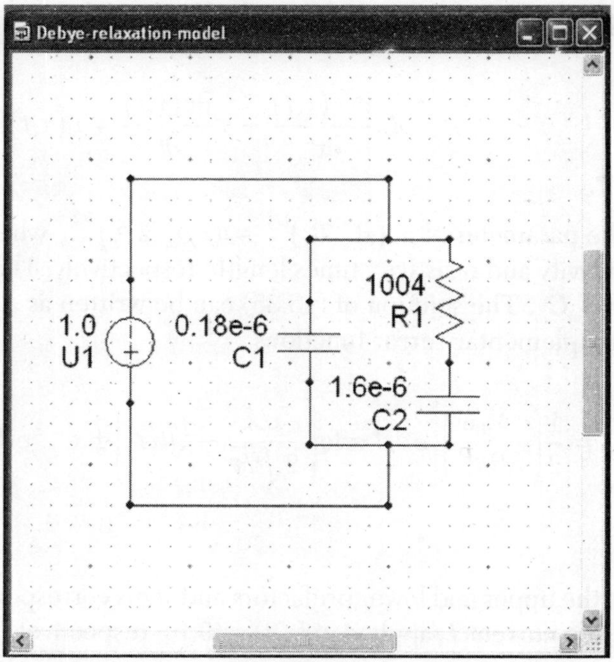

FIGURE 13.16 **Debye-relaxation equivalent circuit model.**

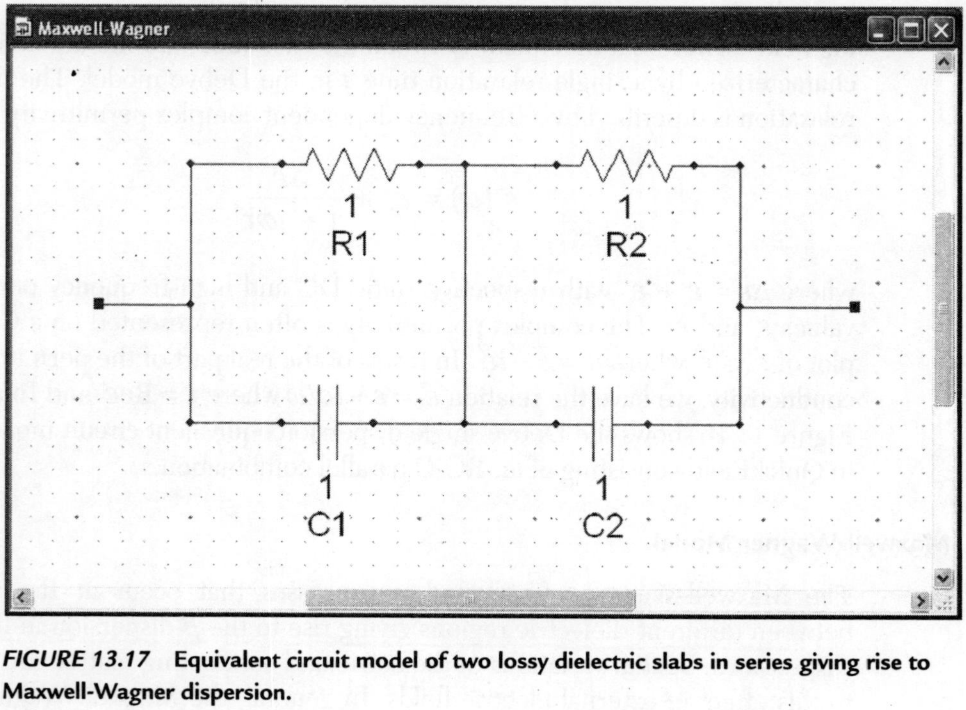

FIGURE 13.17 **Equivalent circuit model of two lossy dielectric slabs in series giving rise to Maxwell-Wagner dispersion.**

depends on the geometry of the dielectric interface with respect to the incident field direction. The simplest electrical model exhibiting this type of dispersion consists of two lossy dielectrics slabs in series. It turns out this effect vanishes in dielectrics with interfaces parallel to the external field. Figure 13.17 shows the circuit model constructed in QuickField equivalent to two lossy dielectric slabs in series.

Modeling Skin Impedance

R. T. Tregear developed a model of skin impedance in 1965. Tregear's electrical circuit model consists of a network of 12 resistors and 12 capacitors, where the capacitors C have equal values and the resistances have decreasing values according to $R_i = R, R/2, R/5, R/10, R/100, R/200,...$

Figure 13.18 shows an extension of Tregear's electrical circuit model of skin impedance with 16 resistors of decreasing value and 16 equal capacitors. This model was rapidly constructed by creating one RC parallel combination. This combination was then duplicated to form four chains of four RC parallel combinations that were connected with wires.

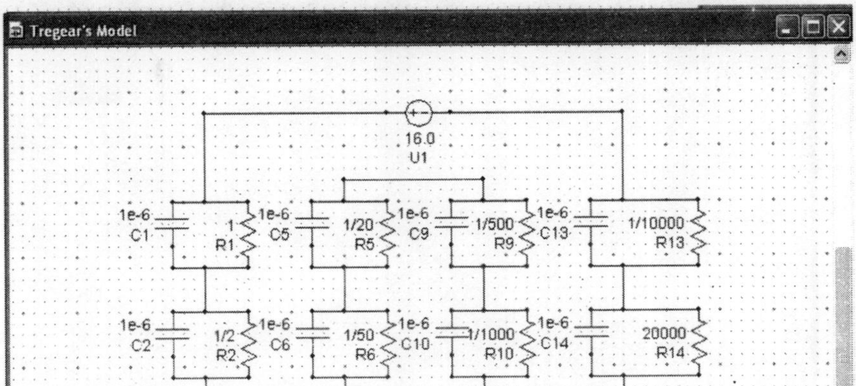

FIGURE 13.18 **Modified Tregear's electrical circuit model of skin impedance consisting of 16 resistors and 16 capacitors, where the capacitors have equal values.**

Modeling Electrode Polarization

Dielectric spectroscopy measurements involving contacting metal electrodes with conducting solutions are subject to large errors at low frequencies due to the polarization effect. Electrode polarization is generated by an electrical double-layer that forms at the solution/electrode interface. The total measured impedance includes contributions from the solution and the interface. The measured low-frequency dielectric response contains this polarization component in addition to Maxwell-Wagner and α-dispersion contributions.

Figure 13.19 illustrates an idealized circuit model of electrode polarization, where C_p and R_p are the polarization capacitance and resistance, respectively. The suspension capacitance and resistance are given by C_2 and R_2. The polarization impedance depends on the type of electrode and electrolyte as well as the excitation frequency. Platinum black electrodes and plate separation techniques can be used to eliminate the polarization impedance, where measurements are made at different electrode separations and then subtracting the common electrode artifact signal. Non-contacting techniques involving torroidal induction coils with torroidal pickup or SQUID magnetometers can also be used to avoid electrode polarization effects.

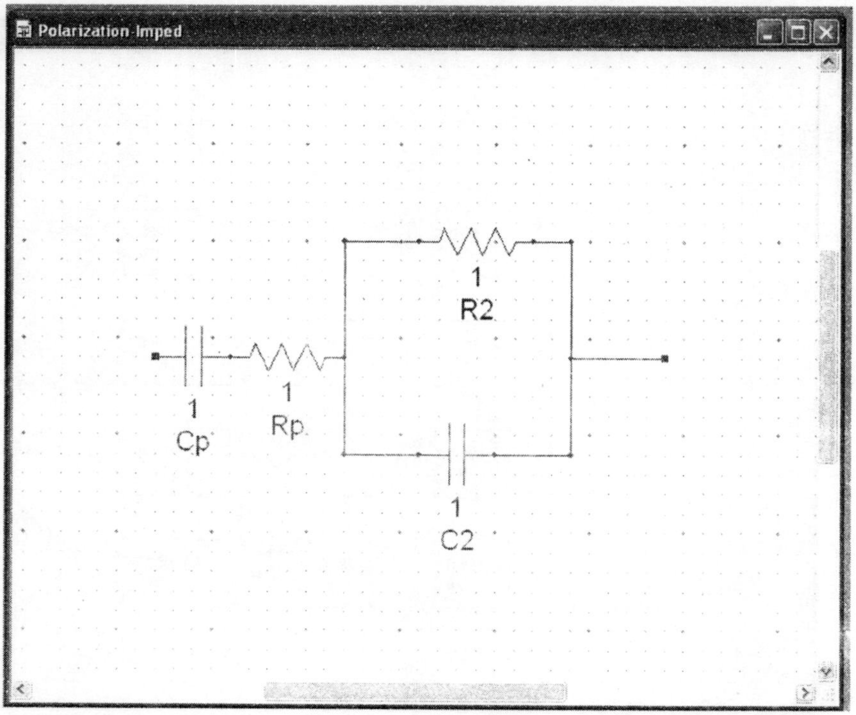

FIGURE 13.19 **Equivalent circuit model of electrode polarization occurring at the electrode and electrolyte interface showing the polarization capacitance C_p and resistance R_p.**

13.10 CHUA'S CHAOTIC CIRCUIT

Finally, we consider an example of an electrical circuit discovered by Chua in 1985 that exhibits chaotic behavior. Chua's circuit shown in Figure 13.20 undergoes complex oscillations that can easily be tested in the laboratory.

Quasiperiodic, chaotic, and intermittent chaotic behaviors have been observed for different resistance and capacitance values in the circuit. The period doubling route to chaos has also been observed in the Chua circuit by varying circuit parameters. The current I and voltages v_1 and v_2 obey the differential equations

$$\begin{pmatrix} RC_1\dot{v}_1 \\ RC_2\dot{v}_2 \\ L\dot{I} \end{pmatrix} = \begin{pmatrix} v_2 - v_1 - U(v_1)/R \\ v_1 - v_2 + IR \\ -v_2 \end{pmatrix}. \tag{13.49}$$

This system may be written in dimensionless form

$$\begin{pmatrix} \dot{x}/\alpha \\ \dot{y} \\ \dot{z} \end{pmatrix} = \begin{pmatrix} y - x - U(x) \\ x - y - z \\ -\beta x - \gamma z \end{pmatrix}, \tag{13.50}$$

where

$$U(x) = bx + \frac{1}{2}(a - b)\left(|x + 1| - |x - 1|\right). \tag{13.51}$$

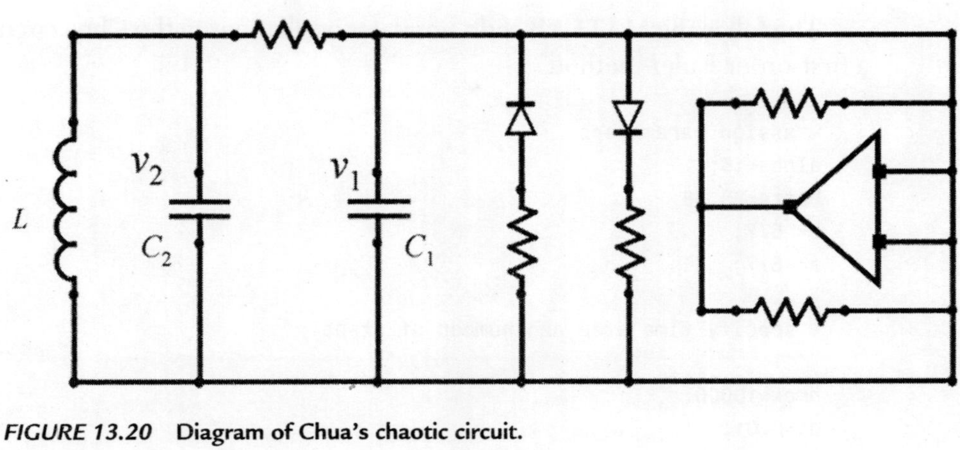

FIGURE 13.20 Diagram of Chua's chaotic circuit.

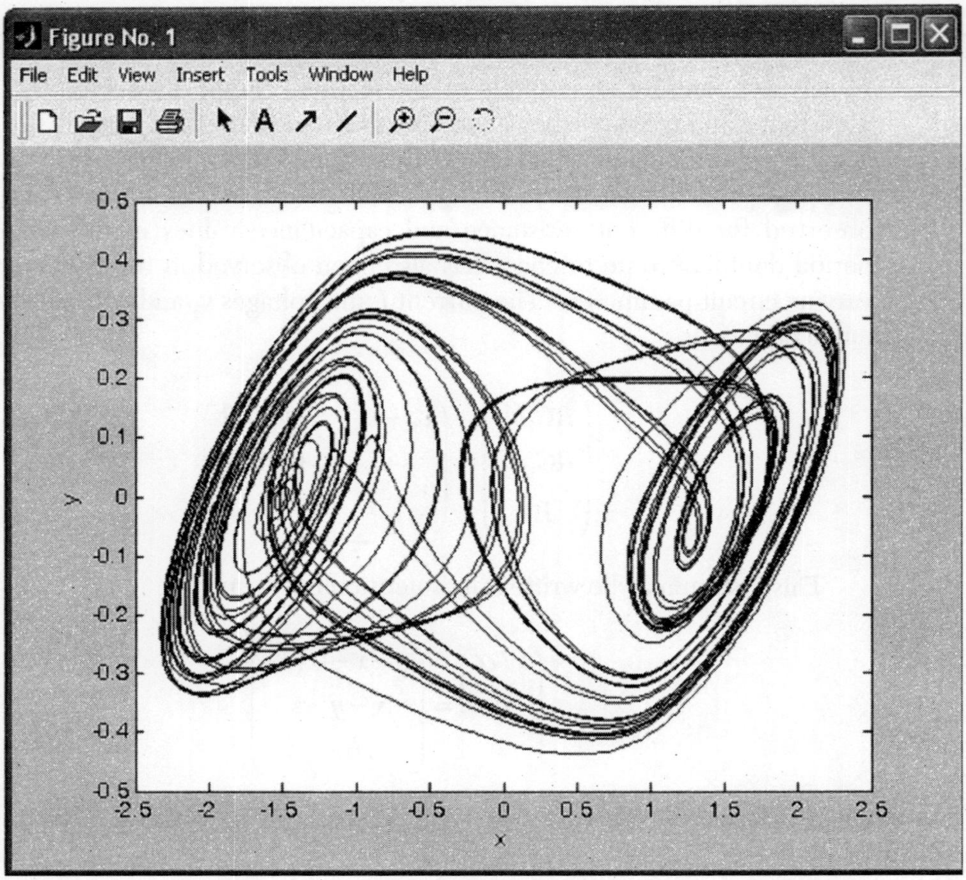

FIGURE 13.21 **Phase space trajectory generated by Chua's circuit simulation.**

The following MATLAB code simulates oscillations in the Chua circuit using a first-order Euler method.

```
% assign parameters
alpha=15;
betta=25.58;
b=-5/7;
a=-8/7;

% specify time step and number of steps

nmax=10000;
dt=0.01;
```

```
% assign time and variable arrays

x=zeros(nmax,1);
y=zeros(nmax,1);
z=zeros(nmax,1);
t=zeros(nmax,1);

% specify initial conditions

x(1)=-1.6;
y(1)=0;
z(1)=1.6;
t(1)=0;

% perform Euler time stepping

for n =1:nmax-1
    f=b*x(n)+0.5*(a-b)*(abs(x(n)+1)-abs(x(n)-1));
    x(n+1) =x(n)+alpha*( y(n)-x(n)-f)*dt;
    y(n+1) =y(n)+(x(n)-y(n)+z(n))*dt;
    z(n+1) =z(n)-betta*y(n)*dt;
    t(n+1) =t(n)+dt;
end

% plot phase trajectory in the z-y plane

plot(x,y)
xlabel('x')
ylabel('y')
```

Figure 13.21 shows the chaotic phase trajectory output by the MATLAB program above. Similar plots may be constructed in the x-y or x-z planes. The time series of each of the three variables may also be plotted.

REFERENCES

QuickField User's Guide: Tera Analysis Ltd. (2007). *Chapter 6: Electric Circuit Definition.*

W. T. Thomson. (1962). *Laplace Transformation, second edition*, Prentice-Hall, New Jersey. *Chapter 3: Electrical Circuits.*

Rodney Cotterill. (2003). *Biophysics An Introduction*, John Wiley & Sons.

Philip Nelson. (2004). *Biological Physics Energy, Information, Life*, W. H. Freeman and Co ıpany.

S. Grimnes and Ø. G. Martinsen. (2000). *Bioimpedance and Bioelectricity Basics*, Academic Press.

R. T. Tregear. (1965). "Interpretation of skin impedance measurements," *Nature* **205**, 600–601.

K. H. Schoenbach, R. P. Joshi, J. F. Kolb, N. Chen, M. Stacey, P. F. Blackmore, E. S. Buescher, S. J. Beebe. (2004). "Ultra short Electrical Pulses Open a New Gateway Into Biological Cells," *Proceedings of the IEEE*, **92** 1122–1137.

J. Gimsa and D. Wachner. (1998). "A Unified Resistor-Capacitor Model for Impedance, Dielectrophoresis, Electrorotation, and Induced Transmembrane Potential," *Biophysical Journal*, 75 1107–1116.

G. Smith, A. P. Duffy, J. Shen, C. J. Olliff. (1995). "Dielectric Relaxation Spectroscopy and Some Applications in the Pharmaceutical Sciences," *Journal of Pharmaceutical Sciences*, **84**, 1029–1044.

R. N. Madan. (1993). *Chua's Circuit: A Paradigm for Chaos*, World Scientific, Singapore.

EXERCISES

13.1 Use the Laplace transformation method to calculate the current flowing in an RC circuit with theta function $\theta(t)$, square pulse $\theta(t) - \theta(t - a)$, and delta function $\delta(t)$ excitations.

13.2 Use the Laplace transformation method to calculate the current flowing in an RL circuit with exponential $\exp(-\lambda t)$, sinusoidal $\sin(\omega t)$, and damped sinusoidal $\exp(-\lambda t)\sin(\omega t)$ excitations.

13.3 Use the Laplace transformation method to calculate the current flowing in an LC circuit with theta function $\theta(t)$, square pulse $\theta(t) - \theta(t - a)$, and delta function $\delta(t)$ excitations.

13.4 Use the Laplace transform method to calculate the current flowing in an RLC circuit with exponential $\exp(-\lambda t)$, sinusoidal $\sin(\omega t)$, and damped sinusoidal $\exp(-\lambda t)\sin(\omega t)$ excitations.

13.5 Use the Laplace transformation method to calculate the current flowing in each segment of the two-loop circuit in Figure 13. 4 consisting of a resistance R in series with an LC parallel combination with theta function $\theta(t)$, square pulse $\theta(t) - \theta(t - a)$, and delta function $\delta(t)$ excitations.

13.6 Use the Laplace transformation method to calculate the current flowing in each segment of the two-loop circuit with the theta function $\theta(t)$, square pulse$\theta(t) - \theta(t - a)$, and delta function $\delta(t)$ excitations in Figure E-13.6.

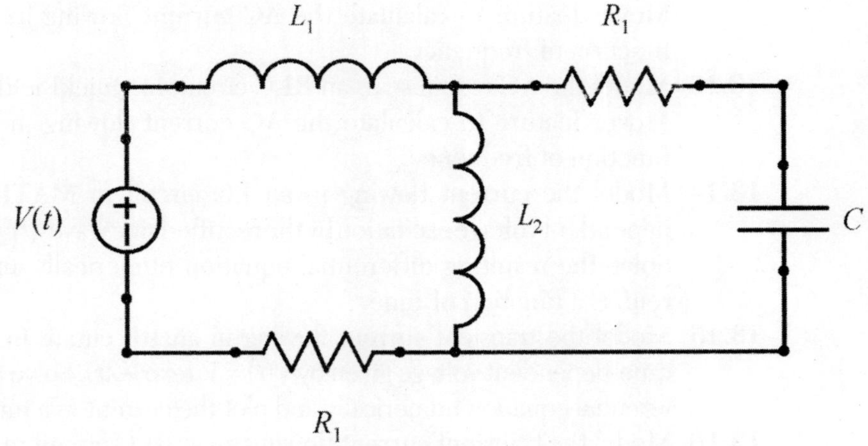

FIGURE E-13.6 **Two loop circuit.**

13.7 Use the Laplace transformation method to calculate the current flowing in each segment of the three-loop circuit with the theta function $\theta(t)$, square pulse $\theta(t) - \theta(t - a)$, and delta function $\delta(t)$ excitations in Figure E-13.7.

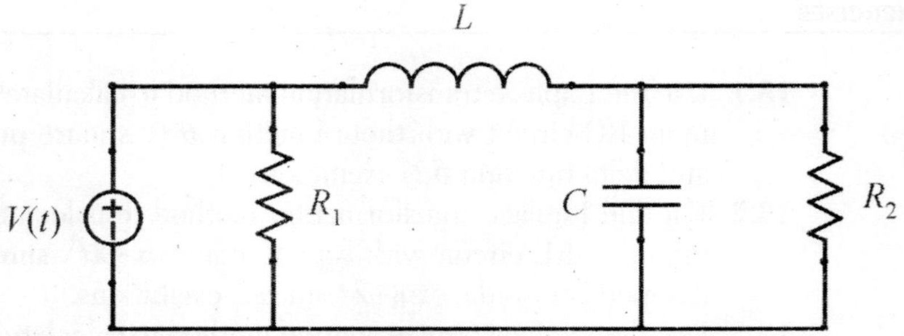

FIGURE E-13.7 **Three loop circuit.**

13.8 Use QuickField to solve for the AC currents flowing in the two- and three-loop circuits shown in Figures E-13.6 and E-13.7.

13.9 Use QuickField to solve for the transient currents flowing in the two- and three-loop circuits shown in Figures E-13.6 and E-13.7 with theta function $\theta(t)$ voltage excitation.

13.10 Model the AC response of an LC circuit in Quickfield. Use the Label Mover feature to calculate the AC current flowing in the circuit as a function of frequency.

13.11 Model the AC response of an RC circuit in QuickField. Use the Label Mover feature to calculate the AC current flowing in the circuit as a function of frequency.

13.12 Model the AC response of an RL circuit in QuickField. Use the Label Mover feature to calculate the AC current flowing in the circuit as a function of frequency.

13.13 Model the AC response of an RLC circuit in QuickField. Use the Label Mover feature to calculate the AC current flowing in the circuit as a function of frequency.

13.14 Model the current flowing in an LC circuit in MATLAB with time-dependent voltage excitation by the rectified sine wave $V(t) = V_0 |\sin(\omega t)|$. Solve the resulting differential equation numerically and plot the current as a function of time.

13.15 Model the transient current flowing in an RL circuit in MATLAB with time-dependent voltage given by $V(t) = V_0 t \exp(-\lambda t)$. Solve the resulting differential equation numerically and plot the current as a function of time.

13.16 Model the transient current flowing in an RLC circuit in MATLAB with time-dependent voltage given by the saw tooth function $V(t) = V_0 \text{saw}(t)$. Solve the resulting differential equation numerically and plot the transient current as a function of time.

13.17 Use QuickField to model the two-stage RC band-pass filter shown in Figure 13.13. Use Label Mover to plot the voltage across the output resistance R_2 as a function of frequency.

13.18 Write a MATLAB code to numerically solve the passive transport equation (13.46) with unit impulse. Make a surface plot of $v(x,t)$. Compare the numerical simulation to the analytical solution (2) at a given location x as a function of time and at a given time as a function of x.

13.19 Write a MATLAB code to numerically solve the passive transport equation (13.46) for a neuron with a square wave excitation. Plot $v(x,t)$ at several x positions as a function of time.

13.20 Create the neuron passive transport model in QuickField and compare the time-dependent voltage across circuit elements both for a step voltage and step current excitations.

13.21 Model the passive neuron response to AC excitation in QuickField. Compare the AC response of models with 5, 10, and 15 cells. Note that individual cells may be copied and pasted for rapid construction of very long models.

13.22 Plot the spatial and temporal solution to the passive transport equation (13.47) in MATLAB given parameter values for crustacean nerve membrane: r = 15 μm, ρ_m = 5000 Ω-cm², ρ_i = 50 Ω-cm, and C_m = 1μF cm^{-2} corresponding to length and time constants of λ = 2.7 mm and τ = 5 ms.

13.23 In 1941 Cole and Cole proposed the empirical formula to model the complex dielectric response of tissue

$$\varepsilon^*(\omega) = \varepsilon_\infty + \frac{\Delta\varepsilon}{1+(i\omega\tau)^{1-\alpha}},$$

where the Cole exponent α can be associated with a distribution of relaxation times. The spread in relaxation times may be due to cellular and molecular interactions, gap junctions, and tissue anisotropy, fractal dimension, and size effect. For water $\alpha \approx 0, \varepsilon_s \approx 80$, $\varepsilon_\infty \approx 6$, and $\tau \approx 10$ ps. Create a Cole-Cole plot of the dielectric response of water in MATLAB potting ε' vs. ε'', where $\varepsilon^* = \varepsilon' - i\varepsilon''$.

13.24 Simulate Tregear's electrical circuit model of skin impedance with 12 resistors of decreasing value and 12 capacitors similar to Figure 13.18. Use the Label Mover feature to calculate the skin impedance as a function of frequency. Compare your results with the extended Tregear's model using 16 capacitors and resistors.

13.25 Calculate the AC-induced membrane potentials across the cell and nuclear membranes using the equivalent shown in Figure E-13.25. Use the Label Mover feature in QuickField to vary the frequency of the excitation voltage.

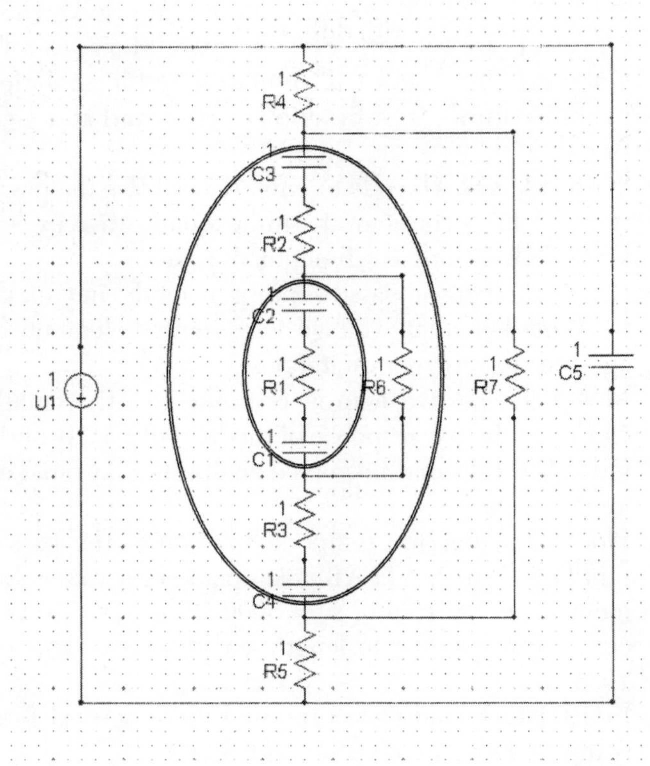

FIGURE E-13.25 **Equivalent circuit model of a cell membrane surrounding cytoplasm with the nuclear membrane surrounding the nucleoplasm.**

13.26 Modify the equivalent circuit in Figure E-13.25 to include membrane resistance in parallel with the membrane capacitance. Use Label Mover to calculate the induced membrane potential as a function of frequency.

13.27 Model Chua's chaotic circuit using Runge-Kutta fourth order integration scheme in MATLAB (See Appendix 1). Plot the time series of one of the variables x, y, or z and make a 3-D parametric plot of the phase space trajectory.

13.28 Investigate the sensitive dependence with respect to small changes in initial conditions in the Chua chaotic circuit in MATLAB by plotting the time series of the same variable together for slightly different initial conditions (e.g., $x_0 = 0.1$ and $x_0 = 0.100001$). The time series should coincide initially and then diverge. Note that this sensitive dependence on initial conditions is a hallmark of chaotic systems.

13.29 Reconstruct the strange attractor in the Chua circuit by creating a time delay plot of $x(t)$ vs. $x(t + \tau)$ in MATLAB, where τ is a suitably chosen time delay. Time delay plots may also be created from the other two variables. Experiment with different values of τ until the plot looks similar to Figure 13.21. Plots with values of τ too small will be a straight line. Values of τ that are too large will produce plots that resemble "pick-up-sticks." Such phase space reconstructions are used to view chaotic phase trajectories from experimental data, where only one time series is recorded.

13.30 Create a Poincare section, or strobe plot, of the Chua circuit in MATLAB plotting points in the x-y plane when z passes through zero. Strobe plots can be made in any of the coordinate planes by plotting a point when the third variable is zero (or by plotting a point periodically in driven systems). It is important to plot points without line connection enabled.

13.31 Create a return map of Chua circuit in MATLAB by plotting subsequent maxima of one of the variables (e.g. z_M vs. z_{M+1}). It is important to plot subsequent maxima without line connection enabled.

13.32 Calculate the current flowing in an LC circuit with square wave excitation using the fourth order Runge-Kutta procedure in MATLAB solving the first-order differential equation

$$\frac{dx}{dt} = f(x,t)$$

according to

$$x_{n+1} = x_n + \frac{1}{6}(F_1 + 2F_2 + 2F_3 + F_4)$$

$$t_{n+1} = t_n + h$$

$$F_1 = hf(x_n, t_n)$$

$$F_2 = hf\left(x_n + \frac{F_1}{2}, t_n + \frac{h}{2}\right)$$

$$F_3 = hf\left(x_n + \frac{F_2}{2}, t_n + \frac{h}{2}\right)$$

$$F_4 = hf(x_n + F_3, t_n + h).$$

Appendix 1

A Walk Through MATLAB

M-Files

M-files are fun and easy to use. In many cases it is better create an M-file when writing code blocks more than a few lines in length. The M-file is saved under a filename such as 'chua_circuit.m' M-files are much easier to edit than code entered in the Command Window. *After code is executed in the Command Window, it is necessary to use the Up Arrow on the keyboard to bring up a given line to modify/correct it.*

To create an M-file select (File → New → M-file) then enter the code and save the file (File → Save as... → filename.m)

As an example, consider the M-file consisting of the code block

```
1+1
```

saved as 'sally.m.' The M-file is executed in the Command Window by typing 'sally' at the Command Prompt

```
>> sally
```

resulting in the output

```
ans =

     2
```

Try creating M-files with some of the graphing examples below.

Basic Math and Syntax

```
% comment

; output suppression
```

```
3+2-4*2-2/3    % addition, subtraction, multiplication, division

3^2            % exponentiation

sqrt(9)        % square root
```

Logical Operators

```
~ % not

| % or

& % and
```

Conditionals

```
if condition
   operations;
end;

if condition
   operations1;
else
   operations2;
end;

if condition1
   operations1;
elseif condition2
   operations2;
else
    operations3;
end;
```

Loops

```
for i=1:n            % i=1, 2, 3,…,n
    operations;
end;

for i=1:2:n            % i=1, 3, 5, 7, 9,…,n
    operations;
end;

for i=1:-3:-10            % i=1, -2, -5, -8
    operations;
end;
```

```
while condition
      operations;
end;
```

Vectors and Matrices

```
v=1:5           % vector v= (1, 2, 3, 4, 5)

u=-1:2:7        % vector u= (-1, 1, 3, 5, 7)

w=1/u.          % vector w = (-1, 1, 1/3, 1/5, 1/7)

A = [1, 2, 3; 4, 5, 6; 7, 8, 9]             % 3 by 3 matrix

A'              % transpose of matrix A

det(A)          % determinate of matrix A

eig(A)          % eigenvalues of A
```

Complex Numbers

```
z= 3 + i        % complex number in form a + bi

real(z)         % real part of z (returns 3)

imag(z)         % imaginary part of z (returns 1)

abs(z)          % magnitude of complex number (returns sqrt(10))
```

Plotting Graphs

Simple Graphs

```
t=0:.01:5;                      %define time variable and increment

plot(t,sin(5*t).*exp(-t))       %plot a single curve
```

Log-log and Semi-log Graphs

```
semilogy(x,y)               % plots x-axis linear and y-axis log

semilogx(x,y)               % plots y-axis linear and x-axis log

loglog(x,y)                 % plots both axes log scale
```

Multiple Curves on One Graph

```
%plot two curves on one graph
plot(t,1./(t-3),t,exp(t)),legend('1/(t-3)','exp(t)') % two curves
```

```
%plot the first six Bessel functions on the same graph
x=0:.1:15;
plot(x,besselj(0,x))
hold on
for n= 1:5
    y=besselj(n,x);
    plot(x,y)
end
hold off
```

Contour Plots of Complex Functions

```
[X,Y]=meshgrid(-3:0.01:3,-3:0.01:3);
H=besselh(0,2,X+i*Y);
contour(X,Y,real(H),-1:.1:1),hold on
contour(X,Y,imag(H),-1:.1:1); hold off
% also try besselj, bessely, besseli, besselk
```

Plots Equipotentials and Vector Field on the Same Graph

```
v=-2:0.2:2;
[x,y]=meshgrid(v);
z=y.*exp(-x.^2-y.^2);
[px, py]=gradient(z,.2,.2);
contour(v,v,z),hold on,quiver(v,v,px,py),hold off
```

Calculating Sums

```
% calculate the sum of the first five natural numbers
sum=0;

for n=1:5
    sum=sum+n;
end
sum          % outputs 15


% vectorized sums eliminates loops and execute faster

n=1:5;
sum(n)         % outputs 15
```

Importing and Exporting Data

MATLAB's Import Wizard can be used to import Excel worksheets (.xls), Text files (.csv, .txt, .dat), and MAT-file array (.mat):
File → Create Time Series from File

For MATLAB workspace arrays:
File → Import from Workspace File → Array Data

This example creates and reads an ASCII data file from the command line

```
>> B= [1 2 3];
>> save c:\docs\jim.dat B-ASCII
>> P=dlmread('c:\docs\jim.dat');
>> P
```

resulting in the output

```
P =

     1   2   3
```

Many more read and write options are available. See also 'Importing and Exporting Data' from the MATLAB Help feature.

Numerical Solution of Ordinary Differential Equations

MATLAB supports an extensive library of ODE solvers. The following example solves the Lorenz system

$$\begin{pmatrix} \dot{x} \\ \dot{y} \\ \dot{z} \end{pmatrix} = \begin{pmatrix} \sigma(y-x) \\ \rho x - y - xz \\ xy - \beta z \end{pmatrix} \tag{A1.1}$$

with initial conditions $(x_0, y_0, z_0) = (0.1, 0.1, 0.1)$ over the time interval $0 \le t \le 100$ using the fourth-order Runge-Kutta scheme.

First, create the following M-file named 'lorenz.m'

```
function yp=lorenz(t,y)

global sigma rho beta

yp= [sigma*(y(2)-y(1));
     rho*y(1)-y(2)-y(1)*y(3);
     y(1)*y(2)-beta*y(3)];
```

The subsequent commands are entered in the Command window

```
>> global sigma rho beta
>> [t,y]=ode45(@lorenz,[0, 100], [.1;.1;.1]);
>> plot3(y(:,1),y(:,2),y(:,3))
```

The output is a 3D parametric curve, where `y (:, 1)`, `y (:, 2)`, and `y (:, 3)` correspond to the x-, y-, and z-columns of the matrix y. The use of global variables allows parameter variation without having to modify the function file. The graph may be further enhanced with axis labels and a title.

```
>> xlabel('x', 'FontSize',16)
>> ylabel('y', 'FontSize',16)
>> zlabel('z', 'FontSize',16)
>> title('Lorenz Model','FontSize',18)
```

A system of two second-order differential equations such as

$$\begin{pmatrix} \ddot{x} \\ \ddot{y} \end{pmatrix} = \begin{pmatrix} f(x,\dot{x},y,\dot{y},t) \\ g(x,\dot{x},y,\dot{y},t) \end{pmatrix} \tag{A1.2}$$

can be converted to a system of four first-order equations

$$\begin{pmatrix} \dot{x} \\ \dot{x}_p \\ \dot{y} \\ \dot{y}_p \end{pmatrix} = \begin{pmatrix} x_p \\ f(x,\dot{x},y,\dot{y},t) \\ y_p \\ g(x,\dot{x},y,\dot{y},t) \end{pmatrix} \tag{A1.3}$$

and solved using a procedure similar to the one above extended to four equations (with four initial conditions). A system of three second-order equations is solved as a system of six first-order equations, and so on.

Symbolic ToolBox

The following examples require the Symbolic Toolbox. These can be executed at the Command Line. Remember to use the Up Arrow key to modify/correct previous operations.

Simple Plots

```
syms x
ezplot(x*sin(1/x), [-.1,.1])
ezplot(besselj(0,x),[0,10])
```

Surface Plots

```
syms x,y
ezmesh(besselj(0,sqrt(x^2+y^2)), [-10,10])
```

Combination Surface and Contour Plots

```
syms x,y
ezmeshc(sin(x)*exp(-x^2-y^2))
% also use the Rotate 3D tool to view different orientations
```

3-D Parametric Curve

```
syms t
ezplot3(sin(cos(t)),cos(sin(t)),t,[0,15])
% also use the Rotate 3D tool to view different orientations
```

Symbolic Integration

```
syms x a
a= 3;
int(dirac(x-a)*besselj(0,x),0,inf)

syms a b positive
syms x
f=exp(-a*x^2+b*x)
int(f,x,0,inf)
pretty(ans)          %use to make the answer more readable
```

Symbolic Differentiation

```
syms x
f=besselj(2,x);
diff(f)
```

Symbolic Solution of Differential Equations

```
dsolve('D2I + 5*DI +6*I = 9*sin(t)', 'I(0)=0', 'DI(0)=0')

dsolve('DX+Y=Z','DY+3*Z=Y','DZ+X=2*Z','X(0)=1','Y(0)=0','Z(0)=1')
ans.X     % returns X(t) where dsolve is evaluated first
ans.Y     % returns Y(t)
ans.Z     % returns Z(t)
```

Symbolic Summation

```
syms x k
symsum(exp(-k*x),0,inf)

syms n
symsum((-1)^n/n^2,inf)
```

```
syms n
symsum(1/2^n,1,inf)
```

Taylor Expansion

```
syms x
taylor(besselj(1,x),8)     % terms up to but not including order 8
```

Symbolic Solution to Systems of Equations

```
syms a b c x y z
S=solve('a*x+b*y=3', 'c*y+b*z=2', 'x+z =1', x,y,z)
S.x      %returns x value
S.y      %returns y value
S.z      %returns z value
```

Factoring Expressions

```
syms x y
factor(x^6-y^6)
```

Simplifying Expressions

```
syms a b positive
f=(a^3+b^3)/(a+b)
simplify(f)
```

Functional Calculator

The Symbolic Toolbox features a functional calculator called funtool. To launch the calculator type funtool at the Command Line

```
>> funtool
```

This calculator displays and performs an operation on functions of one variable. Operations include plotting, differentiation, integration, inverse function, function composition as well as many algebraic functional operations.

REFERENCES

B. Hahn and D. Valentine. (2007). *Essential MATLAB for Engineers and Scientists, 3rd Edition*, Butterworth-Heinermann/Elsevier.

D. J. Higham, N. J. Higham. (2005). *MATLAB Guide, 2nd, edition*, SIAM.

Appendix 2 Laplace Transform of Pulsed and Periodic Functions

Pulsed Functions

Dirac Delta Pulse at $t=0$

$$f(t) = \delta(t) \quad F(s) = 1 \tag{A2.1}$$

Dirac Delta Pulse at $t = a$

$$f(t) = \delta(t - a) \quad F(s) = e^{-as} \tag{A2.2}$$

Heaviside Step Pulse (amplitude 1 $a \le t$)

$$f(t) = \Theta(t - a) \quad F(s) = \frac{e^{-as}}{s} \tag{A2.3}$$

Uniform Pulse Function (amplitude 1 between times $a \le t \le a + b$)

$$F(s) = \frac{e^{-as}\left(1 - e^{-bs}\right)}{s} \tag{A2.4}$$

Sine Pulse Function (amplitude 1 between times $0 \le t \le a$)

$$f(t) = \sin\left(\frac{\pi t}{a}\right) \quad F(s) = \frac{\pi a\left(1 + e^{-as}\right)}{a^2 s^2 + \pi^2} \tag{A2.5}$$

Periodic Functions

The Laplace transform of a periodic function $f(t)$

$$L[f(t)] = \frac{1}{1 - e^{-sz}} \int_0^z e^{-st} f(t)\,dt \tag{A2.6}$$

Square wave (amplitude 1 period $2a$)

$$f(t) = \Theta(t) - 2\Theta(t - a) + 2\Theta(t - 2a) - 2\Theta(t - 3a) + 2\Theta(t - 4a) + \cdots \tag{A2.7}$$

$$F(s) = \frac{1}{s} - 2\frac{e^{-as}}{s} + 2\frac{e^{-2as}}{s} - 2\frac{e^{-3as}}{s} + \cdots \tag{A2.8}$$

$$F(s) = \frac{1}{s}\tanh\frac{as}{2} \tag{A2.9}$$

Rectified square wave (amplitude 1 period $2a$)

$$F(s) = \frac{1}{s\left(1 + e^{-as}\right)} \tag{A2.10}$$

Half-wave rectified sine wave (amplitude 1 period $2a$)

$$F(s) = \frac{a}{\left(s^2 + a^2\right)\left(1 - e^{-\pi s/a}\right)} \tag{A2.11}$$

Full-wave rectified sine wave (amplitude 1 period a)

$$F(s) = \frac{a}{\left(s^2 + a^2\right)}\coth\frac{\pi s}{2a} \tag{A2.12}$$

Saw tooth wave (amplitude 1 period a)

$$F(s) = \frac{1}{as^2} - \frac{e^{-as}}{a\left(1 - e^{-as}\right)} \tag{A2.13}$$

Triangle wave (amplitude 1 period $2a$)

$$F(s) = \frac{1}{s^2}\tanh\frac{as}{2} \tag{A2.14}$$

Appendix 3

TABLE OF CONFORMAL MAPPINGS

CONFORMAL MAPPINGS ON THE HALF PLANE

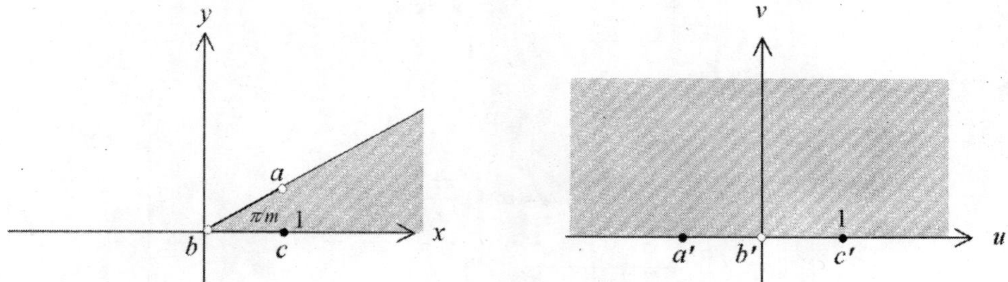

$$w = z^m$$

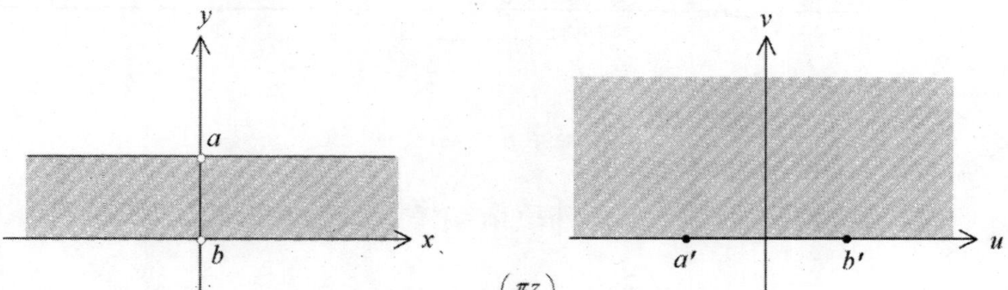

$$w = \exp\left(\frac{\pi z}{a}\right)$$

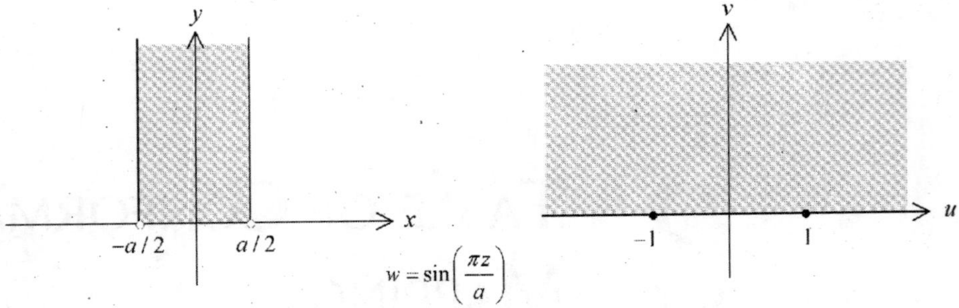
y
x
−a / 2
a / 2
$w = \sin\left(\frac{\pi z}{a}\right)$
v
u
−1
1

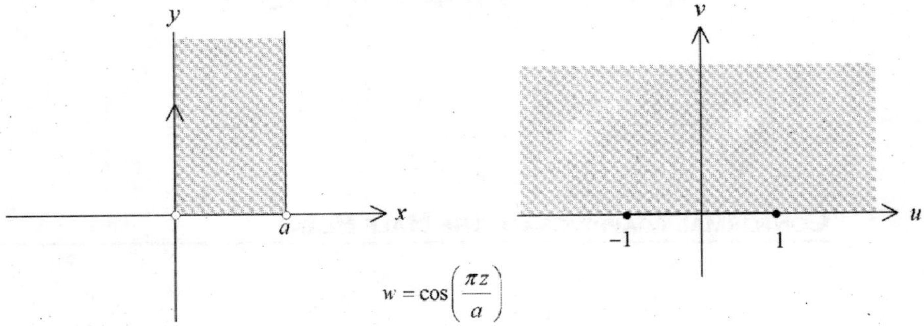
y
x
a
$w = \cos\left(\frac{\pi z}{a}\right)$
v
u
−1
1

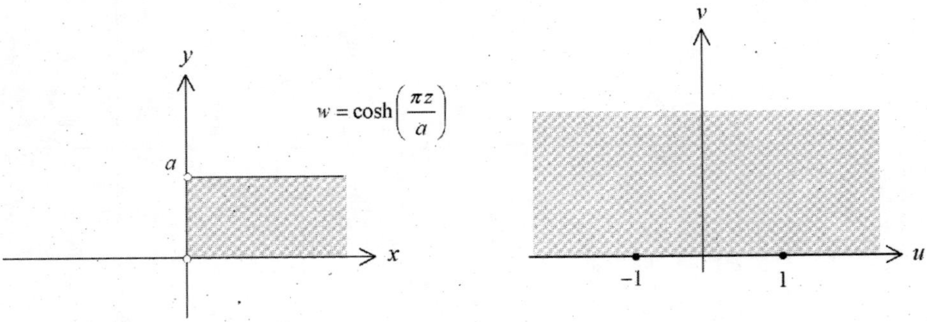
y
x
a
$w = \cosh\left(\frac{\pi z}{a}\right)$
v
u
−1
1

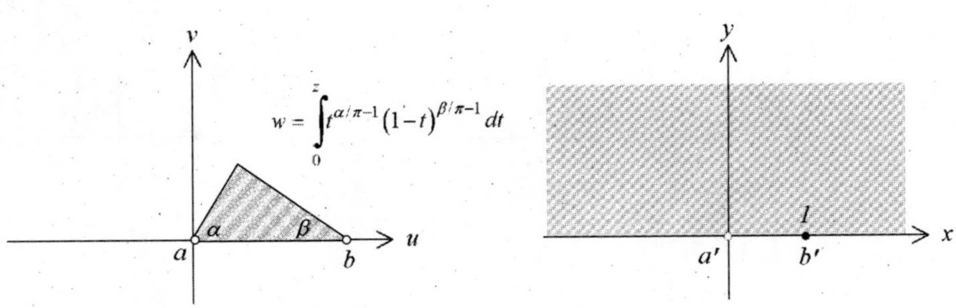
v
u
α
β
a
b
$w = \int_0^z t^{\alpha/\pi-1}(1-t)^{\beta/\pi-1}\,dt$
y
x
1
a′
b′

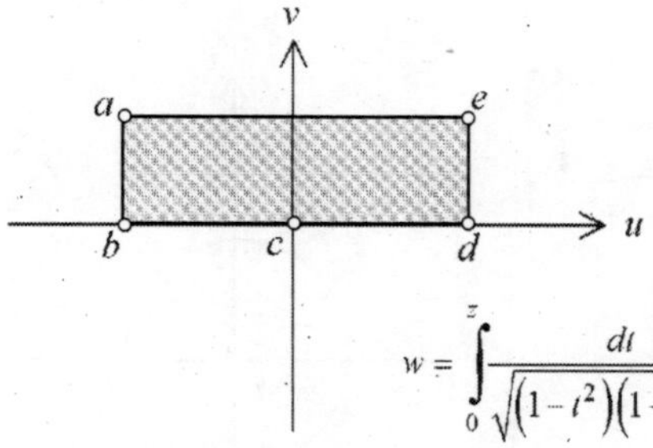

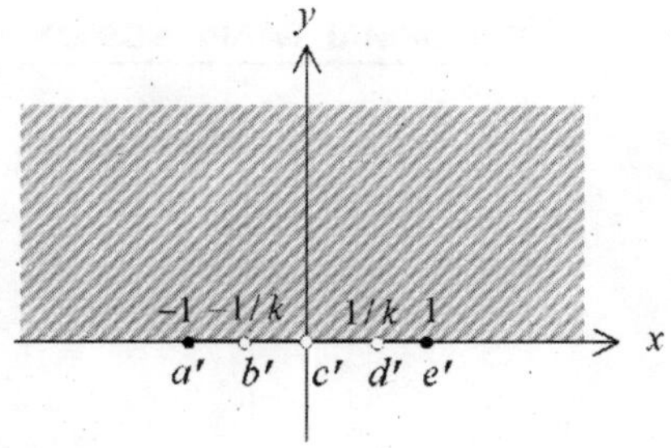

$$w=\int_0^z \frac{dt}{\sqrt{(1-t^2)(1-k^2t^2)}},\quad 0<k<1$$

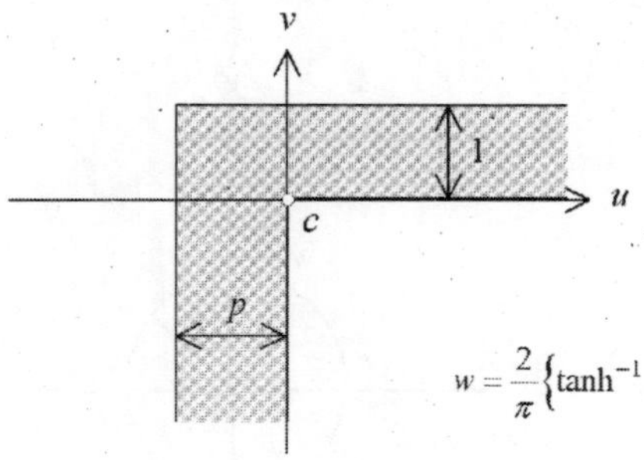

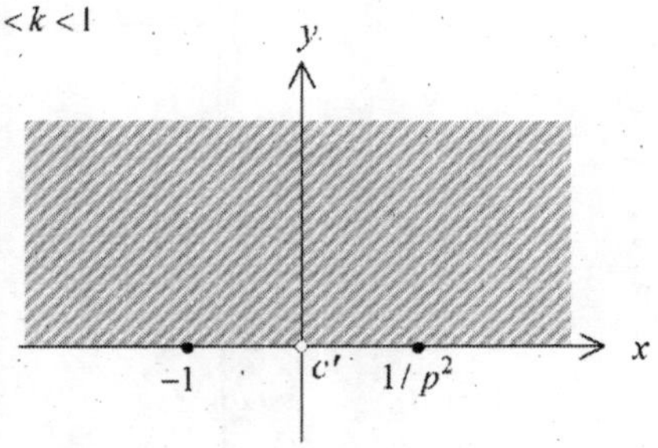

$$w=\frac{2}{\pi}\left\{\tanh^{-1} p\sqrt{z}-p\tan^{-1}\sqrt{z}\right\}$$

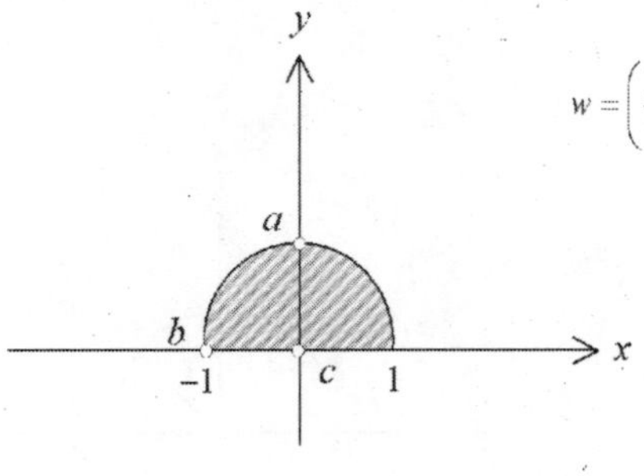

$$w=\left(\frac{1+z}{1-z}\right)^2$$

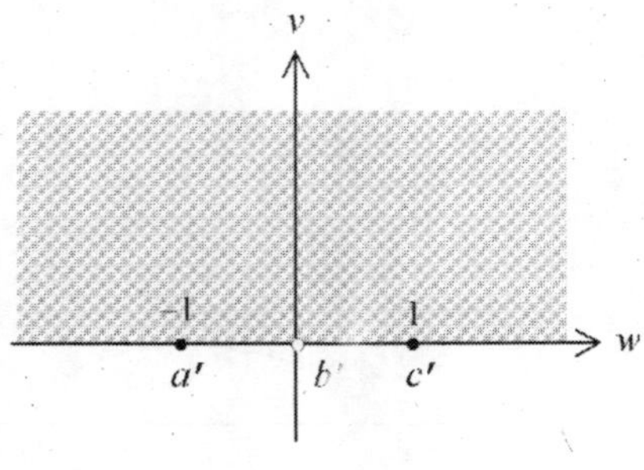

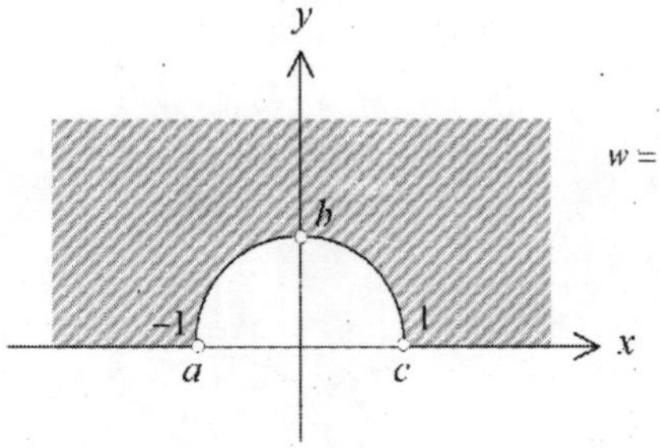

$$w=\frac{a}{2}\left(z+\frac{1}{z}\right)$$

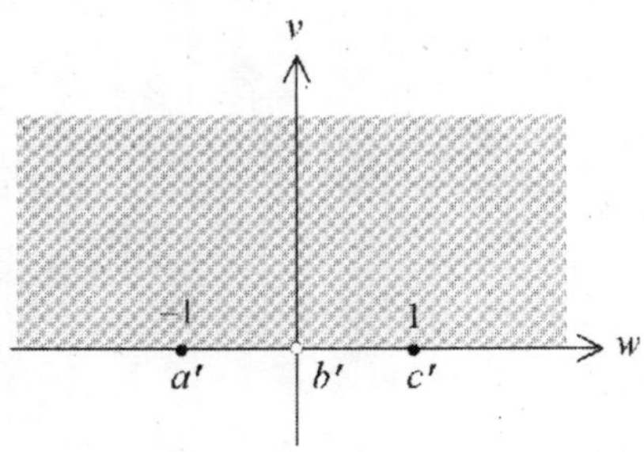

CONFORMAL MAPPINGS ON THE UNIT SPHERE

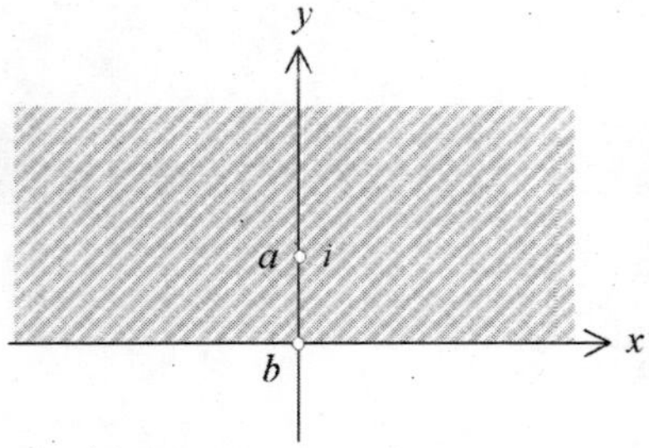

$$w = \frac{i - z}{i + z}$$

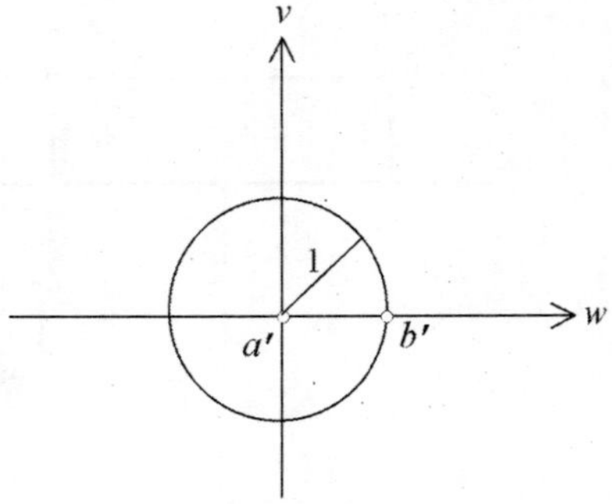

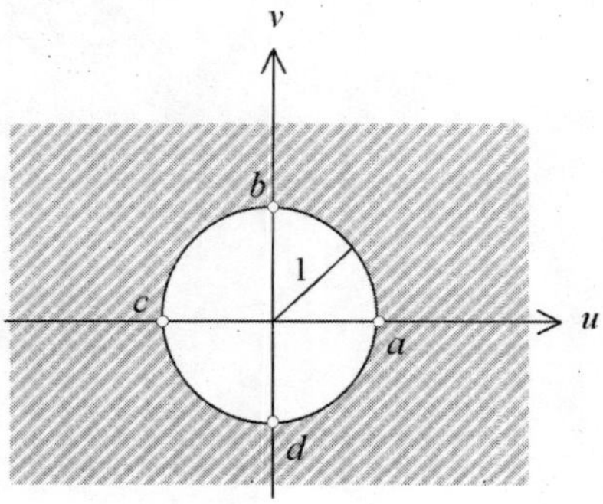

$$w = \frac{1}{z}$$

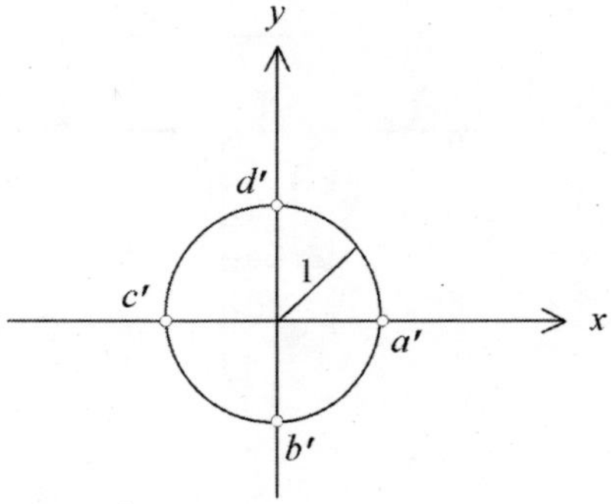

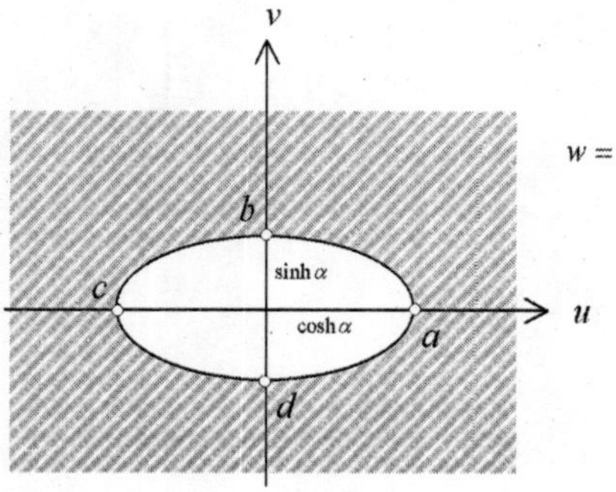

$$w = \frac{1}{2}\left(ze^{-\alpha} + z^{-1}e^{\alpha}\right)$$

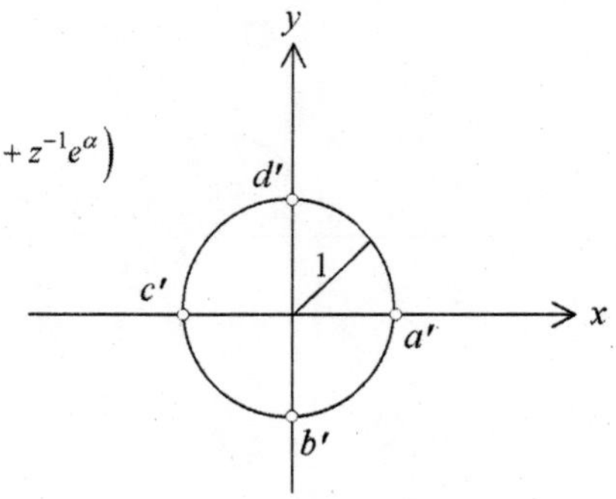

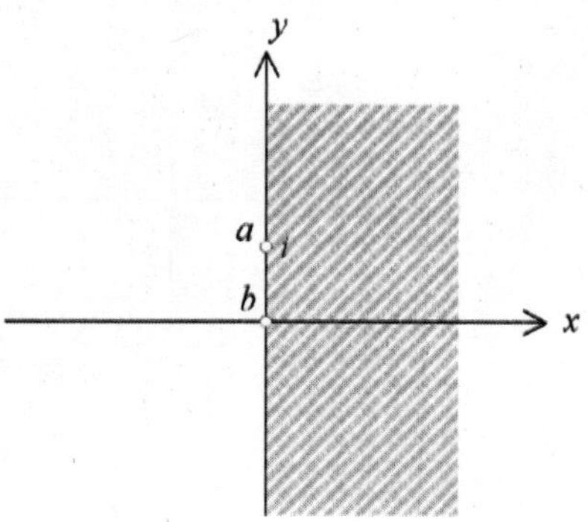

$$w = \frac{z - 1}{z + 1}$$

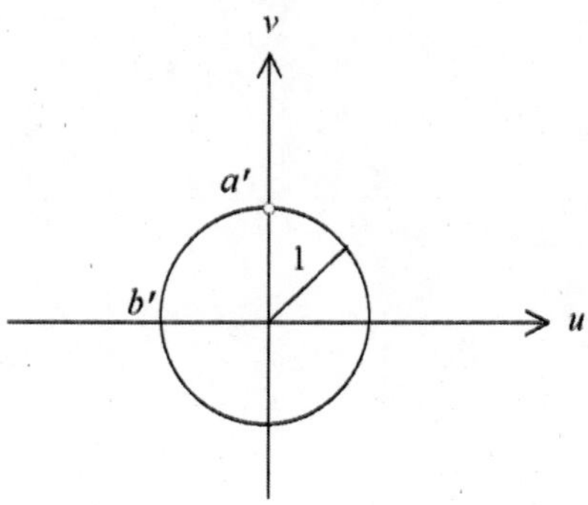

Appendix 4

Engineering Material Properties

Material	Young's Modulus (Gpa)	Shear Modulus
Rubber	0.007–0.035	0.0008–0.0016
Glass	65	8–30
Concrete	25–30	
Marble	50	
Granite	70	
Aluminum	56–77	25–30
Copper	120	50–63
Cast Iron	70–145	50
Wrought Iron	180–210	85
Stainless Steel	190	85
Structural Steel	200	80
Titanium	105–120	41.4
Tungsten	360	150
Diamond	1200	
Carbon Fiber Composite	35–1000	
Carbon nanotube	1000	

Material	Thermal Conductivity (W/m K)	Coefficient of Linear Expansion (10^{-6}/K at 20°C)
Air	0.023	
Water (20°C)	0.59	
Ice	2.22 (0°C)	51 (0°C)
Silver	420	18.8
Carbon Steel	52	11.7
Concrete		12
Stainless Steel	15	17.3
Iron	47	10.5 (cast) 12 (malleable)
Aluminum	237	23
Titanium	180	8.5
Gold	318	14
Brass		19
Diamond		1

Material	Conductivity (S/m)
Silver	$61\ 10^6$
Copper	$58\ 10^6$
Aluminum	$35\ 10^6$
Steel	$2\ 10^6$
Titanium	$2.3\ 10^6$
Graphite	$1\ 10^5$

Earth Resistivity (Ωm)	
Average	100
Dry earth	10^3
Slate	10^7
Sand stone	10^8
Wet soil	10–100
Sea water	1–5

Space Shuttle Tile (composed of SiO_2)	
Density	9 lb/ft^3
Specific Heat	0.15 BTU/lb-°F
Thermal Conductivity 70 F and 1 ATM	0.028 BTU/ft-hr-°F
Thermal Conductivity 2000 F and 10^{-4} ATM	0.073 BTU/ft-hr-°F
1 BTU = 1.054 10^3 J	1 BTU/ft-hr-°F = 1.731 W/m-K

REFERENCES

CRC Handbook of Chemistry and Physics 88th Edition (2007–2008). *Section 4: Properties of the Elements and Inorganic Compounds, Section 5: Thermochemistry, Electrochemistry, and Kinetics, Section 12: Properties of Solids.*

Electrostatic, DC Conduction and Heat Transfer material data library http://www.quickfield.com/free_dat.htm

See also Material Properties at www.EngineeringToolBox.com

Appendix 5 Biological Material Properties

Material	Young's Modulus (Gpa)
Soft Biological Tissue	0.0002
Human Cartilage	0.024
Collagen	0.6
Human Tendon	0.6
Compact Bone in Tension	22
Compact Bone in Compression	10
Elastin	0.0006
Lipid Bilayer (membrane)	.012

Material	Thermal Conductivity (W/m K)
Skin	0.34
Muscle	0.46
Fat (pig)	0.17

Mitochondrion (spherical model)	Dimension	Conductivity (S/m)	Relative Permeability
Outer membrane	7 nm (thickness)	0.95 e-6	12.1
Inner membrane space	0.92 μm	0.4	54.0
Inner membrane	7 nm (thickness)	0.95 e-6	3.4
Matrix	0.832 μm	0.121	54
Cellular plasma membrane	5 nm (thickness)	1.0 e-5–1.0 e-7	8–11.3
Cytoplasm	1~10 μm	0.5	80
Extracellular medium	1~10 μm	1.0	80

Tissue During Cardiac Phase	DC Resistivity (Ohm-m)
Blood	1.38–1.54 Ohm-m (between end-systole and end-diastole phases)
Myocardium	4.2 Ohm-m
Lungs	14.2–15 Ohm-m (between end-systole and end-diastole phases)
Thorax (average)	40 Ohm-m

Cell/Organelle	Dimension (microns)
Mitochondrion	0.1–5 μm
E. coli	2 μm
Liver cell	20 μm
Nerve cell	5 μm (diameter) (1 meter length)
Erythrocyte (red blood cell)	2 μm (thickness) 8 μm (diameter)
Bacteria	0.2–2 μm (diameter) 1–400 μm (length)
Cell membrane	5 nm (thickness)

Tissue	Conductivity (S/m)	Relative Permittivity
Skin	10^{-3} (kHz) 0.1 (MHz)	10^{6} (kHz) 10^{3} (MHz)
Bone	0.01 (dc)	3000–10000 (dc)
Fat	0.03 (10^{2}–10^{6} Hz)	10^{6} (100 Hz) 10 (MHz)
Fat	0.1–0.3 (0.4–5 GHz)	5 (0.4–5 GHz)
Breast Fat	0.2–1 (0.4–5 GHz)	5–50 (0.4–5 GHz)
Breast Tumor	0.7–3 (0.4–5 GHz)	47–67 (0.4–5 GHz)
Brain	0.1 (10^{2}–10^{6} Hz)	10^{7} (100 Hz) 10^{3} (MHz)
Muscle	0.4 (10^{2}–10^{6} Hz)	10^{6} (kHz) 10^{3} (MHz)
Blood	0.7 (dc)	10^{5} (kHz)

References

C. Gabriel, S. Gabriel. (1996). "Compilation of the Dielectric Properties of Body Tissues at RF and Microwave Frequencies," Report Prepared for AFOSR/NL Bolling AFB DC 20332-0001 http://niremf.ifac.cnr.it/docs/DIELECTRIC/Report.html

S. Grimnes and Ø. G. Martinsen. (2000). *Bioimpedance and Bioelectricity Basics*, Academic Press.

K. H. Schoenbach, R. P. Joshi, J. F. Kolb, N. Chen, M. Stacey, P. F. Blackmore, E. S. Buescher, S. J. Beebe. (2004). "Ultra short Electrical Pulses Open a New Gateway Into Biological Cells," *Proceedings of the IEEE*, **92** 1122–1137.

Roland Glaser. (2001). *Biophysics*, Springer-Verlag. *Chapter 2.5.3: Mechanical Properties of Biological Membranes*.

T. Kotnik and D. Miklavčič. (2000). "Theoretical Evaluation of the Distribution Power Dissipation in Biological Cells Exposed to Electric Fields," *Bioelectromagnetics,* **21** 385–394.

Appendix 6 About the CD-ROM

- Included on the CD-ROM are files related to electromagnetism topics
- See the "README" files for any specific information/system requirements related to each file folder, but most files will run on Windows XP or higher

Selected items include:

- All Figures in Microsoft PowerPoint® (color) and *.tiff figure files (B&W)
- QuickField model files
- MATLAB® *.m files
- *QuickField® Student Version* software

Index

D

E

F

G

H